Victorian Science & Imagery

SCIENCE AND CULTURE IN THE NINETEENTH CENTURY

Bernard Lightman, Editor

Victorian Science & Imagery

Representation & Knowledge in Nineteenth-Century Visual Culture

Edited by

Nancy Rose Marshall

University of Pittsburgh Press

Published by the University of Pittsburgh Press, Pittsburgh, Pa., 15260

Manufactured in the United States of America
Printed on acid-free paper
10 9 8 7 6 5 4 3 2 1

Cataloging-in-Publication data is available from the Library of Congress

ISBN 13: 978-0-8229-4653-3
ISBN 10: 0-8229-4653-X

Cover art: *The Peacock Room* (digitally altered details), James Abbott McNeill Whistler
Cover design: Joel W. Coggins

To all the students from whom I have learned, and to the very processes of scholarly debate, conversation, perpetual curiosity, and discovery occurring in classrooms, academic conferences, public lectures, and institutional study days from which these essays originate, I dedicate this volume. And also to Jim, John, Jamie, and Daniele, the ghosts that haunt it.

CONTENTS

ACKNOWLEDGMENTS

As this book originated in a graduate seminar at the University of Wisconsin–Madison, I would like to express appreciation for my home institution's encouragement of instructors to teach subjects both in their field and of their own choosing. When scholarship meshes seamlessly with pedagogy, the resulting alchemy benefits a wide community, and from this one course emerged two panels at the annual conference of the College Art Association and this subsequent volume. Along the way I benefited from a number of conversations with colleagues, and I am grateful as well to the senior scholars who joined the project, no questions asked, when I tapped them for contributions. As a mark of her profound impact on the field of Victorian studies, several individuals in this collection, including myself, have been shaped by their studies with Dr. Caroline Arscott, whose intellectual presence looms large in these pages. Lynn K. Nyhart, Vilas-Bablitch-Kelch Distinguished Achievement Professor in the history of science program at UW–Madison, provided valuable advice and insight at various stages of the collection.

Most instrumental in bringing the collection into focus were editors Abby Collier and Bernard Lightman at the University of Pittsburgh Press, who found perceptive, rigorous readers to shared their expertise and improve the work in countless ways. Also at Pittsburgh, managing editor Amy Sherman shepherded the project into its final state with vigilance and precision. And finally, thanks to Carol Bracewell for her valiant efforts at taming the bibliography.

I would like to express my appreciation for the dedication of the Kohler Art Library team of director Lyn Korenic, reference librarian Linda Duychak, and Soren Schoff. A UW–Madison Vilas Associate Award helped catalyze thought toward this volume, as did a sabbatical leave.

And finally, stay-at-home support for the final stages of this manuscript was delightfully provided by Harry and Ciccina DiPiazza Marshall.

Victorian Science & Imagery

INTRODUCTION

VICTORIAN SCIENCE AND IMAGERY

Nancy Rose Marshall

What happens when we look at an artwork in terms of the ways its depicted subject, its medium, its method of production, or its formal elements correspond with the dominant scientific discourses of the period in which it was produced? This volume suggests that considering how objects from such conventionally separated categories as "empirical research" and "creative works of art" might share epistemological conditions offers fruitful interpretive possibilities. At a crucial moment in the development of the modern world, British and American visual culture helped to produce science, while science in turn newly informed art. Remarkably, in 1875 it was possible for an English scientist to maintain that "science and art . . . act and re-act upon each other with an almost exchangeable importance. Science is theoretical art; art practical science."[1] The art historians in this collection help us see the extent to which nineteenth-century "science" and "art" were not in fact separate fields but instead mutually constitutive.

A focus on the close kinship between the two fields in the nineteenth century is timely and topical, as it may help disrupt the subsequent reification of the idea of two antagonistic cultures, an image that became entrenched by the middle of the twentieth century.[2] Interpretive models emphasizing the ways that cultural knowledge emerges across a wide range of spheres might, in fact, offer us some salient wisdom in facing our current global crises; it will take both imagination and empirical data to solve the issues involved in climate change, for instance.

Foundational to our project is the acknowledgement that art and science are not tightly compartmentalized, separate spheres but rather loosely framed practices and knowledges that emerge from shared

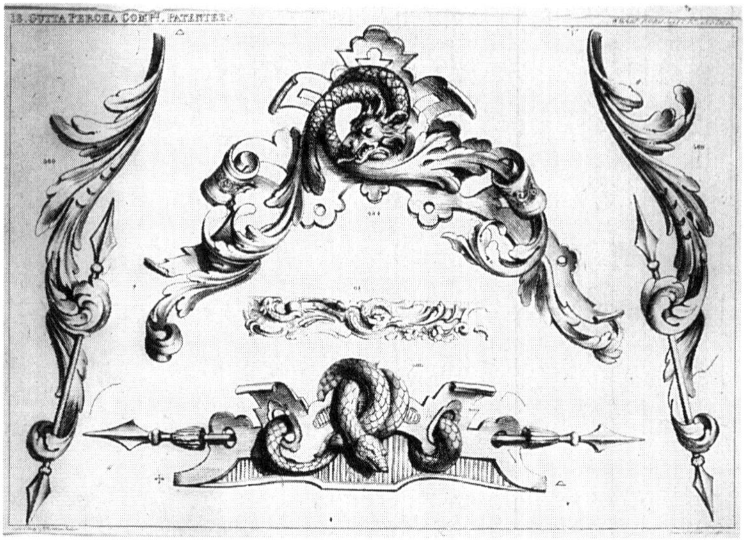

Fig. I.1. Hawkins's ornamental serpent designs reveal his close study of natural history. Benjamin Waterhouse Hawkins, *Decorative Motif in the Form of a Serpent*. Reprinted from *Gutta Percha Company's Pattern Book of Ornaments for the Use of the Trade* (London: Gutta Percha Works, 1850).

beliefs circulating in a society.[3] As Gowan Dawson and Sally Shuttleworth remind us, the same "metaphors, themes, images, and ideological orientations" inform poetry and science; in other words, like science, art and literature are both social products that manifest from and in turn produce the epistemological possibilities at play in any given moment in history.[4] Science does not remain objective, timeless, universal, and stateless while art gets to be imaginative, subjective, and culturally constituted; science is not the straight man to art's funky and unpredictable comedian. In the helpful characterization of Caroline A. Jones and Peter Galison, "neither practice has unique absolute purchase on 'reality,' and neither is as alienated from history as its rhetoric may imply."[5]

The story of one individual serves to illustrate the inseparability of artistic and scientific identities in the Victorian period. Around 1850, naturalist Charles Darwin wrote to Benjamin Waterhouse Hawkins, who had illustrated fish and reptiles for two volumes of *The Zoology of the Voyage of HMS Beagle* (1838–43): "Dear Sir, I have much pleasure in sending my testimony to your skill as an artist . . . , and in several departments in Natural History."[6] The unproblematic promotion of the same man as both scientist and artist suggests the compatibility of these roles in the Victorian period. A fellow of the Linnean and the Geological

Societies and a member of the Society of Arts, Hawkins also received recommendations from Royal Academicians John Rogers Herbert and the vice-president of the Zoological Society William Yarrell. Hawkins, a clubbable sort with a penchant for dramatically illustrated lectures and a gift for public relations, exemplified the possibilities open to Victorian men in possession of both empirical knowledge and artistic skills. He found work in both Britain and the United States due to the shared currency of artistic and scientific thought between the two countries, as well as to their comparable institutional structures. Combining his understanding of sculptural technique acquired from his academically trained teacher, William Behnes, with his natural history and geological knowledge gleaned largely from self-study, Hawkins's projects exemplify the type of hybrid forms seen elsewhere in this book.[7] Hawkins also deployed art to persuade, either in the service of commerce or to direct public opinion regarding scientific and social debates. In designs for the pattern book of the Gutta Percha Company, which fabricated objects from a recently isolated rubberlike material from Malaysian trees, for example, he drew snakes and other creatures featured in mass-produced moldings and decorative elements (see fig. I.1).

In this way Hawkins promoted through his art a new industrial material brought to England by a botanist and a doctor in 1843 and further refined by English scientists of the Joint Committee for Chemistry, Colonies and Trade.[8] Likewise, when commissioned by the planners of the 1854 Sydenham Crystal Palace Park geology display, who were familiar with his sculptural work from Royal Academy exhibitions, Hawkins realized life-size three-dimensional models of the famous paleontologist Richard Owen's visions, including the megalosaurus (fig. I.2). A boisterous dinner party staged by Hawkins for scientists (including Owen), journalists, and investors in the model of the iguanodon was an inspired promotion for what was essentially the world's first Jurassic Park (although Hawkins's dinosaurs remained inanimate, proliferating cartoons and anecdotes of mobile and nightmarish monsters threatening visitors, who sometimes stole their teeth as souvenirs, suggested otherwise).[9] Renowned for his accomplishments at Sydenham, Hawkins next designed a hadrosaur with Joseph Leidy of Philadelphia's Academy of Natural Sciences and then another dinosaur display for Central Park in New York (where it still lies buried, according to popular accounts).[10] These splendid models, undertaken in concert with men on different continents holding differing opinions about evolution and extinction, assured the visibility of both the debate and the debaters. While Leidy supported Darwin's theory of natural selection leading to evolution, Owen rooted his own mechanism for evolution in divine creation.[11]

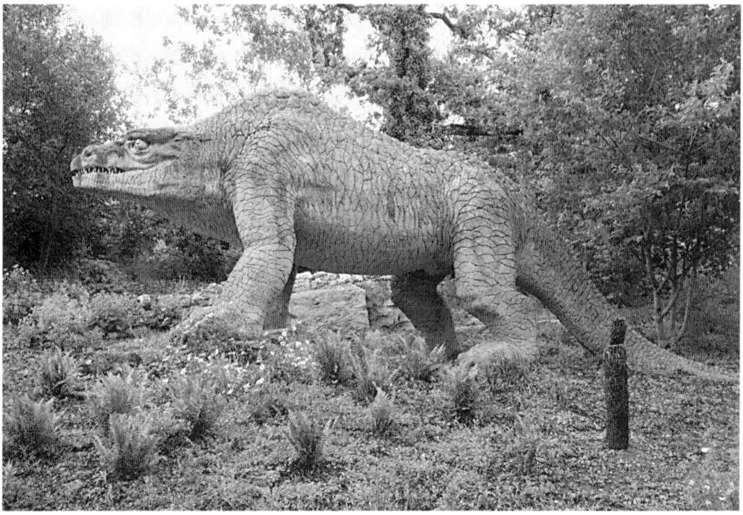

Fig. I.2. Working with paleontologist Richard Owen, Hawkins created the first "Jurassic Park" of life-size dinosaur reconstructions. Benjamin Waterhouse Hawkins, *Megalosaurus*. Concrete, iron, rock, pigment. Sydenham Crystal Palace, London, 1854. Credit: Chris Sampson, Wikimedia Commons.

As a scientist-artist, Hawkins could also design illustrations to express his opinions on the scientific debates of the period. A plate in his 1860 *Comparative Anatomy as Applied to the Purposes of the Artist* (fig. I.3), portraying a white man clasping the hand of a deferential African, simultaneously evoked the iconic "handshake moment" of history painting (suggesting a treaty or agreement) and refuted Darwinian claims about the evolution of man from apes, thereby entering the fray of the acrimonious public dispute between Owen and the biologist Thomas Henry Huxley, "Darwin's bulldog." In the text accompanying the plates, Hawkins emphasized that artists should use these "graphic illustrations . . . to contradict degrading theories in support of the fallacy of the Darwinian Paradox" and his diagram of the various primate species grouped together in harmony was intended to display their fixity rather than their evolution from beast to man.[12] With the central handshake, the Black and white figures appear to accept their positions as no. 2 and no. 1 in Hawkins' hierarchy. In another diagram (fig. I.4), Hawkins represented his interest in the divinely ordained (rather than evolving) hierarchy of man over animal—and of white over black races—by depicting, in skeletal form, a tableau of a standing and a kneeling figure that we can identify only through iconography: The chains and beseeching gesture of the skeleton on the right are part of a large catalogue of abolitionist imagery that worked to

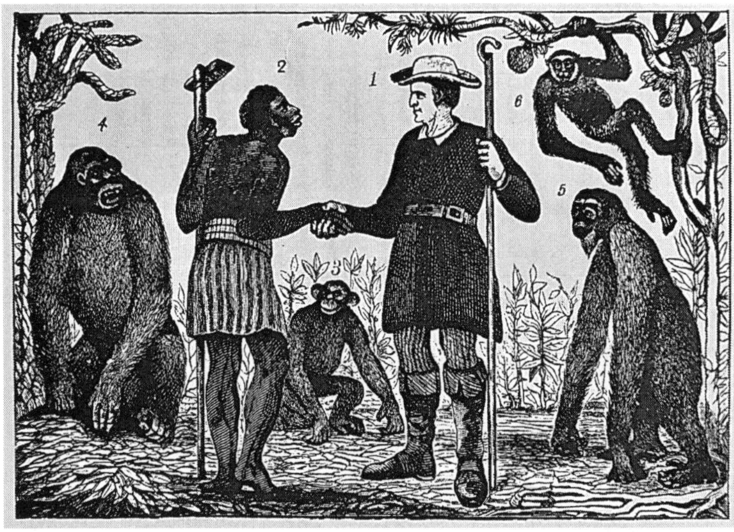

Fig. I.3. By placing the human figures in a numbered array of primates, Hawkins indicated his support for the belief in a divine, rather than evolutionary, natural history hierarchy with white man at its top. *Plate 2*, in Benjamin Waterhouse Hawkins, *Comparative Anatomy as Applied to the Purposes of the Artist*, ed. George Wallis (London: Winsor and Newton, 1860).

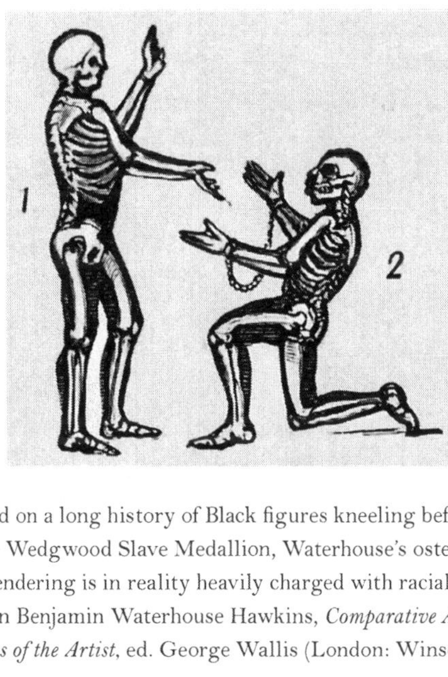

Fig. I.4. Based on a long history of Black figures kneeling before white men, including the Wedgwood Slave Medallion, Waterhouse's ostensibly empirical anatomical rendering is in reality heavily charged with racial meanings. Detail, *Plate 1*, in Benjamin Waterhouse Hawkins, *Comparative Anatomy as Applied to the Purposes of the Artist*, ed. George Wallis (London: Winsor and Newton, 1860).

assert and maintain the superiority of the white man, most influential of which was the medallion crafted by Josiah Wedgwood in 1787 in which the image was inscribed "Am I Not a Man and a Brother?"[13] The fact that racist comparative anatomy was taught to artists is yet another example of the inextricability of the arts and the sciences at this time.

Hawkins's plates, then, demonstrate the complex interweaving of the codes of art and of science in this period. They demonstrate that one man could skillfully deploy art as a weapon in scientific debate, on the one hand, while bringing keen empirical observation to the arts, on the other. Royal Academician Henry Stacy Marks's painting *Science Is Measurement* (plate 1) occasioned unusual scrutiny upon its exhibition at the Royal Academy in 1879, due to just this overt assertion of the relationships between art and science. Portraying a scholar-gentleman in an eighteenth-century periwig and knee-breeches, pencil in mouth and notebook and tape measure in hand, gravely evaluating a skeleton of a stork, the painting was important enough to be selected for Marks's "diploma picture."[14]

Despite the fact that the painting was set about one hundred years prior to its production, the critic for the *Leisure Hour* understood the piece as a paradigmatic and celebratory image of the modern era, proclaiming that it rose to a category of those rare artworks that became "typical representatives of the period of their production." The picture constituted "a doctrine of our more recent times" for making the case that "good could come of measurement applied to organisms, that life and mind and social phenomena could by any possibility be subject to laws, and that these laws could be discovered and made useful to man." Its primary contribution was in its acknowledgement that the biological kingdom had become newly subject to the type of empirical investigation previously reserved for the inanimate world, and that mysteries of organic forms could now be rendered transparent and rational. Everything, art included, now fell under the purview of the coherently calculable: "All art, which is in truth nothing but applied science, depends likewise on measurement."[15]

Marks, though, seems to add a question mark to his positivist title through the puzzled standoff between the stork skeleton and the ornithologist; although the scientist deploys the analytical gaze of the experimental observer, he appears to pause in confusion at what he sees. Moreover, his wig and knee-breeches make the scientist appear quaintly old-fashioned for those not immediately inclined to associate eighteenth-century dress with Enlightenment empiricism. What, in fact, *was* science, if not measurement?[16] For nineteenth-century thinkers and practitioners, and for Marks himself, it was, to a degree, art.

The Nineteenth Century

As we learn from the foundational scholarship of Bernard Lightman, the nineteenth century offers some distinctive contributions to the student of the history of science. A number of significant discoveries in the sciences coincided with an expanding middle class and the rise of commodity production and consumer culture, while a receptive and engaged public eager for entertainment and education in a range of fields drove the proliferation of publications in both science and art. Before scientific knowledge became too arcane for a general audience, journals routinely assumed an educated readership thoroughly absorbed by the cutting-edge debates of the day.[17] The new medium of industrial printing expanded print culture, allowing for cheaper editions of theoretical texts and the spread of debated ideas.[18] Lightman's *Victorian Popularizers of Science* demonstrates that we must look beyond the work of canonical scientific figures of the period to identify and understand the widely varying beliefs of the new middle-class consumers of science, a subject that became both education and entertainment in a remarkably wide range of locations beyond the predictable lecture halls, periodicals, and libraries, including churches, artist's studios, shipyards, and zoos. In an era in which "the lines delineating science from spectacle had not been drawn," scientific display was as likely to be found under the oversight of showmen as in the care of learned dons. To comprehend the complex domains of Victorian science, it is also necessary, as Lightman notes, to explore the discourses and practices of marginalized groups, such as women, people of color, and the working classes, as well as those of alternative sciences such as phrenology or psychical research.[19]

From 1851, the year of the Great Exhibition at the Crystal Palace in London, World's Fairs became important sites for the display and dissemination of new information, and there were countless such locations in London alone, such as the Adelaide Gallery (1832), the Polytechnic Institution (1838), or the Royal Institution (1799).[20] Networks of scientific communication stretched across the Atlantic between individuals and institutions, even as American naturalists fought for acknowledgement as the experts on their own continent.[21] As natural history collections increasingly moved into dedicated spaces, Philadelphia's Academy of Natural Sciences (1812) and the American Natural History Society in New York (1869), followed by London's Natural History Museum in 1881.[22] By 1870 there were 125 scientific societies in Britain and Ireland, including the Royal Astronomical Society (1820), the Institute of Chemistry of Great Britain (1877), and the Geological Society of London (1807).[23] In the United States, the American Association for the Advancement of

Science was formed in 1848, followed by the National Academy of Sciences in 1863.[24]

Like the world of science, the art world became similarly institutionalized and ordered into hierarchies of educational, exhibition, and sales venues. The American Academy of Fine Arts (1802) and the National Academy of Design (1825) in New York and the Pennsylvania Academy of Fine Arts in Philadelphia (1805) were three of the most prominent organizations in the United States, while British art was dominated by the Royal Academy in London (1768).[25] Public art collections expanded in number, scope, and access, prominent among them being the National Gallery in London (1824), the Boston Museum of Fine Arts (1870), the Metropolitan Museum of Art in New York (1872), and the Pennsylvania Museum and School of Industrial Art (1876).[26]

While art and natural history began to be separated in the Enlightenment, they retained connections not fully severed until the twentieth century, if indeed this rupture was ever fully completed. Many scholars have traced the history of both scientific and art display in tandem with changing cultural norms and shifts in how knowledge was organized, particularly the move from *Kunstkammer* (cabinet of curiosities) to museum.[27] Such a scholarly turn of course follows in the wake of *The Order of Things* (*Les mots et les choses*), in which Michel Foucault proposed the concept of the "episteme," a slippery and shifting amalgam of discourses, constructions, institutions, orderings of knowledge, and ideologies that produce conditions of knowing in a given period. Thinking with Foucault illuminates how in fact the arts and the sciences not only can but should be seen as tightly interconnected and cross-fertilizing—as, to use the words of Jones and Galison, "regimes of knowledge, embedded in, but also constitutive of, the broader cultures they inhabit."[28]

Tony Bennett cites Krzysztof Pomian's argument that the fact that cabinets of curiosities gave way to natural history collections points to changing epistemes; focusing on the exceptional or curious object, one that inspired wonder, was well suited to the medieval and early modern worldview in which the notion of seeking for laws in nature was alien: the unique could therefore be exemplary.[29] During the Enlightenment, however, rational logical order, classification, connection and organization became governing principles, pushing art and natural history asunder while retaining less visible threads of attachment.[30] The rise of the scientific focus on the category or type—as opposed to the unique individual or specimen—accompanied the elevation of "objectivity" into an essential scientific moral quality.[31] At the same time, the mutual relationship between artistic and scientific practices continued in a range of practices, however; as Sally Gregory Kohlstedt has found, for instance, natural history illustrations fed into arrangements of specimens, which in turn sometimes were copied by artists.[32]

As science developed, an emphasis on categories of rational classification led to the rise of new disciplines; the Victorian period saw the burgeoning of psychology, chemistry, physics, and many other fields. The term *scientist* was itself a nineteenth-century word, deployed first in 1834 by William Whewell, and it remained a fluid term until later in the century.[33] Dramatic advances drove confidence about the potential ordering and taming of the world through the sheer power and scope of reason, and the list of momentous and transformative discoveries occurring over a very short period of time is formidable. The universe became vaster but more crowded through advances allowing for astronomical observation and calculation, which in turn produced theories such as Whewell's nebular hypothesis imagining the creation of the planets and stars. Simultaneously, our environment grew ever more populated by the invisibly tiny: forty-nine new elements were discovered in the nineteenth century, and the development of the model of the atom as the smallest particle in the universe by John Dalton (1808) was modified at the end of the century by J. J. Thomson's discernment of a subatomic particle, the electron, in 1896.[34] From the mid-eighteenth century chemists proved the idea of the conservation of mass, the principle that nothing could be created or destroyed, followed by the formulation of the law of the conservation of energy.[35] Meanwhile, changing the face of medicine were John Snow's and Louis Pasteur's proof of the germ theory (1850s) for the transmission of disease, along with the deployment of chloroform during surgery (1847) and Joseph Lister's successes with antisepsis (1867).[36]

Even more paradigm-shattering were Charles Darwin's proposals regarding evolution and natural and sexual selection, which radically reoriented many fields, from theology to zoology to sociology, and participated in the gradual relocation of humankind away from its long-accepted position at the center of the universe. People increasingly belonged to the realm of the animal; biology had now "boldly entered the precincts of man's own and special order," as Andrew Wilson, a Scottish follower of Darwin, asserted in 1883.[37] In the wake of German scientists' contributions to the visualization of the workings of the cell and its components in the middle of the century, there blossomed a newly acute sense of the kindred relationship between all living things, based on the concept of protoplasm. Producing such visions as Huxley's fantasy of the transubstantiation of a man into a lobster, an imaginative investment in the exciting idea that there was one common substance shared by all organic creatures refashioned the order of the world.[38] Even plants could become cousins: as Wilson remarked, "the last decade of science has certainly tended to raise the plant as a living, and moreover as a sympathetic and active being."[39]

In light of—and contributing to—the radical changes in structures of belief, traditional attitudes began to crumble. Nature was increasingly perceived as a comprehensible united system rather than an amorphous force inimical to human beings.[40] A view of scientific progress as a gradual mastery of the world also both allowed for and was part of the development of technologies of power deployed to subordinate and discipline all that fell under the category of "nature," relative to the human sphere of "culture." The normalization of values that instrumentalized the natural world in turn served and was served by the growing system of industrial capitalism, reminding us that any definition of nature is itself a product of culture with its own history.[41] Propelled by their growing economies, expanding empires indelibly marked the globe, generating encounters between peoples, species, plants, cultures, and knowledges, which in turn required cataloguing and recording. Indeed, the centralized concentration of raw materials brought together from the far reaches of the earth formed important laboratories. As Julia Voss has concluded, for example, it was Darwin's experience with the extraordinary accumulation of specimens in the British Museum that set him on the path to grasping the mechanisms behind the infinite variety of life.[42]

At the same time that imperial institutions and practices certainly contributed to knowledge, then, the dark side of empire and colonization produced systematic oppression and exploitation of people, animals, and lands, as Rachael DeLue tells us in her essay in this volume—as well as, of course, scientific racism. Art, too, must be situated in all of these contexts.

Victorian Science and Art

The two fields under consideration in this volume are, then, profoundly interconnected. Most obviously, perhaps—certainly to Marks's depicted naturalist—artists and scientists both invested heavily in the faculty of sight.[43] The scientific gaze was an ostensibly objective form of looking increasingly shared by artists who, especially after the forcefully expressed artistic directives of the prominent cultural critic John Ruskin, began to practice close observation of the natural world. Ruskin advocated seeing rather than analyzing: "Do not think, by learning the nature or structure of a thing, that you can learn to draw it," he instructed, continuing, "To draw a man, a flower or a mountain" can be accomplished "only by looking at them; not by cutting them to pieces."[44] Countless scholars have observed the connection between vision, knowledge, and empiricism.[45]

Such a dependence on sight was paradoxically countered by an increasing doubt about the reliability of this sense. Vision, as we have learned, is both flawed and culturally constructed.[46] Feminist and postcolonial theorists have drawn attention to the ways in which the dominance of sight related to rational or scientific knowledge is "phallocentric, colonialist, and calculating," debased by exploitative and oppressive urges.[47] From the Renaissance, the development of one-point perspective, an artistic system for representing three dimensions in two that was perfected in the academic art of the nineteenth century, contributed to the notion that space was generated from the gaze of an omniscient beholder, rigidly defining the gaze as objective, mappable, and immobile. As has been well rehearsed in art history and visual culture, Hal Foster's foundational concept of "visuality"—later defined by historian of science Klaus Hentschel as a "variegated bundle of social factors involved in the process of seeing"—shows us that the optical entertainments and scientific instruments produced by a culture work as models for concepts of vision's function, producing socially contingent "scopic regimes" or "visual regimes."[48]

Art historians have long traced vision as historically conditioned.[49] More recently, Jonathan Crary, Christopher Otter, and others have generatively examined the historiography of the viewer, tracing how the Renaissance model of an objective eye passively reflecting the world had, by the nineteenth century, shifted to one in which vision was embodied and subjective, dependent on the responses of individual sensoria.[50] The discovery of binocular vision proved that sight was processed in the brain, and the phenomenon of after-images likewise demonstrated the extent to which it could be subject to misprision; by the mid-nineteenth century physicist Hermann von Helmholtz's experiments had definitively demonstrated the unreliability of the eye. Vision became less a universal, objective function and more an embodied, individualized experience; perception was, in Crary's words, transferred to "the thickness of the body."[51]

In chapters 4 and 7 in this volume, respectively, I and Barbara Larson take up these issues in the context of the study of hallucinations and optical physiology. The fact that our vision is not to be trusted, as Barbara Stafford notes, in turn informed anxious or dismissive responses to the image itself: "Visual evidence [came to be seen as] synonymous with legerdemain."[52] The denigration of the visual arguably continues in our time, a situation in which the contributors to this volume make an intervention through the vigorous assertion of the unique work performed by art and visual culture.

Ways of representing linked to ways of knowing, which in turn generated frameworks of power and control. Svetlana Alpers's *Art of*

Describing famously relates the Dutch seventeenth-century capitalist focus on possession of objects and land to acutely observed and obsessively rendered surfaces of their art.[53] Ann Shelby Blum follows Alpers to observe that the detailed, descriptive, mapping style of northern art was intertwined with—both produced and produced by—"the empirical agenda of scientific revolution."[54] And, as Daniela Bleichmar further suggests, the project of making visible was intimately bound up in making "imperial nature moveable, knowable, and ideally, governable."[55] Images allowed for notional possession.

With all due respect to Ruskin and Alpers, in the nineteenth century the dominant regime of knowing came increasingly to be based on the probing of invisible interiors rather than visible surfaces. New technologies such as X-rays retooled cognition to imagine what lay beneath, and in turn to conquer those depths.[56] For nineteenth-century scientists, such scientifically generated images became bound up in the pursuit of the holy grail of "objectivity," as Lorraine Daston and Peter Galison famously articulated: "The image, standard bearer of objectivity, is inextricably tied to a relentless search to replace individual volition and discretion in depiction by the invariable routines of mechanical reproduction."[57]

As Naomi Slipp recounts in chapter 5 of this volume, Foucault's analysis of the history of medicine found that patients were increasingly diagnosed by internal signs apparent only to the trained physician, rather than by external symptoms. Such a pattern corresponds with Foucault's larger argument that nineteenth-century science moved away from classification by visible markers to more arbitrary links based on abstractions. In her chapter on photographs of early operations on etherized patients demonstrating both the effects of the drug and the lack of pain, Slipp addresses the challenges of representing these invisible truths.

Indeed, the invisible—and the need to make visible—drove much science and art in this period, as we find also in my own essay on the ghostly and DeLue's chapter exploring the representation of the subterranean. "Imponderable matter," an etheric unmeasurable force or substance, became a leading subject for scientific scrutiny and experiment, sweeping up other practices such as mesmerism and spiritualism.[58] Some scholars have attributed the new concentration on the invisible as a response to the proliferation of things in the Victorian mass-produced, overstuffed world. Alice Jenkins, for example, points to a "profoundly dematerializing tendency in nineteenth-century physical science."[59] Observing this dissolving, dissipating materiality, Pre-Raphaelite associate John Tupper wrote a facetious poem titled "Progress of the Species," about the newly and disturbingly invisible. Citing "galvanism, and mesmerism,

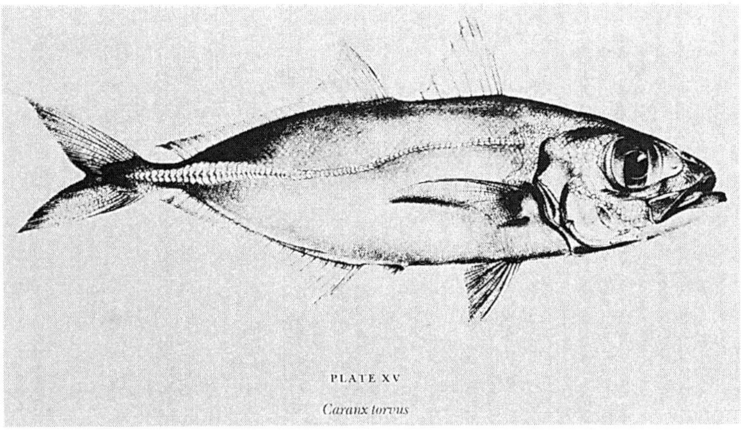

Fig. I.5. Hawkins produced the illustrations for Darwin's volume on fish by relying on a combination of careful observation of specimens and creative rendering. Benjamin Waterhouse Hawkins, *Goggle Eyed Scad,* engraving, plate 15 in Charles Darwin, ed., *Voyage of H.M.S. Beagle, Part 4 (Fish)* (London: Smith, Elder, 1840–1842).

and steam, / And gas, and anæsthetics!" as well as photography, Tupper suggested the ways in which Victorians were faced with disappearing things, exclaiming, "Dear man, you must have hated tangibles."[60]

Many of the images in this book, such as Southworth & Hawes's daguerreotypes (see figs. 5.1, 5.2, 5.4) or Joseph Maclise's anatomical drawings (see plate 5; figs. 6.1, 6.3, 6.5), demonstrate the overlaps and blurred boundaries between what was at the time understood as a subjective artistic practice and the objective, mechanical and documentary mode used by the sciences; between realism and classicism; and between scientific "objectivity" and nineteenth-century academicism. To produce the illustration of the delightfully dubbed "goggle-eyed scad" (fig. I.5) for Darwin, for instance, Hawkins first worked from preserved specimens to create an outline, which was approved by a zoologist; Hawkins then detailed the rendering on his own and copied the result to a lithography stone.[61] Imagination often complemented data and dry facts, as Martin J. S. Rudwick reminds us in *Scenes from Deep Time: Early Pictorial Representations of the Pre-historic World.* To recreate models of "antediluvian monsters" from—in some cases—extremely partial fossil evidence for the Crystal Palace Park at Sydenham, Hawkins drew on historically embedded artistic conventions used to illustrate deep time. In the "intersection of two traditions," dinosaurs were born.[62]

Some writers insisted that science itself was a source of imaginative inspiration, as did, for instance, influential early psychiatrist Henry

Maudsley, who declared, "No, science has not destroyed poetry, nor expelled the Divine from nature, but has furnished the materials, and given the presages, of a higher poetry and a mightier philosophy than the world has yet seen."[63] Likewise, John Tyndall maintained that scientists as diverse as Darwin, Isaac Newton, and Michael Faraday understood that "nourished by knowledge patiently won; bounded and conditioned by cooperant Reason, Imagination becomes the mightiest instrument of the physical discoverer."[64]

As scholars have demonstrated, however, others feared the world would become depleted of its mystery given too great an understanding of its mechanisms.[65] As a way of reenchanting subjects that seemed to have been rendered dull and bare through too much analysis, some nineteenth-century writers viewed science as sharing origins, models, and publics with the magical and supernatural. In fact, much recent writing has tracked the interconnection and inseparability of magic and science in this period, at a time when the pursuit of inexplicable phenomena could just as easily send investigators down paths that today might be dismissed as pseudoscience as on trails that ended in transformative empirical discoveries. As I discuss in chapter 4 in this volume, spiritualism and psychology were closely connected; physiologist and antispiritualist William Carpenter acknowledged that contemporary technological feats such as the Atlantic telegraphic cable strained the bounds of credulity in their seeming miraculousness.[66] Public educational spectacles often took on aspects of magicians' showmanship, as in the demonstration of the necessity of oxygen for respiration through the removal of air from a glass vessel in which a fluttering bird demonstrated a corresponding distress.[67]

Science and art also shared certain assumptions propelling their practices and pedagogy. Both participated in the nineteenth-century quest for universal principles uniting many complex, seemingly separate phenomena. Heat, light, electricity, magnetism, and gravity, for instance, were now understood as different manifestations of the same force. A drive toward wholeness marked many publications of the period, such as Alexander von Humboldt's influential *Kosmos* (1845), as scholars sought a single sublime law to explain the functions of the universe.[68] In the new world of Victorian matter, the same rules applied to the macro and the micro, making the minute or the detail itself a whole world, one that fulfilled, albeit in an empirical mode, the Romantic promise of William Blake "to see a world in a grain of sand / and a heaven in a wildflower."[69] "Though conversant with the minute forms of things," chemist Humphry Davy remarked concerning modern thinkers in 1830, "they have for their ultimate end the great and magnificent objects of nature.

They regard the formation of a crystal, the structure of a pebble, the nature of a clay or earth; and they apply to the causes of the diversity of our mountain chains, the appearance of the winds, thunder-storms, meteors, the earthquake, the volcano, and all those phenomena which offer the most striking images to the poet and the painter."[70] Confidence in the human ability to lay bare all spheres of existence through the intelligibility of laws applicable on the scale of both the pebble and the volcano was a hallmark of the Victorian ideology of progress. Human capacity to perceive expanded into the infinite space of the universe, on the one hand, and the impossibly tiny world of the atom and the subatomic particle, on the other.[71]

Nature, then, became a unity in which each part was vitally interrelated with all others. In chemistry, for instance, determining the relation of parts to the whole arose as a dominant experimental mechanism, allowing chemists to embrace perpetual change in the framework of constancy.[72] This was also true in other fields, such as medicine and anatomy; Galenic practice had treated the whole person, but increasingly the body became a system first of interrelated parts and then, eventually, of individual cells. As anatomist Joseph Maclise, discussed in Keren Hammerschlag's chapter in this volume, pronounced, "The womb of anatomical science is pregnant of the true interpretation of the law of *unity in variety*."[73]

The nineteenth-century tendency to establish unities and continuities, in a quest for single great laws to comprehend all the functions of the universe, corresponded to the aesthetic doctrines of the day, which advocated above all for unity and harmony. As in nature, all parts of an artwork should relate to the whole; Ruskin inimitably intoned: "A pure or holy state of anything . . . is that in which all parts are consistent. The highest or organic purities are composed of many elements in an entirely helpful state. The highest and first law of the universe, and the other name of life, is, therefore, 'help.' The other name of death is 'separation.' Government and cooperation are in all things, and eternally the laws of life; anarchy and competition eternally, and in all things, the laws of death."[74] That is, all parts should contribute to the whole. Such a grandiose set of claims helps us realize the seriousness and complex interrelationship of aesthetics and political and social structures in this period. For Ruskin, socialism generated vitality, while capitalist competition (or Darwin's and sociologist Herbert Spencer's survival of the fittest) was fatal for the social organism. Drawing manuals repeatedly characterized the aesthetic goals of variety in unity, coloration, lighting, and expression, signaling the operation of an epistemological framework extending across spheres of science and art. Yet, as one advised:

> This variety must not run riot, and be introduced merely for its own sake, but under certain circumstances and relationships, so as to allow—or, perhaps, rather to suggest—the idea of Unity. This idea of Unity, or oneness, which is the subordination of all the parts to the completeness of the whole is . . . essential in a work of pictorial art. We find it, like a cord, running through and tying together all nature; and it seems to have been a divine idea in the creation of all things, binding organic forms, from the highest to the lowest, into one complete cycle.[75]

As suggested even in art instruction, then, the Victorian era was a period of a complete reconsideration of the categories of matter and form, body and mind, and body and spirit. In the wake of German philosopher Immanuel Kant, British philosophers, physiologists, and psychologists shared a common preoccupation with solving the age-old debate over whether there was a world apart from the mind's perceptions. Materialists insisted that things existed regardless of the presence of human beings to perceive them, while idealists argued that there was no reality outside of human perception. Huxley neatly formulated one solution relevant to both artists and scientists: "In itself it is of little moment whether we express the phenomena of matter in terms of spirit, or the phenomena of spirit in terms of matter; matter may be regarded as a form of thought; thought may be regarded as a property of matter."[76] Art critic Philip Gilbert Hamerton could therefore similarly argue that "artistic aesthesis is rooted in physical sensation."[77] The materiality of artworks themselves was therefore newly relevant, Hamerton observed: "In the Graphic Arts you cannot get rid of matter. Every drawing is *in* a substance and on a substance. Every substance used in drawing has its own special and peculiar relations both to nature and to the human mind."[78]

The place of the material body in the nineteenth century, however, was famously vexed. Dawson's work on the "Fleshly School" of poetry, a moniker coined in 1871 by a critic hostile to the sensual verse of Dante Gabriel Rossetti, Algernon Charles Swinburne, and others, demonstrates the extent to which the experience of the senses might be read through an ethical lens that could equate materialism with atheism.[79] Larson astutely points out connections between Rossetti's first "fleshly painting" of 1859 and Darwin's *On the Origin of Species*, noting that the so-called New Painting, or aestheticism, with its disinterest in moral truth twinned with an engagement in material pleasures, took on the same position in its field as materialist, apparently godless evolutionists marked out in theirs. Elsewhere in her work on the Victorian understanding of the body's relationship with its environment, Larson draws on German philosopher Robert Vischer's idea of empathy relating to vibrations of the nerves caused by our sensory engagement with objects.[80] For decades now, readings of the novel, from D. A. Miller's *The Novel and the Police* (1988) to Nicholas Dames's *Physiology of the Novel* (2007), have likewise been based on understanding the way in which the sensations produced by certain types of prose engage the physiology of the reader.[81]

This type of interpretive gambit, based on interlaced discourses of humanities and science, seems uniquely suited to illuminating this period of radical change. Literary critic Katherine Boehm notes how industrialization, paired with discoveries in physiology, encouraged a collapse of subject/object relationships, in which "subjects and objects constitute one another in a relationship characterized by 'reversibility' and 'intertwining'"; she cites Maurice Merleau-Ponty's contention that the body "is double-belongingness to the order of the 'object' and to the order of the 'subject.'"[82] The subject was reconfigured, becoming so interconnected with its environment that boundaries of selfhood could break down. Similarly, just at this time, the turn toward "Brunoism" in medicine, a practice named for the Scottish physician John Brown, insisted that "the body was the site of active interchange of forces with the external world."[83] Comprehending the body's shifting definitions and boundaries in the nineteenth century is instrumental in shaping theories regarding the intersection of representation and knowledge.

Methods for Histories of Art and Science

Until recently the interpretation of Victorian culture in light of the science of the period has largely emerged from the minds of literary critics and historians of science; this volume, in contrast and in complement, makes a case for its distinctive foregrounding of the practices, tools and contributions of art history. Models abound for the challenging project of interdisciplinary treatment of humanities fields and science, and this book is therefore in a privileged position in the wake of exceptional and wide-ranging academic work.

A major focus for this blended field has been the way that art was transformed by scientific shifts. Martin Kemp's early work in this area, *The Science of Art* (1990), concentrated on finding evidence that artists were drawing on scientific practices or knowledges. Kemp, like others, used the methodological approach of intellectual history largely adopted by historians of science by tracing personal networks, libraries, accounts of processes or intentions, and communities of practice and reception.[84] Scholars of this bent often attempt to establish the monodirectional impact of one discipline or set of ideas on another; for example, Phillip Prodger argues that Darwin "forever changed the way pictures are seen and made," extending a "decisive influence on the history of art."[85] Darwinian theory in particular has indeed proved a fertile area to consider regarding its consequences on visual culture, and in the last decade, several exhibitions and their catalogues have demonstrated a range of ways artists deliberately drew on concepts such as the survival of the fittest in ways that allowed them to disrupt conventional knowledges.[86]

Many direct responses by artists to innovations in thought and materials are indeed visible in the historical record. After the discovery and isolation of magnesium came the development of efficient and effective flash photography, for instance.[87] Quick to take up new pigments produced by innovations and experiments in chemistry, Joseph Mallord William Turner and the Pre-Raphaelites altered the face of British art.[88] Some nineteenth-century artists believed that aesthetics were generated by natural laws; optical experiments and color theory produced the concept of complementary colors and also yielded new meanings for pigments, which could now convey emotional or symbolic qualities.[89] A scientific basis emerged for imaginative interpretations of the world and of art, such as the anthropomorphizing ideas of the poet, novelist, and natural historian Johann Wolfgang von Goethe, who argued that blue was cold and male, yellow warm and female.[90]

More recently, cultural historians have moved away from the practice of uncovering precise proof regarding the influence of science on art to draw instead on models such as intersection or association. In his study of the Pre-Raphaelites, for instance, John Holmes is not concerned with retrieving historical records of the exact knowledge possessed by these artists; instead, as he notes, "we don't know what they knew but we can be sure of the significance and visibility of science within the wider culture of the 1840s and 50s."[91] He finds both direct and indirect evidence for the Pre-Raphaelites' incorporation of empirical discoveries, and, in turn, argues that "Pre-Raphaelitism came to be the visual language of science" via the architecture of the University Museum at Oxford and the Natural History Museum in London.[92] In another relevant volume, *Nature's Truth: Photography, Painting, and Science in Victorian Britain*, Anne Helmreich asserts that at the core of her project is the belief that "artworks can be persuasively *associated with* changes in nineteenth-century paradigms."[93] This more capacious model allows for art historians to contemplate objects rather than biographies, meaning rather than means of production. Matthew C. Hunter's *Wicked Intelligence: Visual Art and the Science of Experiment in Restoration London* offers an especially sophisticated model for parsing the interrelationships of emergent "experimentalist" and artistic communities in early modern

Britain, reading the ways in which the "didactic dimensions" of texts like Robert Hooke's *Micrographia* "are embedded in form and content alike." Hunter examines "how scientific images turn out to be scientific objects and vice versa." Rather than mere "illustrations or models," diagrams, drawings, and other images are "the residue of actual thought processes" that "reveal what it was to think through scientific questions because it was itself part of the thought."[94] From an art historian's perspective, likewise, we can see how certain paintings are only possible in the wake of altered scientific paradigms. Martin Meisel and Michel Serres, for instance, have both considered Turner's amorphous energetic vortexes of paint in relation to how the "science of heat based on the study of energy and entropy led to new channels of thought and feeling."[95]

One reason for the relative paucity of any serious analysis of scientific imagery is that science itself has only recently embraced the image, having suffered both from a fundamental distrust of the reliability of the visual and, paradoxically, from an overinvestment in its transparency.[96] Citing Steve Woolgar's *Science: The Very Idea*, visual sociologist Luc Pauwels notes that the act of representation itself is linked to the idea that "there are, or were, discrete objects 'out there' that exist independently of our perception . . . [they are then] reflected *truthfully* through various forms of representational devices."[97] The belief that the eyeball simply recorded what it saw, Blum relates, allowed for the "confidence in the possibility of unmediated representation."[98] But, as Pauwels continues, this is never possible: "There is no state in which things are perceived in an unbiased form."[99] Scientific imagery is therefore never about reproducing reality.[100] Similarly, Bruno Latour observes that for many, illustrations in a scientific text "are the world itself," so that "to call them image, inscription, representation, to have them exposed in an exhibition" is to reveal them as fabricated.[101] Visual culturists like Pauwels remind us that the "issue of *representation* touches upon the very essence of all scientific activity": "What is known and passed on as science is the result of a series of representational practices."[102]

While historians of science originally shared their field of study's initial disdain for images, neglecting to attend to them with much sensitivity, this is no longer the case.[103] Alex Pang helpfully summarized the state of this field in 1997, citing Bernard Smith's *Imagining the Pacific: European Vision and the South Pacific* (1959) as an exemplary account of how the scientific renderings of the artists who accompanied explorer James Cook dramatized connections between the climate and the ecology.[104] As noted, the foundational work of Daston and Galison established the importance of image-producing recording devices in the march toward ostensibly "objective" experiments.[105] Pang's project, however, is ultimately concerned with contributing to and explaining the history of science, producing a "more detailed and complete picture of how scientific practices and ideas work," rather than interpreting art.[106]

It is crucial, then, to bring art historians' methods to the field of the history of science, considering the medium and the formal elements of scientific imagery. Ann Shteir and Bernard Lightman engage in some of the most comprehensive thinking about the intersections between history of science and history of art in their introduction to *Figuring It Out: Science, Gender and Visual Culture.* They ask, "How can a study of visual features help us to understand better the subtle and complex ways that the modern Western culture of science developed?" while pointing out the pitfalls of too blithe an approach of this kind, noting that "using images to make explicit the often hidden cultural connotations in the work of science is inevitably problematic."[107] Lightman and Shteir observe that the meaning of any given picture is unstable and must be understood in light of the history of its chain of quotations and references, urging the necessity of contextualizing an image in its means and moment of production.[108] Lightman, for instance, notes how the visual and material properties of colorfully illustrated popular science books, particularly those by women, contributed to their particular form of natural history.[109] The goal of scrutinizing the "ornamental, rhetorical, and authoritative use of visual images" is shared by the authors of this volume.[110]

Other writers, including several assembled here, are engaged in considering more general metaphorical or epistemological links between the fields. It is therefore as useful to question the ways in which literary metaphors or artistic forms might have formed scientific thought as it is to discern how scientific concepts produced new literary tropes.[111] Concepts like "transmission," for instance, applied both to social ideas and to natural forces, and the electrical technologies involved in new forms of transmission gave rise to literary metaphors of influence and creativity.[112] Catalyzed by Gillian Beer's pioneering *Darwin's Plots* (1983), science writing has been dissected as literature, often due to its authors' saturation in aesthetic theory—in the case of Darwin, his absorption of the categories of the sublime, the picturesque, and the beautiful was generative for the models he produced, while the variety and changeableness characteristic of the picturesque can be related to his interest in species differentiation.[113]

Voss suggests that Darwin's own drawings—as well as the images that appeared in his publications—made meaning in specific ways related to their visual forms. Observing that scientists in search of methods by which to portray processes that occurred too slowly for measurement developed new kinds of representations, such as the geological cross section or the series suggestive of embryonic development, Voss finds that geologists' "translation of form and time into a symbolic system of rows, lines, angles, or points opened up space for Darwin to conceptualize his theory of evolution."[114] Blum underscores the fact that scientific illustrations are not only about their depicted subjects but also about the mechanisms by which those subjects were represented on the page.[115]

Art history, with its affinity for form and style, as well as for history, is clearly ideally positioned to contribute to this conversation. While formalism in literary studies has returned to the fold with productive results, it has never left the art historian's tool box.[116] Just as literary historians have produced illuminating new work by evaluating the ways in which Victorian texts drew on the language and forms of science, this collection calls on art historians to probe their subjects in this light. The creative, thoughtful essays here assembled remind us that science and art shared forms, which we see manifest as waves, layers, lines, or geometries; and of course, that both fields invest in the idea of the evolution of form as well as generate surprisingly kindred responses: pain, pleasure, empathy, sympathy, to name just a few. These writers are alert to the work that artistic media themselves do, as for instance in Slipp's analysis of the uncertainties "regarding the purposes and technological limitations of photography" or Larson's interest in the facture of Rossetti's paintings. Moreover, they also take into account the material and experiential contributions of the technologies and empirical disciplines they invoke. This collection, in short, rewrites and regenerates the histories of both science and art.

Art history is clearly a productive partner in this investigation of nineteenth-century science, although its proffered assistance has been often oddly overlooked even in works purporting to be about visual culture. Jonathan Smith rightly admonishes "sociologists and historians" for treating images as if they illustrate an a priori event, text, or object, and, on the other hand, "literary and cultural historians" for ignoring the original contexts of scientific illustrations.[117] Yet, strikingly, he leaves out art historians altogether. Too often images are seen as passive mirrors of preexisting ideas by historians of science.[118]

As Diana Donald reminds us, however, images are in reality "constitutive of meaning, giving form to concepts, perceptions, reflections and emotions that could not be expressed in words."[119] As Barri Gold puts it, "Art, literature and science work *together* to form and reform how we understand the world."[120] Art is active, "working" through its mediations of the world, not merely passively reflecting a truth or reality "out there."

Chapter Synopsis

Crucial to the writers in this collection, then, is clarity regarding interpretive processes, hermeneutic armatures, and methodological goals. These are not essays in the history of science but rather in the history of art; their focus is to illuminate conditions and configurations of knowledge shared by science and art in the nineteenth century. Treating British and American visual culture as they relate in terms of common cultures of science and imagery, the volume contributes to an excitingly expanding way of seeing. The authors assembled here "investigate the dialogical relationship between artists and scientists,"[121] ways in which artists may have been "attuned," to use Alison Syme's word, to the science of the period. Syme examines "tropes that cut across the different contexts and representational regimes with which the illustration engages" with a focus on "the figure of metamorphosis in particular—transformation or translation from one state to another, material, linguistic, or otherwise." George Levine characterizes this practice as an interest in discerning distinctive cultural narratives, a seeking of "how the culture tells stories, that is, imagines its life, subtly informs the way science asks questions, arrives at the theories that reshape the culture that informed them."[122]

In some instances, artworks are interpreted via popular scientific theories (such as Darwinian evolution and sexual selection); other authors investigate how images functioned when read in light of broader contexts of science. Linking the different chapters is a shared attentiveness to the ways in which the visual produced new notions about the place of the human in the natural world during a period of burgeoning theories and discoveries that challenged longstanding boundaries between animal, plant, and human. Artists and scientists studied here both struggled to come to terms with the issues of imperial modernity, in a world in which dawning recognition of entropy and climate change suggested a potential dissolution of the man-centered world order—and even of civilization altogether, as Syme observes.

In chapter 1, "Measuring Native America: Early American Archaeology and the Politics of Time," Rachael DeLue explores the geological underground as a subject for artists and other image makers in the nineteenth century in the United States by exploring how these practitioners confronted the paradox of picturing what exists beneath the earth's visible surface. DeLue illuminates the intellectual and historical stakes of envisioning the subterranean, here understood broadly as a location or space as well as an object, concept, phenomenon, or metaphor, and elucidates a sustained attempt in the nineteenth century to compel images to take measure of what lies beneath. The intertwined fields of archaeology, paleontology, and geology, she contends, produced new temporal configurations that worked to write Native Americans out of history.

Continuing to mine the geological vein in chapter 2, "'All That Is Solid Melts Into Air': Burne-Jones, Glaciation, and the Matter of History," Syme argues that commercial and glacial allusions in art offer a glimpse of the very material forces that have shaped our natural and cultural histories in a way that resonantly suggests our evaporating futures. In this chapter, Syme contextualizes one of Edward Coley Burne-Jones's illustrations for the Kelmscott Chaucer in terms of nineteenth-century writing on and imaging of glaciers by Ruskin and others, and its depiction of a glassy mountain of inscribed ice in terms of Victorian consumer culture, glassworlds, and celebrity. Blurring the line between nature and culture, the illustration suggests the erosions and reshapings to which human history—as exemplified, made, and marked by its famous men—is subject. Syme contends that the evocative conflation of the geological and historical records in the illustration encapsulates Burne-Jones's concerns about the erosion of history, substance, and meaning in modernity: it pictures the transience of all human words and deeds, the material substrate and visual history of which seemed to be fading at an accelerating rate in the artist's commodified Victorian modernity.

Moving to the surface of the earth, Carey Gibbons considers the process of forecasting the weather during the Victorian period, in chapter 3, "Grasping the Elusive: Victorian Weather Forecasting and Arthur Hughes's Illustrations for George MacDonald's *At the Back of the North Wind*," revealing a similarity between the thought processes and representational techniques developed within the field of meteorology and those employed by Hughes in his illustrations for MacDonald's 1871 novel. The wave expresses a tension between the desire to grasp multiple conditions concurrently and the realization that mental mastery and control are continually challenged by unpredictable, inexplicable forces.

In chapter 4, "A Haunting Picture, in Light of Victorian Science: John Everett Millais's *Speak! Speak!*" a chapter bridging the volume's interests in the physical and biological sciences, I analyze a painting in light of how the Victorian understanding of candle flame related to debates about materialism, human perception, and the nature of reality. Millais's *Speak! Speak!* (1895) takes up these themes in ways that were also shared by contemporary debates in the intertwined fields of chemistry and psychology. I argue that this painting deploys its forms, particularly its light effects, to represent a complex meditation on life, death—including the artist's own—and art. *Speak! Speak!* is, then, a painting alight with fire's metamorphic potentiality.

Shifting to medicoscientific representations of the human body, in chapter 5, "Photographing Ether, Documenting Pain: Representing the Chemical Invisible in the Daguerreotypes of Southworth & Hawes," Naomi Slipp discusses how the production of surgical daguerreotypes linked medicine with technological and chemical progressiveness, relating the presumed documentary effects of the photographic medium to medical objectivity early in the professionalization of both fields. These daguerreotypes have typically been viewed as transparent records of early surgery, but Slipp's study repositions them within the history of medical professionalization, period theories on corporeal pain, and a comparison of the perceived "magical" effects of ether and daguerreotypy. Slipp further reveals the professional agendas of doctors and photographers who consciously used the visual arts to advertise medical advancements that were, ironically, difficult to communicate visually. Slipp underscores the relationship between photography and power, following scholars such as Daniel Fox and Christopher Lawrence, whose 1988 *Photographing Medicine* proposed that "photos have a history and pictures like this have been among the means by which medical power has been legitimated and extended."[123]

Moving from the image of the actual, if anaesthetized, body to the ideal body of anatomical treatises, Keren Rosa Hammerschlag explores the complex and contested relationship between anatomy and race in the developing discourses of evolution in mid-nineteenth-century Britain and America in chapter 6, "Drawing Racial Comparisons in Nineteenth-Century Surgical Anatomies." The visually arresting mid-nineteenth-century surgical anatomy by Joseph Maclise (surgeon and brother of the successful Royal Academy artist Daniel Maclise) included an illustration of a black man who is rendered white in the American edition of the same atlas. Hammerschlag inquires what this "whitewashing" means for the histories of medicine, art, and the representation of race.

Taking up the discussion of evolution and art in chapter 7, "The Post-Darwinian Eye, Physiological Aesthetics, and the Early Years of Aestheticism, 1860–1876," Barbara Larson examines ways in which the development of psychophysiology and evolutionism in Great Britain of the 1860s and early 1870s helped produce the theory connected with the art movement aestheticism. Both Darwin and psychophysiologists emphasized the significance of a coordinated sensorium that responded to the environment in a holistic manner. Larson ties together two usually unrelated phenomena, the rise of physiological theory and the transition in art practice away from a focus on detail toward an interest in obscurity, the atmospheric, or materiality.

Finally, functioning as a case study of some of Larson's themes, Caitlin Silberman's chapter, "Darwinian Aesthetics and Aestheticism in James McNeill Whistler's Peacock Room," investigates the web of connections between the decorative program of James McNeill Whistler's celebrated Peacock Room (1876–1877) and discourses of evolution and aestheticism. In filling the room with anthropomorphic peacocks, Whistler employed two ostensibly distinct but overlapping ways of thinking through the origins and instrumental value of humans' sense of beauty: the Darwinian approach and the aesthetic mode. In its claim that beauty needed no moral purpose, the threat of Whistler's aestheticism differed little from Darwin's materialism.

Together our authors illuminate the world in which Hawkins was able, in his purposefully titled *The Science of Drawing Simplified* (1843), to argue for the importance of artistic training for the scientist: "To the naturalist . . . the study of form and facility in drawing are indispensable: the anatomist, botanist, ornithologist, entomologist, and geologist observe generic distinctions so minute as to escape the vulgar eye; and to delineate these accurately, requires not only correctness and dexterity but scientific knowledge."[124] Hawkins also reasoned, conversely, that the empirical observation skills taught by drawing produced moral benefits: "By quickening the child's perception of the external characteristics of objects, it creates a lively interest in them. . . . It promotes habits of consideration both for living creatures and inanimate objects, and of care in handling them; both which are influential in preventing the practice of cruelty and the tendency to destroy; a knowledge of the structures of animals induces humane feeling, and promotes reflection; for the child will not injure what it has so much pleasure in observing."[125] We might well take a lesson from Hawkins in his equation of looking closely with discovering in ourselves care and empathy for the observed object. In light of centuries of environmental havoc wrought by the innovations of human civilization, we are in need of just such nurturing ways of seeing our place in the world in relation to other forms of life. It is to be hoped that art, science, and the sort of visual historical consciousness found within this volume will help us do just this.

CHAPTER 1

MEASURING NATIVE AMERICA

EARLY AMERICAN ARCHAEOLOGY AND THE POLITICS OF TIME

Rachael Z. DeLue

The Copernican Revolution of the sixteenth century displaced the earth from the center of the universe, and Darwin's theory of evolution remade the human as just another animal among countless other beasts. In analogous fashion, geology in the eighteenth and nineteenth centuries radically expanded the timescale of the earth while drastically reducing humanity's share in its chronology.[1] Geology's scrutiny of the earth's crust also revealed ancient civilizations and the fossilized records of extinct organic life, fueling the disciplines of archaeology and paleontology and producing a revolution in knowledge about the relationship between the present and the past, including the status of modern humans in relation to earlier human cultures and to the animal kingdom.

Revolutions in thought inevitably produce problems for representation. In Europe and North America from the sixteenth century forward, visual images played a substantial, often essential role in scholarly inquiry and the production and dissemination of new knowledge.[2] In the nineteenth century, geologists on both sides of the Atlantic, confronted with radically new theories about the earth's history and temporality, found existing conventions and strategies of visualization severely lacking. Established models of scientific illustration, predicated on observation and mimesis—as with a botanical illustration or a drawing of a dissected corpse—could not manage the task of delineating what the new science described: abstract, extraempirical concepts and unobservable phenomena. Consequently, scientists and image makers sought new modes of picturing, seeking to visualize within the limits of a static, two-dimensional format the animate, ever-changing, and newly immeasurable ancient terrain of the earth's crust and its interior depths.

The development of new image types in the sciences went hand in hand with the formulation of new knowledge, the one catalyzing the other. As did their colleagues in geology, paleontologists and archaeologists contended with the problem of picturing the distant past, not least because relatively little of that past, including its life forms, remained to be observed. More problematic, however, was the fact that the temporal and biological relationship of past life, human and otherwise, to organic life in the present had yet to be fully fathomed. Linnaean taxonomic systems accounted for morphological and physiological relationships among living species, but the knowledge and the technology for determining how one species might relate to another *across time* were still developing in the period. Moreover, time itself would not submit to measure. The inability to establish anything but a relative rather than absolute timeline for existence—to assign an actual *date* to a rock layer or a fossil specimen—made the attempt to cognize and render the flow of time in relation to the present all the more vexing.[3]

In the case of archaeology in the United States, such calculations came equipped with immensely high social and political stakes, as exemplified by a particularly striking episode in the early history of American archaeology to which the remainder of this chapter is devoted. The most visible product of this episode, *Panorama of the Monumental Grandeur of the Mississippi Valley*, a moving panorama commissioned by the physician and amateur archaeologist Montroville Wilson Dickeson to entertain and edify the American public, embodies the difficulties and desires that attended any attempt to delineate the past and measure time in light of the modern human. Examining and analyzing the panorama's particular visual and narrative strategies, which drew equally on scientific and popular contexts and colluded in the formulation of knowledge about America's past and present, elucidates in especial fashion the intellectual and political formations that shaped the actions of American institutions and individuals at midcentury while illuminating the active role played by visual culture therein. As an image-*cum*-apparatus, the moving panorama shared with the visual culture of geology the attempt to render multiple temporalities, including deep time, within a single, always present-tense pictorial plane. Unlike geology, which persistently endeavored to understand all things and all histories as existing as part of a single and comprehensive timeline, Dickeson's panorama, for reasons equally scientific and political, postulated multiple and contradictory chronologies for the span between then and now.

The first major landmark of archaeology's eventual professionalization in the United States in the late nineteenth century was the establishment in 1812 of the American Antiquarian Society. That name of course connotes an interest in the distant past, and the moniker also served to align American scholars with their counterparts in Europe. The society's founders took as their model the Society of Antiquaries in England as well as other such organizations in Europe and the United Kingdom.[4] The very existence of such a society presumed an *American* antiquity on par with the ancient civilizations of Europe, including ancient Greece and Rome. The rediscovery of classical antiquity in the eighteenth century, including the celebrated excavation of the Roman towns of Pompeii and Herculaneum, led to the rise of archaeology in nineteenth-century Europe as a full-fledged scholarly and professional discipline, one that functioned as a national and imperial endeavor through the amassing of national collections, often by way of conquest and plunder.[5] Those individuals in the United States eager to dig up America's ancient past could only follow so far in the footsteps of their European colleagues, in part because they lacked the training and institutional support to do so, but also, of course, because the so-called New World possessed no classical Greek or Roman ruins. As the Reverend Abiel Holmes stated in an address to the American Antiquarian Society in 1814, on the occasion of its second anniversary, "Our antiquities . . . do not extend back beyond the distance of two centuries. . . . In vain, then, do we make pretensions to an ancient origin," thus associating "antiquity" with Anglo-European history and white settlement of the North American continent in the early seventeenth century.[6] Yet members of America's intellectual and political elite, driven by scientific rivalry and an aggressive nationalism, desired an American antiquity to rival Europe's ancient past.

The study of America's past, including its visual and material culture, would highlight America's many noble accomplishments as well as corroborate the nation's exceptional promise. "Who is not gratified," asked Holmes, "in tracing our salutary institutions to their origin, and in discovering the causes of our rapid progress in population, wealth and refinement, of our freedom and independence? Who can behold the portraits of the first settlers of New England, without mingled delight and admiration? . . . Who can step on that rock, on which the pilgrim fathers from Leyden first stepped, or even survey its fragments without grateful emotion?"[7] Scientific inquiry abetted this nationalist project. Geology became such a celebrated and popular science in nineteenth-century America in part because its revelations made the New World as old as deep time, enabling the United States to claim a storied history, albeit a natural rather than a human one.[8] Archaeology took a somewhat different turn: again, because it had to, but also because of its fundamental connection in the United States to the discipline of ethnology and to the project of territorial expansion, and because what lay buried beneath the ground presented myriad problems of identity, visibility, and time for science and for the American nation-state.

Few primary visual documents illustrate this better than the *Panorama of the Monumental Grandeur of the Mississippi Valley*, created around 1850 by John J. Egan at the behest of Dickeson in the thick of territorial reconnaissance and the interlinked professionalization of archaeology and ethnology, which in the United States served as intellectual arms of westward expansion.[9] Based in Philadelphia, Dickeson left his medical practice in 1837 to head west to study Native American earthworks in the Mississippi and Ohio River Valleys, becoming one of the first European Americans to excavate and document a large number of Native American burial mounds.[10] Dickeson's fieldwork belongs to the preprofessional phase of archaeology in the United States, which originated in the late eighteenth century in the form of curiosity about Native American earthworks—mounds in particular—and professionalized largely *as* the study of Native America. Dickeson was thus equal parts archaeologist and antiquarian, curious about America's past and bent on amassing as many artifacts for his collection as he could. His travels overlapped with the rapid expansion of both ethnology and archaeology in the 1840s, signaled by the foundation of the American Ethnological Society in 1842, which Dickeson joined in 1847, and abetted by numerous expeditions launched to explore the territory west of the Mississippi River.

Dickeson's own findings played no small role in stimulating interest in indigenous peoples and North America's ancient history.[11] He spent nearly a decade in the field, returning to Philadelphia with a sizable collection of artifacts, including human and animal bones, pottery, arrow points, stone tools, metal ornaments, beads, rock and mineral specimens, and fossils, many of which he deposited at the Academy of the Natural Sciences in Philadelphia.[12] Dickeson published a series of detailed accounts of his travels, collectively titled "American Antiquities," in the Philadelphia journal the *Lotus*. He also delivered lectures on "Indian Antiquities," illustrated with objects from his extensive collection, and staged several exhibitions, including an "Indian Cabinet" that opened at Philadelphia's City Museum in 1854. In 1876, visitors to the main hall at Philadelphia's Centennial Exhibition could view a selection of Dickeson's artifacts, which were displayed again in 1885 in Fairmount Park's Memorial Hall.[13]

Many others preceded Dickeson in heading west to view Native American earthworks, prompted, as was he, by increasing fascination with America's so-called prehistory and what scholars have dubbed "mound mania."[14] Observers in the eighteenth century and earlier had reported the existence of monumental earthworks in the Southeast and the Mississippi and Ohio River Valleys, part of a much larger network of indigenously produced and enormously varied earthen structures throughout the Americas, including structures that continue to play a vital role in Native American culture today. Largely dating between 1000 BCE and 1700 CE, these earthworks represent what specialists have described as two main traditions, the Woodland and Mississippian, each of which cuts across different Native American social groups. Construction decreased with the arrival of European explorers and colonization.[15]

Nonindigenous period observers from the late eighteenth century onward almost always commented on the remarkable character of these earthworks, and most accounts noted the mystery of their existence as well as their probable ancient origins. Such fascination with potentially ancient American monuments spurred tourism as well as scientific study—or what counted as scientific study at the time. From the point of view of indigenous cultures then and now, archaeology functioned as a glorified version of grave robbing, a violent act of desecration rather than enlightenment. Feeding the mania was the alluring hypothesis that the mounds were built by a lost, possibly non–Native American race, one far more ancient and advanced than living Native American peoples, and thus more in line with the citizens of ancient Greece and Rome, or even the denizens of biblical history. It was proposed, for example, that the ancient builders of the mounds were Israelites, Mesoamericans, Vikings, or Danes.[16] As one observer wrote in 1791, "Who those inhabitants were, who have left such traces; from whence they came, and where they are now; are queries to which we never, perhaps, can find any other than conjectural answers." Another, writing in 1794, similarly characterized the "ancient fortifications and stupendous mounds of earth" as "vestiges of immense population and infinitely greater share of science than is possessed by the present 'red people.' Who then were the authors, is the question that is proposed—and *when*, and *where* did *they* migrate."[17]

This majority view persisted into the 1840s. The first major publication on the Mississippi River Valley mounds, Ephraim George Squier and Edwin Hamilton Davis's *Ancient Monuments of the Mississippi Valley* (1848), postulated that the builders of the earthworks were an extinct race, likely Mesoamerican or Peruvian in origin, a view reflecting what by midcentury had become conventional wisdom. Such hypotheses granted America an ancient, storied past while also advancing the white supremacist racial theories that buttressed the political agenda of the settler colonial state, including territorial expansion and the removal or extermination of contemporaneous indigenous populations.[18]

Dickeson's thinking fell right in line. Recalling in the *Lotus* his excavation of the earthworks on the plantation of a Major General Quitman,

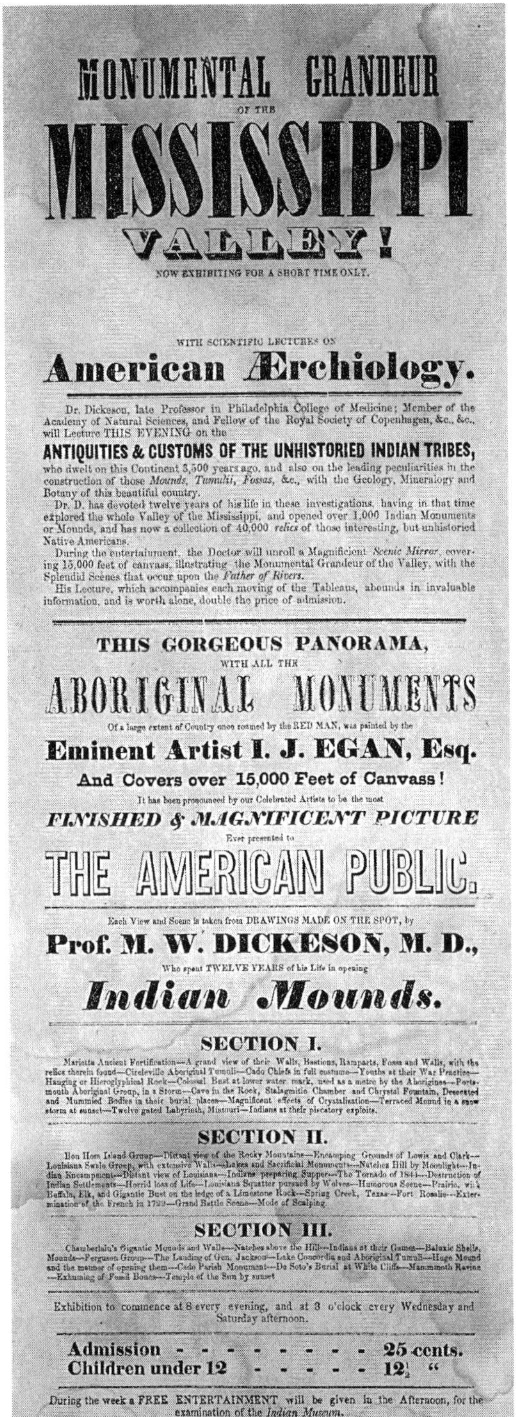

MONUMENTAL GRANDEUR
OF THE
MISSISSIPPI
VALLEY!
NOW EXHIBITING FOR A SHORT TIME ONLY.

WITH SCIENTIFIC LECTURES ON
American Ærchiology.

Dr. Dickeson, late Professor in Philadelphia College of Medicine; Member of the Academy of Natural Sciences, and Fellow of the Royal Society of Copenhagen, &c., &c., will Lecture THIS EVENING on the

ANTIQUITIES & CUSTOMS OF THE UNHISTORIED INDIAN TRIBES,

who dwelt on this Continent 3,500 years ago, and also on the leading peculiarities in the construction of those *Mounds, Tumulii, Fossas,* &c., with the Geology, Mineralogy and Botany of this beautiful country.

Dr. D. has devoted twelve years of his life in these investigations, having in that time explored the whole Valley of the Mississippi, and opened over 1,000 Indian Monuments or Mounds, and has now a collection of 40,000 *relics* of those interesting, but unhistoried Native Americans.

During the entertainment, the Doctor will unroll a Magnificient *Scenic Mirror*, covering 15,000 feet of canvass, illustrating the Monumental Grandeur of the Valley, with the Splendid Scenes that occur upon the *Father of Rivers.*

His Lecture, which accompanies each moving of the Tableaus, abounds in invaluable information, and is worth alone, double the price of admission.

THIS GORGEOUS PANORAMA,
WITH ALL THE
ABORIGINAL MONUMENTS
Of a large extent of Country once roamed by the RED MAN, was painted by the
Eminent Artist I. J. EGAN, Esq.
And Covers over 15,000 Feet of Canvass!
It has been pronounced by our Celebrated Artists to be the most
FINISHED & MAGNIFICENT PICTURE
Ever presented to
THE AMERICAN PUBLIC.

Each View and Scene is taken from DRAWINGS MADE ON THE SPOT, by
Prof. M. W. DICKESON, M. D.,
Who spent TWELVE YEARS of his Life in opening
Indian Mounds.

SECTION I.

Marietta Ancient Fortification—A grand view of their Walls, Bastions, Ramparts, Fossa and Walls, with the relics therein found—Circleville Aboriginal Tumuli—Cado Chiefs in full costume—Youths at their War Practice—Hanging or Hieroglyphical Rock—Colossal Bust at lower water mark, used as a metre by the Aborigines—Portsmouth Aboriginal Group, in a Storm—Cave in the Rock, Stalagmitic Chamber and Chrystal Fountain, Desecated and Mummied Bodies in their burial places—Magnificent effects of Crystalization—Terraced Mound in a snow storm at sunset—Twelve gated Labyrinth, Missouri—Indians at their piscatory exploits.

SECTION II.

Bon Hom Island Group—Distant view of the Rocky Mountains—Encamping Grounds of Lewis and Clark—Louisiana Swale Group, with extensive Walls—Lakes and Sacrificial Monuments—Natchez Hill by Moonlight—Indian Encampment—Distant view of Louisiana—Indians preparing Supper—The Tornado of 1844—Destruction of Indian Settlements—Horrid loss of Life—Louisiana Squatter pursued by Wolves—Humorous Scene—Prairie, with Buffalo, Elk, and Gigantic Bust on the ledge of a Limestone Rock—Spring Creek, Texas—Fort Rosalie—Extermination of the French in 1729—Grand Battle Scene—Mode of Scalping.

SECTION III.

Chamberlain's Gigantic Mounds and Walls—Natchez above the Hill—Indians at their Games—Baluxie Shells, Mounds—Ferguson Group—The Landing of Gen. Jackson—Lake Concordia and Aboriginal Tumuli—Huge Mound and the manner of opening them—Cado Parish Monument—De Soto's Burial at White Cliffs—Mammoth Ravine—Exhuming of Fossil Bones—Temple of the Sun by sunset

Exhibition to commence at 8 every evening, and at 3 o'clock every Wednesday and Saturday afternoon.

Admission - - - - - - - -	25 cents.
Children under 12 - - - - -	12½ "

During the week a FREE ENTERTAINMENT will be given in the Afternoon, for the examination of the *Indian Museum.*

PRINTED AT THE MERCURY OFFICE—NEWARK, N. J.

Fig. 1.1. Dickeson produced a broadside advertisement to entice the public. Montroville Wilson Dickeson, "Monumental Grandeur of the Mississippi River Valley! Now Exhibiting for a Short Time Only, With Scientific Lectures on American Aerchiology [*sic*]" (Newark, NJ: Mercury Office, ca. 1851), printed broadside, 6 × 7 1/2 in (15 × 19 cm). Call number E78. M75 M65e, Western Americana Collection, Rare Books and Special Collections, Firestone Library, Princeton University, Princeton, New Jersey.

near Natchez, Mississippi, Dickeson lamented the disappearance of the "by-gone people" who had once populated the area. "Where are now the thousands that once thronged these glorious heights," he wondered, "girded in their warlike armor, and who fought, even unto death, for the defence of these, their native hills? They have long since gone to the bar of God, where neither the victor's wreath nor the largest conquest of earth will avail them aught."[19] For Dickeson, it was the unknown, ancient origins of the mounds—"the mystery, the impenetrable mystery veiling these aged sepulchres, which gives them an interest for the traveller's eye"—along with their status as *the* object of inquiry and locus of debate for early American archaeology that motivated his expedition to the Mississippi Valley in the first place.[20] And it was his sense of the public's own fascination with America's earthworks that inspired him, on returning to Philadelphia, to commission John J. Egan to create a moving panorama, which Dickeson took on tour in the 1850s.

In commissioning Egan to create a moving panorama, Dickeson capitalized on the "panoramania" of the 1840s in the United States. Like many of the most popular moving panoramas in the period, Egan's *Monumental Grandeur of the Mississippi River Valley* featured the region of the Mississippi River, including locales Dickeson visited during his travels.[21] A broadside advertisement indicated that Egan used sketches made by Dickeson "on the spot" to ensure accuracy, and promised, in the inflated rhetoric typical of the industry, "over 15,000 feet of canvas" (fig. 1.1).[22] In actuality, the painted canvas, now in the collection of the Saint Louis Art Museum, measured around 350 feet long, and featured over two dozen scenes that unspooled one by one before the audience. They depicted in no particular order a variety of subjects and periods: hillsides and waterways at sunset or in snowstorms, steamboats alighting at riverside towns, crystal-lined caves, a devastating tornado, the exhumation of two mastodon skeletons, and a humorous vignette depicting a man chased by a wolf, as well as a view of the Rocky Mountains, scenes of Native American life, the burial of Hernando de Soto, the camping grounds of Lewis and Clark, and scenes from the War of 1812 and other battles, including the 1729 revolt of the Natchez people against the French at Fort Rosalie, the center of France's colonial operations in the region. As one would expect, Native American earthworks featured prominently throughout.

Dickeson billed the panorama as an education in the "antiquities and customs of the unhistoried Indian tribes," touting himself as an expert who had explored the "whole Valley of the Mississippi" while excavating "over 1000 Indian Monuments or Mounds."[23] Appropriately enough, one of the panorama's most elaborate scenes, labeled on the broadside "Huge Mound and the manner of opening them," features Dickeson's excavation

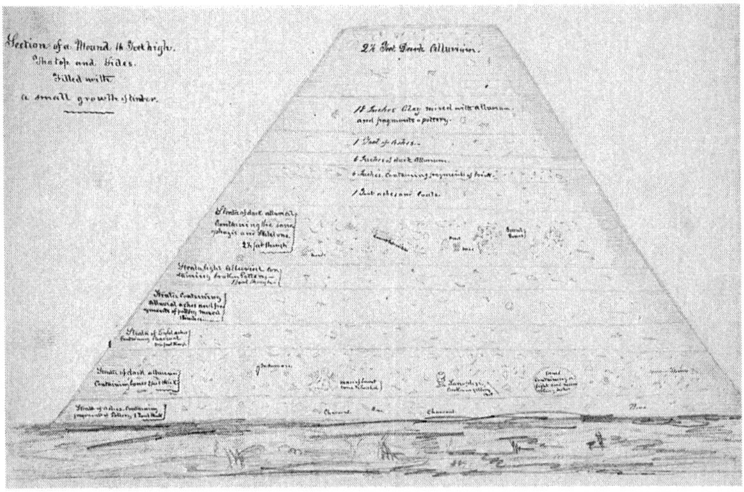

Fig. 1.2. Dickeson regularly made drawings in the field. Montroville Wilson Dickeson, *Cross-Section Drawing*, Montroville Wilson Dickeson Collection, University of Pennsylvania, Penn Museum Archives. Courtesy of the Penn Museum, image no. 229176.

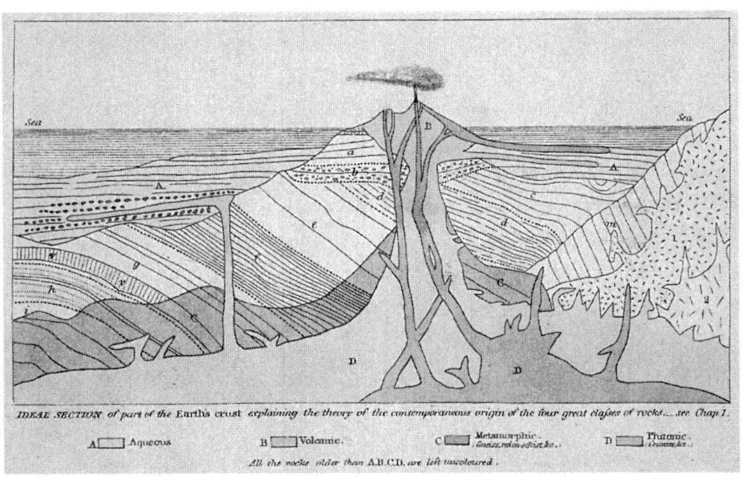

Fig. 1.3. Charles Lyell's illustration features a typical visualization of geological stratification. "Ideal section of part of the Earth's crust explaining the theory of the contemporaneous origin of the four great classes of rocks." Reprinted from Charles Lyell, *Elements of Geology*, 1st American ed., from the 1st London ed. (Philadelphia: James Kay, Jun. and Brother; Pittsburgh: C.H. Kay and Co., 1839), frontispiece. Photography: John J. Blazejewski.

in January 1844 of a Native American mound located on the Louisiana plantation of William Feriday, one of the largest plantations in Concordia Parish (plate 2).[24] As Dickeson noted in his published account of the excavation, he opened several mounds on Feriday's property. In some cases, he extracted bones and artifacts to add to his collection, and in others, he sketched the specimens in situ when they proved too fragile to remove.[25]

Egan's excavation scene matches closely Dickeson's detailed description of the third mound he excavated on Feriday's property, including the composition and color of the strata and the skeleton buried in a seated position toward the top, which Dickeson identified as "Choctaw" and reported finding adorned with beads and accompanied by several objects, including a vase and a hatchet.[26] In the image, enslaved Black men loaned to Dickeson by Feriday have removed approximately half of the total mass of the mound, exposing a cross section of horizontal layers containing human remains and pottery bisected at the top by a vertical burial shaft. At the base of the excavated mound, a pot decorated with rectangular motifs contains a stack of bones, while more bones lie scattered on the ground amid a variety of objects exhumed from the mound, including arrow points and stone tools. Several decorated pots and baskets sit atop piles of excavated earth. Two white men, one wearing a hat and the other holding a notebook, stand amid the enslaved laborers and survey the site, perhaps conversing about the progress of the excavation. Another similarly dressed figure toward the left, adjacent to a man wielding a pickaxe, makes a cross-sectional drawing of the excavation, a detail probably meant, along with the neat and regular cuts of earth, to suggest the systematic and expert nature of the undertaking. This figure represents Dickeson himself, at work in the field, and the drawing resembles one of Dickeson's own (fig. 1.2). Other figures, all of them white, wander about the landscape marveling at the mounds, their presence transforming the earthen structures into sights befitting a picturesque tour.[27] Egan also included in the scene a small Native American encampment, along the right edge, peopled by a pair of men conversing and two seated women tending to an infant.

Both the method of extraction featured in the scene and the manner of Egan's depiction of the mound as a stack of strata contained by a single, emphatic contour line underline its kinship with the banded layers of the earth's crust as identified and visualized through the physical and visual methods of geology, in particular stratigraphy, a staple of geology by the early nineteenth century for use in determining the order and relative position of rock layers in relation to the geological time scale. Stratigraphy, as a description of the distinctive layers that formed the separate performance, and concluded in a quartette, in the manner in which they kept time, was truly astonishing. We retired to the parlor, and enjoyed some delightful music, of a higher order, on the piano and flute. Mr. Feriday favored us with an exquisite performance on the eight key flute. In his early days, he was a pupil of the celebrated Nicholson. Whilst the family was retiring, Mr. Mitchell and company repaired once more to the margin of the lake, and opposite the house, blew several airs on the bugle and flute, which reverberated enchantingly upon the broad face of the water.

At eight o'clock, on the following morning, we mounted our fleet horses, and repaired to the contemplated groups of mounds. It was the rainy season, and the contiguous swamps were filled with water, which saturates the roads, and renders them almost impassible. At every step the horses would sink almost to their breast; and it appeared, to one unaccustomed to this swamp-riding, as if the next plunge would surely bring him down. To a northerner it is often extremely alarming; but, to the Mississippian or Louisianian it is unworthy of notice, and both male and female sit their horses with the greatest ease and elegance.

Nine o'clock brought us to the system of mounds. They were situated in an angle formed by the junction of Cocodria and Black Bayou. There are eight in number, and there is no very perceptible difference in their dimensions, except the one which faces the Cocodria. This measures as follows: Base, 52 feet, height, 11 feet; and it has seven angles. We divided it, by extending a ditch entirely through, giving the order of stratification. (See Fig. 1.)

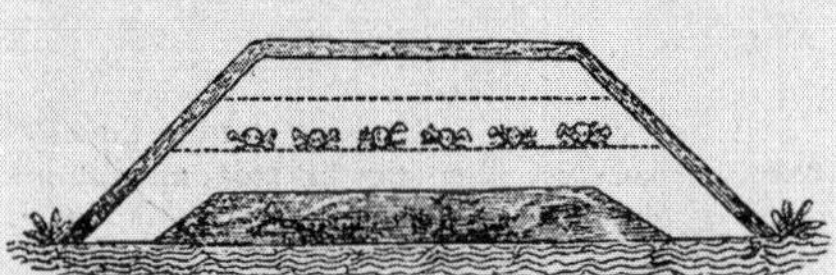

(Fig. 1.)

Commencing at its base, the first formation was composed of a conglomerate of ashes, coals, burnt bones, and fragments of bricks and pottery, two feet thick. This our force had great difficulty to penetrate. No particular form could be given to it. It emitted a caustic smell, not unlike that from ashes while being slacked. We found nothing of importance, save a few arrow points, of beautiful navaculite, and of elaborate workmanship. Several fragments of mica, well burnt, and two fragments of terra cotta pipes lay near the top surface of this stratum. Above this one, the formation was of a dark rich loam, filled with broken pottery, of an unusual fine texture, and in some instances painted. This stratum varied in thickness, from eighteen inches to two feet. The next stratum contained the skeletons, it was a light yellow loam, three feet through. The bones lay in small heaps; and apparently;

Fig. 1.4. Dickeson's sketch of the Feriday mound evokes the visual methods of geology. Montroville Wilson Dickeson, detail, "Fig. 1," in "Opening of Mounds upon the Plantation of William Feriday, Louisiana," *Lotus* 1, no. 10 (June 3, 1848): 146. Courtesy of the American Philosophical Society, Philadelphia.

earth's crust as well as a conceptual tool for aligning the physical and the temporal, provided the template for an abundance of images that featured in geology's scientific as well as popular iterations (fig. 1.3). Dickeson's relatively systematic approach to excavation has been noted in the scholarly literature, and his notes from the field as well as his essays in the *Lotus* make clear that he took care to document his step-by-step process, which included describing the basic features of a given site, identifying and measuring distinct strata, recording the location of finds, and making drawings and diagrams on site. His approach, then, as well as the imagery that resulted, including the schematic images of excavated mounds published in the *Lotus*, closely resemble the methods and the visual culture of geology (fig. 1.4). But a major difference looms, one that distinguishes Egan's image of descent, survey, and extraction from the ways of knowing of geology writ large. Geology as a scientific practice generated a need for novel image types because scientists finally got a handle on the deep time of the earth. The Egan painting, on the other hand, purposely gets time wrong, installing deep time where it does not belong.

Archaeology in Europe trained attention on long-lost civilizations, with an eye to reinstating them within the flow of history and creating a vital link between the present and the past.[28] Dickeson drew a direct connection between this aspect of Europe's study of classical antiquity and the research he conducted in the United States. He titled the lecture accompanying the display of the panorama "American Aerchiology," referring specifically to the study of ancient things as pursued on the other side of the Atlantic, and nascently in the United States, and he promised to illuminate for audiences America's ancient past—the "unhistoried" Native Americans who "dwelt on this Continent 3,500 years ago." Dickeson claimed to have collected forty thousand "antiquities" and "relics" during his travels, a choice selection of which he displayed alongside the panorama. Together with the panorama's visual survey of "all the aboriginal monuments" that remained in a territory "once roamed by the RED MAN," these artifacts promised to paint a "magnificent picture" of a time long gone.[29] Such rhetoric appears throughout Dickeson's writing, and he constantly drew analogies between classical antiquity and American prehistory, likening the "aged sepulchres" of the Mississippi River Valley to the amphitheaters of ancient Greece, the remains of Herculaneum and Pompeii, and the tombs of Egypt, and lamenting the failure of Americans to "preserve these venerable piles" in order to save them from oblivion.[30]

But, of course, the paradigm of archaeology in Europe as the recovery of a lost but desired past hardly applied to the American context, and to Native America in particular. It goes without saying that Native Americans in the nineteenth century were not ancient things, artifacts or relics that hovered outside of history, awaiting posthumous reintegration into the chronicle of humanity. Far from being a race that "once" roamed the continent, Native Americans, of course, existed in the present, and some of the mound builders overlapped with European explorers and settlers into the eighteenth century and beyond.[31] Those Europeans and European Americans who explored, surveyed, settled, and dug up the American continent encountered artifacts from the past *and* interacted with living civilizations. Yet more often than not, archaeology in America suppressed the latter reality in favor of the myth of a vanished race, a civilization existing outside of history and lost to time. Put another way, archaeology in America, premised from the outset on claims of the inferiority of indigenous cultures, endeavored to transform the living present into a distant and ancient past, discontinuous with the modern. The reasons for this are well known, but they bear repeating: Native Americans were widely understood from a Euro-American standpoint as obstacles to American progress, and archaeology, along with the rhetoric surrounding it, abetted the supposedly preordained project of territorial expansion, an overspreading of the continent premised on belief in the essential superiority of whites and perceived to be as natural and inevitable as the demise of the "savage" races. This attitude fueled legislation, scientific research, industry, westward migration, homesteading, genocide, and war, all underpinned by a pervasive, pressing need to envision the American Indian as an artifact, the vestige of a static, unchanging culture—a denizen of the *then* rather than the *now*.[32]

Thomas Jefferson's *Notes on the State of Virginia* offers perhaps the most well-known account of an excavation of a Native American burial mound like the one depicted in Dickeson's panorama, and the rhetoric of Jefferson's text, despite its earlier date, closely resembles that of Dickeson's project overall. Jefferson, in fact, set the tone for midcentury discourse on Native America and the mission of territorial expansion writ large, as accelerated by his Louisiana Purchase in 1803 and by the subsequent Lewis and Clark Expedition, itself a perfect example of the collusion of scientific inquiry and expansionist politics in the period, with dire consequences for Native Americans inhabiting the lands in question. As the sponsor of a scientific expedition and a venerated Enlightenment thinker, Jefferson was a touchstone for scholars and entrepreneurs like Dickeson who hoped to contribute equally substantially to the intellectual development of the American republic, as reflected by the frequent invocations of Jefferson's work in guidebooks accompanying Mississippi River panoramas in the 1840s.[33]

Circulated initially in manuscript form, and then in printed editions in 1785 and 1787, Jefferson's *Notes* included a detailed chapter on Virginia's "savage Americans" that fell in line with what was fast becoming a rote account of native populations, with antecedents as early as Thomas Hariot's sixteenth-century account of England's efforts to colonize the Virginia territory, and persisting throughout the nineteenth century: Indians lacked a system of governance; as a people they were fragmented rather than unified; their ritual practices possessed little logic or order; they were ill equipped for the wages of civilization; their innate vulnerability necessitated federal protection and oversight; their numbers were rapidly dwindling; those remaining consisted of a mere shadow of a once great and pure race; only the "last vestiges" of their languages and customs survived; they produced only rudimentary tools and wares and no architecture, art, or other cultural monuments of which to speak; and so on.[34] Jefferson's study of Native America was made urgent by his belief in the imminence of Indian extinction—a conviction, true or not, vital to his aggressive expansionist stance and to the Indian removal policies he crafted to abet his territorial ambitions.[35]

Much like the language of the broadside for Dickeson's moving panorama, and pervasive in the antebellum United States, the rhetoric of loss pervades Jefferson's text, especially in the section that recounts his experience excavating a burial mound located on the South Fork of the Rivanna River, near Charlottesville.[36] When he wrote the *Notes*, Jefferson did not consider Native American burial structures to be works of architecture. "I know of no such thing existing as an Indian Monument," he wrote, only "arrow points, stone hatchets, stone pipes and half-shapen images," yet he treated the mound as if it were a ruin all the same.[37] The passage detailing his dig is worth quoting at length for the sense it gives of Jefferson's approach to Native America, which mirrored that of American archaeology in the late eighteenth century and the first half of the nineteenth century more broadly:

> I first dug superficially in several parts of it, and came to collections of human bones, at different depths, from six inches to three feet below the surface. These were lying in the utmost confusion, some vertical, some oblique, some horizontal, and directed to every point of the compass, entangled, and held together in clusters by the earth. Bones of the most distant parts were found together, as, for instance, the small bones of the foot in the hollow of a scull, many sculls would sometimes be in contact, lying on the face, on the side, on the back, top or bottom, so as on the whole to give the idea of bones emptied promiscuously from a bag or basket, and covered over with earth, without any attention to their order. The bones of which the greatest

> numbers remained were sculls, jaw bones, teeth, the bones of the arms, thighs, legs, feet, and hands. A few ribs remained, some vertebrae of the neck and spine. . . . There were some teeth which were judged to be smaller than those of an adult; a scull, which, on a slight view, appeared to be that of an infant, but it fell to pieces on being taken out, so as to prevent satisfactory examination; a rib, and a fragment of the under jaw of a person about half grown; another rib of an infant, and part of the jaw of a child, which had not yet cut its teeth. . . . It was part of the right half of the under jaw. The processes, by which it was articulated to the temporal bones, were entire, and the bone itself firm to where it had been broken off, which, as nearly as I could judge, was about the place of the eye-tooth. Its upper edge, wherein would have been the sockets of the teeth, was perfectly smooth. Measuring it with that of an adult, by placing their hinder processes together, its broken end extended to the penultimate grinder of the adult. This bone was white, all the others of a sand color.[38]

Jefferson denigrates the people who built the mound by dismissing their burial practices as promiscuous and disorderly. In addition, he characterizes the tomb and its contents as would a clinician, or, more aptly, a professor of gross anatomy narrating a dissection, albeit a less than systematic one: a disassembling and accounting of pieces and parts that also calls to mind the visual dissection typical of a botanical or zoological illustration—disarticulated anatomy arrayed on a blank field—or the sorting of specimens into the multiple drawers of a collecting cabinet, a classificatory impulse visualized in the Egan scene by the large pot at the base of the mound that houses a stack of similarly sized bones. Jefferson identified some of the bones he dug up as those of infants or children, but he hardly envisioned the remains he unearthed as human beings that once lived and breathed. They are, to him, exactly what he described finding in the tomb: parts and rubble tossed in a pit, not people in the round. Jefferson's account makes clear that for him Native America consisted of brute matter and data—enumeration, measurement, and, in the *Notes* chapter, six pages of charts.

The anthropologist and curator Chip Colwell has described how archaeology in nineteenth-century America transformed the Native American subject into an inert specimen for scientific study, a lifeless object for use in the production of knowledge and, ultimately, the exercise of power. Jefferson's *Notes* offers a perfect example of this transformation, as does the Dickeson panorama. Both share with the surgical anatomies that Keren Rosa Hammerschlag discusses in chapter 6 of this volume the racialization of dissection as well as the deployment of dissection as a marker or metaphor of racial difference. And both resemble

TABLE OF ANATOMICAL MEASUREMENTS.

NATIONS OR TRIBES.	No. in Catalogue.	Plate.	Longitud. diameter.	Parietal diameter.	Frontal diameter.	Vertical diameter.	Inter-mast. arch.	Inter-mast. line.	Occipito-frontal arch.	Horizontal periphery.	Length of head and face.	Zygomatic diameter.	Facial angle.	Internal capacity in cubic inches.	Capacity of anterior chamber.	Capacity of posterior chamber.	Capacity of coronal region.	Sub-coronal region.
* Peruvian.	75		6.6	6.	4.6	5.1	15.5	4.1	13.5	19.8	7.8	5.5	72°	83.5	36.5	47.	14.5	69.
* Peruvian.	76		6.5	5.	4.1	4.9	14.4	4.	13.5	18.6	7.8	4.7	73°	64.	23.	41.	11.9	11.9
* Peruvian.	77		6.6	5.7	4.2	5.2	15.5	4.4	13.	19.4	7.8	5.5	75°	75.	35.	30.	12.25	62.75
Peruvian.	79		6.6	5.7	4.5	5.3	15.	4.4	13.9	19.3	7.9	5.3	74°	74.5	31.5	43.	11.4	63.1
Peruvian.	81		5.8	5.7	4.4	5.3	15.2	4.3	13.3	18.3	7.9	5.1	76°	79.	32.25	46.75	18.	61.
Peruvian.	82		6.6	5.6	4.5	5.4	15.5	4.	13.8	19.2	8.	5.4	79°	75.	31.5	43.50	16.1	58.9
* Peruvian.	85	11–B	6.3	5.8	4.5	5.3	15.	4.	13.2	19.	7.2	5.5	80°	76.5	30.	46.5	12.25	64.25
* Peruvian.	86	11	6.1	6.	4.7	5.5	16.	4.5	14.1	19.5	7.8	5.4	81°	83.	33.5	49.5	15.75	67.25
* Peruvian.	87	8 & 9	5.8	5.7	4.4	5.1	14.5	4.1	12.7	18.4	7.4	5.	75°	71.75	28.75	43.	11.4	60.35
* Peruvian.	699		6.6	5.6	4.4	5.2	14.8	4.	13.6	19.2	7.8	5.2	74°	72.	26.5	45.5	9.25	62.75
* Peruvian.	90		6.3	5.5	4.2	5.	14.5	3.7	13.2	18.5	7.6	4.9	75°	70.	28.5	41.5	12.6	57.4
Peruvian.	91		6.2	5.8	4.3	4.9	14.5	4.1	12.6	18.7	7.4	5.2	73°	66.5	25.	41.5	9.7	58.6
* Peruvian.	92		6.8	5.4	4.5	5.3	14.7	4.2	14.	19.5	7.8	5.7	75°	74.5	34.50	40.	13.5	61.
* Peruvian.	93		6.1	5.9	4.6	5.2	15.2	4.1	13.2	19.2	7.7	5.2	75°	76.5	34.25	42.25	14.2	62.30
* Peruvian.	95	11–A	6.7	6.	4.5	5.6	16.2	4.5	14.5	20.2	8.	5.6	80°	89.5	34.	55.5	20.5	69.
* Peruvian.	96		6.4	5.9	4.5	5.	14.1	4.2	13.4	19.4	7.5	4.9	73°	68.	31.5	36.5	12.2	55.8
* Peruvian.	97	11–D	6.5	5.5	4.6	5.6	14.8	4.5	13.6	19.5	7.9	5.6	75°	68.5	33.	35.5		
* Peruvian.	100		6.2	5.5	4.4	5.	13.6	3.8	12.6	18.7	7.3	5.	70°	60.	27.5	32.5	12.5	47.5
* Peruvian.	400		6.5	5.7	4.4	5.2	14.7	4.3	13.4	19.5	7.8	5.5	76°	70.	30.5	39.5	13.1	56.9
* Peruvian.	697		6.7	5.5	4.6	4.9	14.1	4.	13.4	19.5	8.	5.5	73°	71.	32.5	38.5	10.5	60.5
* Peruvian.	402		6.9	5.6	4.4	5.3	15.2	4.1	14.	20.2	8.	5.5	77°	78.5	34.	44.5	11.8	66.2
* Peruvian.	403		6.6	5.6	4.3	5.3	14.9	3.9	13.8	19.5	7.6	5.2	74°	79.	31.	48.	15.5	63.5
* Peruvian.	405		5.9	5.6	4.3	4.9	14.5	3.9	12.9	18.4	7.4	4.9	75°	62.	23.5	38.5	10.1	51.9
* Peruvian.	406		6.3	5.7	4.3	5.3	15.	3.9	13.7	19.	7.5	4.8	76°	75.5	29.5	46.	12.4	63.5
* Peruvian.	446	11–C	6.	5.9	4.4	5.	15.5	4.	13.2	19.	7.6	5.4	80°	77.	28.	49.	11.3	65.7
Peruvian.	447		6.5	5.9	4.6	5.1	14.9	4.	13.2	19.2	7.8	5.4	74°	71.	29.	42.	10.3	60.7
Peruvian.	448		6.1	5.7	4.5	5.	15.	4.	12.9	18.7	7.9	5.3	74°	74.	32.	42.	14.9	59.1
Peruvian.	449		6.3	6.1	4.9	5.3	16.	4.4	13.2	19.5	8.	5.5	77°	83.5	35.5	48.	19.5	64.
* Peruvian.	450		5.9	6.	4.7	5.	15.	4.1	12.6	19.1	7.6	5.3	75°	74.5	35.	39.5	14.3	60.2
Peruvian.	451		6.8	5.4	4.3	5.6	14.6	4.3	14.3	19.5	7.8	5.2	78°	87.	31.	56.	19.75	67.25
Peruvian.	452		6.3	5.8	4.4	5.2	14.4	4.2	13.2	19.2	7.8	5.4	69°	83.	34.	49.	11.5	71.5
* Peruvian.	685		6.3	5.3	4.4	4.6	14.	3.9	13.	18.7				69.	29.	40.	12.3	56.7
Peruvian.	686		6.4	5.5	4.3	5.2	14.8	4.	13.2	19.	7.6			71.	38.	33.	14.4	56.6
Chimuyan.	11	6	6.5	5.4	4.2	5.2	14.3	3.8	13.4	18.8	7.6	4.9	76°	67.5	28.5	39.	10.25	57.25
Quichua.	637		7.	5.2	4.4	5.2	14.6	4.	14.4	19.5	8.2	5.2	70°	79.	33.	46.	17.5	61.5
Atacames.	651		6.6	5.3	3.8	5.2	14.5	3.9	14.	19.2	8.	5.1	73°	75.	29.	46.	14.5	60.75
Atacames.	652		6.6	5.4	4.3	4.8	14.2	4.	13.1	19.5	7.8	5.	74°	74.	27.5	46.5	8.75	65.25
Atacames.	653		7.2	5.5	4.4	5.1	14.8	4.1	13.7	20.2	8.2	5.4	76°	80.	34.25	46.75	13.3	66.7
Araucanian.	654	68	6.7	5.4	4.7	4.9	14.2	4.1	13.4	19.5	7.8	5.	72°	77.	32.	45.	11.9	65.1
Araucanian.	655	66	6.9	5.4	4.1	5.4	15.	4.1	14.2	19.5	8.	5.	76°	84.5	32.5	52.	19.	65.5
Araucanian.	656		6.6	5.3	4.2	5.	14.2	4.	13.8	19.	7.9	4.9	76°	75.	26.	49.	15.25	59.75
Mexican.	714	61	7.1	5.6	4.6	5.5	15.5	4.1	15.	20.2	8.3	5.2	80°	87.				
Mexican.	559	17	6.8	5.5	4.6	6.	15.6	4.4	14.6	19.9	8.1	5.3	80°	89.5	33.5	56.	19.5	70.
Mexican.	715	59	6.3	5.3	4.4	5.4	14.3	4.2	13.5	19.2	7.7		76°	74.				
Mexican.	499		7.1	5.6	4.5	5.4	15.2	4.3	14.2	20.	8.4	5.3	73°	87.5	36.25	51.25	10.25	68.25
Mexican.	716	60	6.6	5.3	4.4	5.4	14.	4.	14.	19.3	7.8		77°	76.				
Mexican.	A.N.S.	16	7.1	5.7	4.4	5.2	15.9	4.	14.	20.5	8.4	5.5	72°	83.	39.	44.	17.5	65.5
Mexican.	717		7.	5.3	4.3	5.3	14.5	4.1	14.	20.			78°	77.				
Mexican.	681	17–A	6.6	5.3	4.3	5.2	14.6	4.1	13.6	19.	7.5	5.4	77°	74.	28.	46.	11.5	62.5
Mexican.	718		6.8	5.5	4.8	5.7	15.9	4.	14.6	19.9	8.	5.4	76°	77.				
Mexican.	682		7.	5.4	4.3	5.3	15.	4.1	14.	19.8	8.	5.4	76°	80.5	36.5	44.	15.5	64.5
Mexican.	720		6.9	5.3	4.5	5.6	14.7	3.9	14.6	19.8	8.1		80°	82.				
Mexican.	A.P.S.	18	6.4	5.7	4.5	5.4	14.6	4.5	13.5	20.2	7.2	5.2	78°	77.	30.	47.		
Mexican.	34	18–A	6.9	5.2	4.2	5.4	14.5	4.1	14.	19.2	8.	5.1	76°	78.	30.	48.	14.25	63.75
Chetimaches.	43		6.5	5.7	4.3	5.9	15.5	4.1	14.	19.1	8.	5.1	77°	80.	30.5	49.5	16.25	63.75
Chetimaches.	70	19	6.9	5.6	4.2	5.9	15.5	4.3	14.	20.	8.5	5.7	71°	85.	39.25	45.75	13.25	71.75
Seminole.	C.		6.9	5.6	4.6	5.3	15.	4.2	13.6	19.8			75°	80.				
Seminole.	707	23	7.1	5.6	4.7	5.5	15.	4.1	14.8	20.3	8.	5.3	78°	89.				
Seminole.	604	22	7.3	5.9	4.6	5.8	15.9	4.4	15.3	20.7	8.4	5.3	72°	93.	35.5	57.5	25.	68.
Seminole.	708		7.	5.5	4.4	5.4	14.9	4.2	14.6	20.1	8.1	5.1	73°	86.				
Seminole.	C.		7.3	5.6	4.2	5.6	15.2	4.7	15.	20.4			73°	82.5				
Seminole.	456	24	7.	5.9	4.5	5.8	14.7	4.6	14.2	20.5	7.9	5.6	81°	91.5	44.	47.5	18.1	73.4

65

Fig. 1.5. Samuel George Morton used quantitative data to shore up his specious racial theories. Samuel George Morton, "Table of Anatomical Measurements," *Crania Americana, or, A Comparative View of the Skulls of Various Aboriginal Nations of North and South America* (Philadelphia: John Pennington; London: James Madden, 1839), 258. Call number ReCAP Oversize GN71.M8f, Rare Books and Special Collections, Firestone Library, Princeton University, Princeton, New Jersey. Courtesy of Princeton University Library.

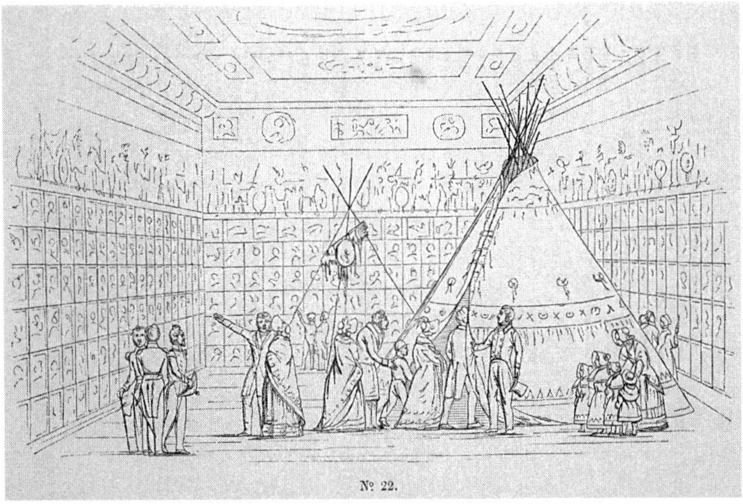

Fig. 1.6. George Catlin displayed his paintings of indigenous people alongside artifacts he collected in the field for audiences in the United States and Europe. George Catlin, "No. 22," *Catlin's Notes of Eight Years' Travels and Residence in Europe, with His North American Indian Collection*, vol. 2 (London: Printed by the author, 1848), plate no. 22, between 292 and 293. Rare Book Division, Special Collections, Princeton University Library, Princeton, New Jersey. Courtesy of Princeton University Library.

other quantitative aggregations from the period intended to serve as evidence for the essential difference and inferiority of Native Americans, among them the teeming statistical tables of cranial measurements that Dickeson's Philadelphia associate Samuel G. Morton used to illustrate his 1839 *Crania Americana, or, A Comparative View of the Skulls of Various Aboriginal Nations of North and South America*, a classic of pseudoscientific racism published in the midst of the forced removal of thousands of indigenous people to federal territory west of the Mississippi pursuant to the Indian Removal Act, signed into law by President Andrew Jackson in 1830 (fig. 1.5). Jefferson's dismissive characterization of Native American cultural objects as "half-shapen images" thus proves a fitting descriptor for Native American peoples as they fared in period accounts.

European archaeologists desired to revive antiquity in order to mourn the loss while preserving the memory of great civilizations. Jefferson offered an analogous lament, but he did so prematurely, *before* the fact of loss. He and others in the early years of the republic and the decades of nation building that followed, Dickeson included, deployed archaeology and its rhetoric of artifacts and ancient history to transform living cultures into dust and bones. In the words of Henry Rowe

Schoolcraft, a prominent midcentury ethnologist and the author of an early federally funded study on America's indigenous cultures, "America is the tomb of the Red man," a sentiment echoed by many at the time, including an 1854 visitor to a mound in Tennessee, who wrote that he had come face to face with the Red Man's "last resting place," and by the artist George Catlin, who assembled his touring "Indian Gallery" in the late 1830s in order to preserve the memory of a "dying nation" that was "rapidly passing away from the face of the earth" (fig. 1.6). The guidebook that accompanied John Banvard's wildly popular *Panorama of the Mississippi River*, which debuted in 1846, similarly characterized Native American earthworks as memorials to a vanished race, "the affecting and imperishable traces of generations that are gone."[39]

Archaeology's chief methods, excavation and preservation, thus yielded in the United States a *prophecy* of destruction rather than an account of a past demise. Jefferson's characterization of Native America as lacking the hallmarks of an advanced civilization only shored up his configuration of contemporaneous indigenous peoples as nothing more than remains. Such narratives proffer a mind-bending negotiation among competing desires: for the pastness of Native America as well as for the presence of Native America as evidence of an *American* past, one on par with Europe's classical roots and analogous to the positing of natural history as American history, geology as the American nation's deep time. But Jefferson of course misappropriates geology's vision of a vastly ancient earth, for his account demands that deep time exist both in the past and in the present, in the form of a civilization at once ancient and vanished *and* presently doomed to perish. Jefferson offered a similarly convoluted argument in the *Notes* when comparing the ancientness of Native America to the populations of other continents. The "red men of America" are of "greater antiquity than those of Asia," he claimed, because they speak so many languages, the result of linguistic divergences that could have arisen only over a vast length of time. This misappropriation of temporality embodies the temporal acrobatics necessary for Jefferson to have his Indians and extinguish them too, a classic case of imperialist nostalgia, to be sure, but here incarnated as historical science.[40]

Dickeson adopted a more systematic approach than Jefferson when excavating and reporting his finds, reflecting the ongoing professionalization of archaeology in the first half of the nineteenth century. Although he used methods that archaeologists now consider crude, it is clear from his field notes and published reports that he usually took pains to excavate in regular sections and to document his work.[41] Yet Dickeson, too, envisioned himself exhuming the remnants of a long-gone culture. Earthworks represented, in his words, "the last and only

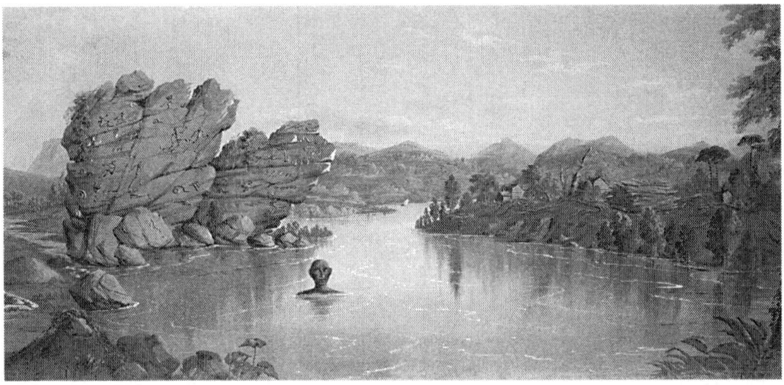

Fig. 1.7. In Egan's panorama, a partially submerged statue and rock-face petroglyphs signify a long-lost past. John J. Egan, "Hanging or Hieroglyphical Rock—Colossal Bust at lower water mark, used as a metre by the Aborigines," *Panorama of the Monumental Grandeur of the Mississippi Valley*, ca. 1850, tempera on lightweight fabric, 90 in × 348 ft (228.6 × 10607 cm). Eliza McMillan Trust, 34:1953, Saint Louis Art Museum, Saint Louis, Missouri.

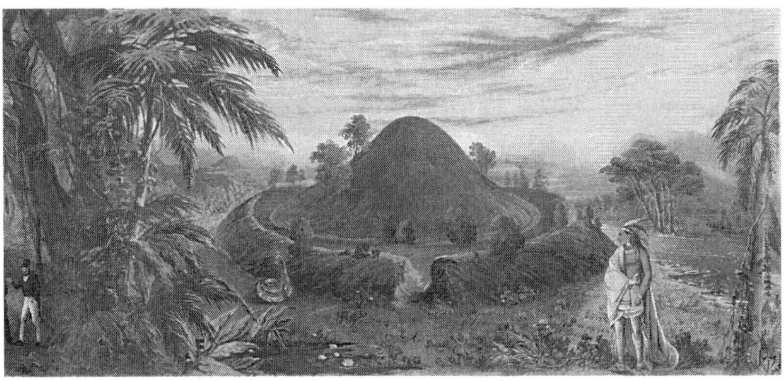

Fig. 1.8. In Egan's panorama, the sun setting over a Native American burial mound signals the passing of indigenous civilizations. John J. Egan, "Temple of the Sun by sunset," *Panorama of the Monumental Grandeur of the Mississippi Valley*, ca. 1850, tempera on lightweight fabric, 90 in × 348 ft (228.6 × 10607 cm). Eliza McMillan Trust, 34:1953, Saint Louis Art Museum, Saint Louis, Missouri.

relic of the past." They existed outside of Euro-American narratives of history as a process of transformation and progress and apart from standard measures of human time—unaccountable by the gauge of chronology and alien to historical narrative. "Who were the mighty masters of these broad lands?" he queried in the *Lotus*. "No history can reveal their earthly sojourn; the same blue sky, the same green fields, the same wide waters remain, but the living mass of beings that once animated this scene, are all gone—gone to the grave!" The monuments of this long-gone race, he wrote, are "land marks in the lapse of age, beneath whose shadows generations have mouldered, and around whose summits a gone eternity plays!" Deep, immeasurable time—lapsing ages, eternity—separates us from them, Dickenson's remarks implied, and their existence could not be squared with known temporal models. "All traces of a historical nature of its origin have long since departed," he lamented of an artifact he collected in Pascagoula, Mississippi, in 1842, and "the testimonials of antiquity lend us no aid." One searches, he wrote, "for some tradition, some time-stained chronicle, but all is silent."[42]

In the image of Dickeson's dig on the Feriday plantation featured in the panorama, Egan relegated Native America to the margins of the scene. The inward turn of the Native figures, away from the progress represented by science, tourism, and the plantation economy, disassociates indigeneity from contemporaneity while preserving it, artifactlike, for posterity. The panorama as a whole underwrites this gesture, positing in image after image Native America as essentially and innately past, and immeasurably so: partially submerged colossi, setting suns, blasted trees, snow-covered earthworks, and time-worn petroglyphs (figs. 1.7, 1.8). In the excavation scene, Egan depicts the enslaved Black men as sheer physicality in relation to the intellectual work of their white counterparts. Their downturned gazes along with the rhythmic correspondence of their bodies and pickaxes with the piling dirt, curving mound, and excavated pots imply their connection to the brute matter of the earth. Yet even these enslaved persons retain a specific and necessary role in the endeavor. The Native Americans idling in the wings of Egan's painting play no such part, save for adornment. Only Native American bones compel the attention of the excavators, while living Native Americans serve simply to decorate the scene, cartouchelike.

Of course, it is possible that Egan, by including this foursome, meant to indicate the actual presence of Native Americans at some of the sites excavated by Dickeson. Dickeson and his contemporaries at midcentury inevitably came into contact with members of indigenous groups, including through the souvenir trade, which involved indigenous production of wares for sale to nonindigenous consumers, including travelers and tourists. It is probable that Dickeson procured some of the objects he brought back to Philadelphia through such trade, which is evoked by the tourists and onlookers that populate many of the scenes of Egan's panorama. Such wares were desirable for their aesthetic qualities as well as for their value as "authentic" Indian art, and what Ruth B. Phillips has characterized as the "dynamic aspects" of object exchanges between indigenous and nonindigenous cultures suggests that Americans held a range of attitudes toward indigenous people in the period, including interest, empathy, and a desire for coexistence.[43] But Dickeson, like the bulk of quasi-professional artifact accumulators in the period, Catlin among them, amassed his collection in anticipation of what he perceived to be Native America's imminent and inevitable demise. "I have flown to their rescue," wrote Catlin, "not of their lives or of their race (for they are '*doomed*' and must perish), but to the rescue of their looks and their modes" in order that "phoenix-like, they may rise from the 'stain on a painter's palette,' and live again upon canvass, and stand forth for centuries yet to come, the living monuments of a noble race."[44] Observers in the period, sympathetic or not, often described the indigenous people they encountered while visiting burial mounds as present at the mounds in order to grieve over the unfortunate plight of their race, erroneously attributing to them a version of the imperialist nostalgia typical of Dickeson, Catlin, and settler colonial rhetoric more generally. The figural group of Native Americans that adorns Egan's excavation scene, *as* decoration, shores up just such a narrative of loss.[45]

It is worth noting that the founders of the American Antiquarian Society classed Indian artifacts alongside a host of other items, among them books, pamphlets, newspapers, maps, manuscripts, and fossil specimens. "The chief objects of the enquiries and researches of this society," wrote one of its founders, Isaiah Thomas, "will be American Antiquities, natural, artificial, and literary," including what he termed "handicrafts of the Aborigines." As Holmes put it, the American antiquarian studied all manner of American antiques, from graveyard inscriptions, musty books, ancient manuscripts, and long-buried edifices to "any misshapen stone, which may haply prove to be some Indian relic, if not even an idol."[46] The records of the early years of the society reflect the capaciousness of its approach and, consequently, of its collection, which housed, for example, an assortment of books printed in the sixteenth, seventeenth, and eighteenth centuries; nine hundred volumes from the Mather family library; a tobacco box given by Sir Walter Raleigh to Increase Mather, which descended through several generations of Mathers; a collection of land titles from the Worcester, Massachusetts, Registrar of Deeds; an axe of French manufacture excavated in Maine; an Indian arrowhead; a stone pestle, gouge, fishing weight, and chisel said to be of "Indian manufacture"; and a pamphlet titled *A Sketch of All the Invasions or Descents upon the British Islands, from the Landing of William the Conqueror.*[47]

In this way, American scholars followed the example of their European counterparts. Antiquarianism came into being in the seventeenth and eighteenth centuries as a supplement to historical scholarship, which

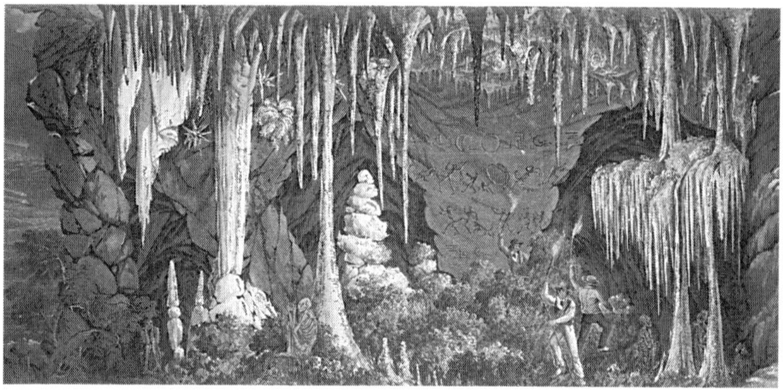

Fig. 1.9. Egan's panorama included a scene of Dickeson exploring a cave containing petroglyphs and Native American remains. John J. Egan, "Cave in the Rock, Stalagmite Chamber and Chrystal Fountain, Desecated and Mummied Bodies in their burial places—Magnificent effects of Crystalization," detail, *Panorama of the Monumental Grandeur of the Mississippi Valley*, ca. 1850; distemper on cotton muslin, 90 in × 348 ft (228.6 × 10607 cm). Eliza McMillan Trust, 34:1953, Saint Louis Art Museum, Saint Louis, Missouri.

derived knowledge about the past from documents and other forms of verbal expression. The antiquarian's study of a diversity of material remains aimed to augment the historian's scrutiny of written records and oral traditions, providing a whole new class of evidence for the scholar. This context in turn abetted the rise of archaeology as a discipline, and the two—antiquarianism and archaeology—were closely linked, often overlapping, in Europe and, later, the United States.[48]

Given the absence of the sort of written records that facilitated the study of ancient Greece and Rome, and bearing in mind the presumed inability or unwillingness of nonindigenous scholars in the first half of the nineteenth century to fathom indigenous forms of record-keeping, the study of material remains, including those excavated from burial mounds, provided the American antiquarian a tangible substitute for the proverbial "time-stained chronicle." Just as the scrutiny of a text from the fifth century might shed light on the civilization of ancient Greece, the study of Native American artifacts could provide knowledge about the people that made and used them. But the nineteenth-century scholar could only learn so much from an object unaccompanied by any discernible explanation, textual or otherwise, as reflected in Holmes's suggestion that what to him looked like misshapen stone could just as easily be "be some Indian relic, if not even an idol." The absence of information was thus a hindrance, but it was also a boon, for in yielding minimal data, Native American material remains offered themselves up to wide-ranging interpretations or fantasies on the part of those studying them. Consequently, the meaning of indigenous remains was especially malleable. As objects extracted from their physical and cultural contexts, they could be mobilized as if in a vacuum, made to signify anything that their interlocutors wished them to, including the essential inferiority and pastness of Native America. In repeatedly using the terms *relic*, *antique*, or *antiquity* to describe indigenous material remains, Dickeson not only consigned Native America to the past but also characterized indigenous peoples as more akin to the object world than the human one, more artifact or remnant than cultural producer, while also evacuating any sense of reverence that might be associated, for instance, with the relic of a Christian saint or, in the United States, a relic of the founding fathers—George Washington's war tent, say, or his snuffbox or a lock of his hair.[49]

Many of the images in the Dickeson panorama that depict Native America feature a dynamic tension between vitality and lifelessness that explicitly fashions Native America as a temporal paradox: a people existing out of time yet lodged in the deep past *and* perpetually present as an idea or image necessary to sustain America's national myth. Take, for example, Egan's scene of cave exploration (fig. 1.9). Into a breathtaking array of stalactites, stalagmites, columns, calcite curtains, flowstones, and crystal formations that together connote inconceivable depths of time, Egan has tucked three iterations of Native America: skeletal remains, a mummified corpse, and petroglyphs. As lifeless bodies and stick figures, this trio exists in purposeful contrast to the energetic and gesticulating explorers pictured in the cave. Moreover, the Native bodies are made to appear far more lifeless and inert than the actual inanimate matter on view in the scene; namely, the elaborately rendered geological formations adorning the cave, enlivened and set into shimmering motion by the light of several brightly burning torches, an animating effect that belies their immeasurable age as well as their mineral nature, and perfectly in line with period scientific rhetoric of an animate, even willful earth possessed of its own, uncanny "inner life."[50] The explorer figures, one of whom is likely Dickeson, play a double role: their vitality underlines the lifelessness of Native America, represented in the scene by corpses and line drawings, and their presence suggests that indigenous bodies are only at home in or "native" to spaces that exist apart from or before human history.

Plotlines involving cave descents like the one depicted by Egan appeared in a wide variety of narrative formats in nineteenth-century America, from melodramatic tales of murder, kidnapping, and moldering

Indian bones, such as Burton Saxe's *The Cave Secret, or, The Mystery of Night Island* (1865), to memoirs of sheltering in dugout caves during the Civil War, as in *My Cave Life in Vicksburg, with Letters of Trial and Travels* (1864), written by "A Lady." Almost without fail these texts advance, first, the underground for white people as merely a way station, a space of temporary habitation followed by an inevitable ascent, and, second, the realm below the earth's surface as a natural habitat for other persons, phenomena, and things, including everything from eyeless fish, ghostly plants, inexplicable noises, hermits, madmen, and alien races to the bones of prehistoric giants and all manner of Native Americans, living or dead.[51] In *The Cave Secret*, white settlers descend into a series of caves only out of necessity, to search for a missing person, while an elderly Indian chief inhabits a cavern called the "Chamber of the Dead" in order to commune with the bones of his ancestors and sing hymns to them "in a strange unearthly tone," as if already and naturally a resident of the underworld.[52] Likewise, Egan presents the explorers in the crystal-ridden cave as not belonging below ground—they exist in full animation, acting in space and time against a backdrop of lifeless matter, Native bodies included. They enter the cave to make observations, and return to the surface to transform what they observed into knowledge for use in the world above.

Because Dickeson's sketches from the field and the cross-sectioned mound in Egan's panorama closely resemble the images of rock strata created by geologists to illustrate the relationship between space and time, situating a physical feature of the earth in temporal terms, both the sketches and the panorama scene suggest an attempt to locate Native America within a comprehensive chronology, one that included the formation of the earth's topography and the appearance of humans. But for Dickeson, and for American archaeology in the period, strata of the sort visible in his excavated mound were in fact not signifiers of time. Although stratigraphic methods were commonly used by geologists in Europe and England by the early nineteenth century, it was not until much later that American archaeologists adopted them to assign dates to sites and to shed light on the timelines of cultures. Previously, in large part because archaeologists believed that North America had been inhabited by humans for such a relatively short time, considerations of humans in relation to geological time, deep or not, remained moot.[53]

The nonsignifying nature of strata within archaeology thus left plenty of room for someone like Dickeson to invest Native American earthworks with incongruous temporalities, designating Native America as America's ancient, ennobled past *and* outside of time, belonging to no time at all. In the advertisement for his "Indian Antiquities" lecture series, Dickeson promised to educate the audience about "the most wonderful Antiquities in the World . . . coeval with the earliest works of man." Their "primordial History," he wrote, has remained "buried in the earth, and obscured by the hand of time." In other words, Native America belongs to antiquity, and thus to the category of great civilizations including Egypt, Greece, and Rome. But, by the same formulation, Native America also belongs to the primordial murk, extant apart from time and from history, *buried* by both, and as alien to Euro-American temporality as the mummified Indian corpses in Egan's crystal-filled cave are to life at the surface. The Native America that arises from such a paradox of claims cannot be accounted for by any modern measure. Incommensurate with the gauges of the historian and the archaeologist, indigeneity in Dickeson's telling falls outside the scope of all manner of chronicle or record. Native America is left to be its own measure, native only to itself and to the fantastical, ahistorical time conjured for it by others.

Dickeson ends his broadside pitch for the lecture series by stating that the Indian antiquities on exhibit with the panorama might justly be considered a source of "true Chronology."[54] This is a telling statement, for it makes clear that for Dickeson and others in the United States the archaeology of Native America was not simply a matter of studying the past. It was also an out-and-out search for time, for a way to conceive of the temporality of existence in the absence of incontrovertible evidence or a definitive model of duration, all with an eye to making time into a servant of expansionist ideology. Many archaeologists in the period between the Louisiana Purchase and the height of mound mania characterized their encounters with Native American earthworks as occurring in a kind of temporal vacuum, as had Dickeson, and they described the study of these structures as a quest for chronology.

Exemplary of this are the remarks of Bishop James Madison, who in a letter published in the *Transactions of the American Philosophical Society* argued against the claim that the mounds he had observed on his travels along the Kanawha, a tributary of the Ohio River in West Virginia, had been designed as fortifications. "What was the real design of them may be left to future inquiry," he concluded. "It is true, that we want here a compass to guide us, and are left to find our way through this night of time, in the best manner we can."[55] Madison's remarks refer to the paucity of knowledge about burial mounds in the early nineteenth century, while also suggesting that indigenous cultures might not belong, ultimately, to knowledge as such. The metaphors of directionless and darkness he employs make a claim for indigeneity's essential incommensurability with any discipline or system of order, such as anatomy, geology, and history, those practices of knowing by which one converts the data of the physical earth and the exploits of the human into understanding and truth. Dickeson's digging and its subsequent reincarnation as scenes in a moving panorama aimed to compensate for Madison's missing compass for use in navigating the obscure reaches of the ancient past. Dickeson strove to excavate time from the depths and lay it bare, but he did so in the form of bespoke temporalities, including, for Native America, a tailor-made combination of ancientness and a kind of atemporal presence that, ultimately, configured as historical absence.

Dickeson's *Monumental Grandeur of the Mississippi Valley* brilliantly incarnated such made-to-measure timelines. As the length of painted canvas unspooled over the course of an hour or an hour and a half, the (presumably white) viewer traveled through space and time, aided by the surrogate explorers and tourists populating Egan's vignettes and by Dickeson's verbal annotations, which interlinked the panorama's sequence of scenes and directed the viewer's imaginative exploration of them. This sonically stimulated viewer in virtual motion would thus have encountered things *in place* along the way, including the bodies and bones of Native Americans, trapped by Egan's paint, like insects in amber, and entombed in the overall logic of the panorama as relics of a premodern, extrahistorical past. But Native America remained silent. Looking at the panorama thus entailed looking into a deep past and encountering a long-dead civilization, one put to rest by the very act of picturing that past in the present.

In refusing a continuous narrative that unfolded linearly in time and space, with the Mississippi River as a throughline, Dickeson's panorama differed from other river panoramas on tour in the 1840s, all of which used the fiction of travel on the waterway as a spatial and temporal binder. Instead, *Monumental Grandeur of the Mississippi Valley* offered only visual continuity. Each scene is linked to the next by a shared landscape element and all exist at roughly the same scale. But rather than following the river from start to finish, and then returning along the opposite bank, as was standard, Dickeson's spectator bounced around a patchwork of geography and chronology, shifting, for instance, directly from a scene of Native Americans fishing in Missouri to a view of the Rocky Mountains, and then to an image of earthworks in Louisiana, and after that back through time to a view of an eighteenth-century battle.

The moving panorama format itself more than accommodated such temporal and spatial acrobatics. What one scholar has called the "virtual voyaging" induced by moving panoramas in the 1840s United States became in Dickeson's hands a kind of time travel as well, with the result that only a fraudulent continuity arose from the formal interlinking of each scene and the linear motion of the moving canvas as it unrolled.[56]

This false pretense doubled down on what scholars have described as the moving panorama's particular reality effect. Visitors to these entertainments expected to be transported, subject to an illusion, while bearing in mind at every instant that they beheld a fiction and, also, understanding that this conscious being in the know was part and parcel of the entertainment. Hence the unique pleasure of this type of early moving image, and the distinctness of its operations from those of the stationary panorama, which trafficked in the experience of total immersion by conjuring an illusion of continuity between the embodied perspective of the viewer and the expansive prospect depicted in the round.[57]

The moving panorama, however, produced multiple and varied subject positions for the viewer to occupy, none of them truly immersive, and each one possessed its own, distinct spatial and temporal coordinates. Panorama operators anticipated this multiplying effect, even banked on it, as reflected in the written instructions they supplied to help viewers half-dupe themselves into occupying various and contradictory perspectives. John Banvard, for example, instructed viewers of his *Panorama of the Mississippi River* that "as the painting alternately ascends on one representation, and descends the next, it will be necessary for the reader, when ascending, to commence at the end of this description, and follow it back to the beginning." Banvard's note refers to the fact that any presentation of a moving panorama began where the last presentation had left off, to spare the canvas the wear of extra rolling and unrolling. But his directions also prepared the viewer for an experience of reading against the grain; that is, from the end of the published explanatory pamphlet rather than the beginning, while compelling every reader to imagine the text as addressing and embodying at least two distinct points of view.

In the guidebook to his *Leviathan Panorama of the Mississippi River*, John R. Smith similarly conjured multiple subject positions by conjoining in his narrative the perspective of actual travel with that of a person viewing a picture. In instructing his viewer that "we must necessarily pass many small towns, as Bloomington, Oquawka, Clarksville, Quincy, Warsaw, &c., as they would prove uninteresting to the mass, and besides, they present no peculiarities," he supposed that viewer to be at once a traveler on the river, passing by characterless towns, and a member of the audience desiring to be shown pictures of the most interesting sites. Smith included in his guidebook a "table of distances" so that visitors could calibrate the relationship between the actual terrain and painted images of it, further underlining the coexistence of disparate points of view. Smith similarly wrote multiple subject positions into his narrative description of his *Tour of Europe* panorama, speaking alternately as a traveler in Europe, a spectator viewing the panorama, and the creator of the painting.[58]

Moving panoramas created by the likes of Banvard and Smith thus configured the spectator as a split subject, distributed rather than singular, in sync with no single perspective as proffered by the painted canvas and its accompanying verbiage, and as a result existing multiply in relation to the painted scenes. Dickeson does not appear to have authored a guidebook akin to those that accompanied the panoramas of his peers, and one can only surmise what instructions he might have given his viewers had he done so. But on their own, the irregular sequencing of subject matter and the discordant chronologies proffered by *Monumental Grandeur of the Mississippi Valley* would have produced for the viewer a decidedly inconstant experience as shaped by a distinctly nonlinear narrative. And the panorama's known paratexts, including the broadside Dickeson authored, his verbal annotations, and the music that accompanied the presentation (not to mention the noise of the apparatus itself, including the labored cranking of two massive spools), further multiplied the points of view and subject positions offered up by the spectacle in toto, thus leaving ample room for choreographed as well as serendipitous flights of spatial and temporal fancy.[59]

Dickeson clearly capitalized on these multiplying capacities, conjuring with their aid multiple American temporalities, some continuous and progressing, others decidedly not, that together delineated a history of America as a continent and as a nation state. In this way he crafted a timeline for the so-called New World, and for the United States in particular, that was distinct from the global temporal unity imagined by geology—the earth and all its inhabitants arrayed in neatly progressing chronological order—and distinct from the all-encompassing, immersive temporality of the stationary panorama. Dickeson's broadside characterized the *Monumental Grandeur of the Mississippi Valley* as a "Magnificent Scenic Mirror," a phrase intended to convey the fidelity and accuracy of Egan's images, in keeping with the claims of lifelikeness that pervaded panorama publicity at the time. But the locution is apt for another reason. A mirror represents the real through a false incarnation of it, in the form of reflected light. In the same fashion, Dickeson's moving panorama, in its capacity as a visual medium, presents an illusion; namely, an illusion of time in the form of an illogical and impossible temporal model masking as a scientific likeness of duration and extinction that reflected the desires of one group in the form of an anachronistic image of another group's demise.

Advertisements for moving panoramas in the period without fail touted their edifying function, and Dickeson took every opportunity to publicize the educational value of his presentations, which he characterized as offering important, highly interesting, and expert information.[60] A spectacle, then, but also a promiscuous timepiece, Dickeson's *Monumental Grandeur of the Mississippi Valley* did its work by *telling* time, presenting a multidisciplinary and multimedia object lesson in chronology in a manner that a hole dug in a mound or an artifact exhumed from a grave—archaeology on its own—could not manage. Together, the spatial disruptions of archaeology and the visual and temporal disruptions of the moving panorama, the former physical and the latter fantastical, plotted a distinctive course for American time.

CHAPTER 2

"ALL THAT IS SOLID MELTS INTO AIR"

BURNE-JONES, GLACIATION, AND THE MATTER OF HISTORY

Alison Syme

What Edward Burne-Jones and the Kelmscott Chaucer have to do with the history of science is not immediately obvious. Inspired by medieval manuscripts and early printed books, William Morris founded the Kelmscott Press in 1891 to revitalize the craft of bookmaking, as he considered most contemporary, machine-printed books shoddily made and aesthetically alienating. In 1896, the press issued *The Works of Geoffrey Chaucer*, an edition of the fourteenth-century poet illustrated by Burne-Jones; ornamented with initials, borders, and type designed by Morris; and hand-printed on linen paper or vellum—a book famously described by Burne-Jones as a kind of "pocket cathedral."[1] An enormous undertaking born of love of the past, the Kelmscott Chaucer seems, in both form and content, far removed from the practice, discoveries, and debates of Victorian science. Chaucer, however, was popularly understood, for much of the Victorian period, as a "nature poet," and Morris and Burne-Jones saw his work "through the grimy filter of that age's encounter with the modern city in its raw . . . onslaught on green and pleasant England," as one scholar has put it.[2] Through a detailed reading of one of the illustrations for the unfinished allegorical poem *The House of Fame*, I argue not only that Burne-Jones was attuned to the scientific discourse of his time but also that the Chaucer provided the context and impetus to visualize, in distilled form, some of the complex relationships between natural change and human activity that his contemporaries were beginning to consider.

The visual and verbal strategies and allusions employed in the illustration take time to unfold. Burne-Jones solicits of the viewer, as does the Kelmscott Chaucer as a whole, patience. My argument is thus built

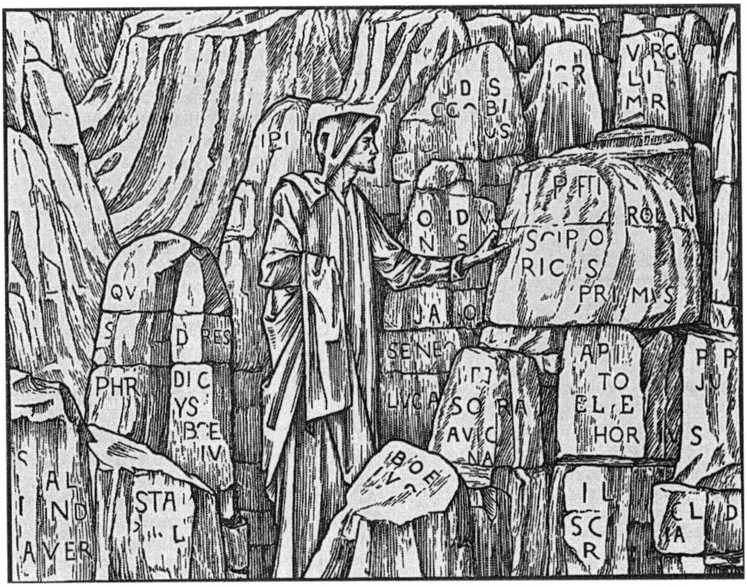

Fig. 2.1. Burne-Jones's illustration foregrounds the indeterminacy of the mountain's substance. Edward Burne-Jones, *Names of the Famous Engraved in Ice*, illustration for *The House of Fame* in Geoffrey Chaucer, *The Works of Geoffrey Chaucer*, ed. F. S. Ellis (printed by William Morris at the Kelmscott Press, 1896), 459. Courtesy of the Thomas Fisher Rare Book Library, University of Toronto, Toronto, Canada.

on "slow looking," the practice of a decelerated critical attention that allows perceptions to accrete over time.[3] This approach seems appropriate, given that the Kelmscott Press was an exercise in what Elizabeth Carolyn Miller has called "slow printing," meaning that in practice and principle it opposed the accelerated pace of "literary and journalistic mass production."[4] In this case, it allows me to gradually draw out the illustration's connections to science, history, and contemporary print and visual culture. Taking seriously Bruce Clarke's argument that the "application of allegorical models across scientific, imaginative, and critical constructions has generated a cultural network of formal and thematic resonances linking the discourses of literature, art, and science within a common matrix,"[5] I pay careful attention to tropes that cut across the different contexts and representational regimes with which the illustration engages. The figure of metamorphosis in particular—transformation or translation from one state to another, material, linguistic, or otherwise—proves key to understanding how the illustration slowly renders perceptible states of change, including the accelerated pace of modern life and its effects.

The House of Fame describes a dream in which the narrator, Chaucer himself, visits three fragile edifices in which human fame of all sorts is inscribed and/or broadcast: a glass temple of Venus; the mountaintop palace of the goddess Fame, made of beryl; and a flying wickerwork House of Rumour. He learns, in the course of his journey, about the transience and arbitrariness of earthly fame and authority. Burne-Jones designed seven illustrations for this poem between 1891 and 1895, and I will concentrate on the one that depicts the narrator's encounter with the high rock (higher than any in Spain, he says) on which the palace of Fame stands (fig. 2.1). The illustration encapsulates the Victorian artist's concerns about the erosion of history, substance, and meaning in modernity.

In Chaucer's poem, as the narrator climbs the rock to the palace above, he tries to discern "What maner stoon" it is, "for hit was lyk a thing of glas, / But that hit shoon ful more clere"—its kind of "congeled matere" puzzles him until he realizes that it is a "roche of yse."[6] Names of the famous, or once famous, are engraved on its frozen surface. Some letters have melted away, rendering the names increasingly illegible, while others, which lie in the shade of Fame's castle above, are still in sharp relief. In Burne-Jones's illustration, the dreaming narrator's uncertainty is conveyed to the viewer by what read, at least at first glance, as quite rocky rocks. The cloaked figure reaches out with one bare hand to feel the substance of Fame's foundation and peers intently at it, scrutinizing both the half-legible names and the ostensible stone's strange matter. Some fragments of names from the classical and medieval past, including Virgil, Ovid, Seneca, Socrates, and Boethius—foundational figures in Western literature and culture—are still partially legible, while others are too reduced to signify at all.

In Burne-Jones's rendering, the stacked, stony blocks inscribed with half-vanished names of the once famous evoke both natural and human-made forms, such as menhirs and tombstones, and are juxtaposed against, or are part of, what appears to be a landscape of metamorphic rock, a term for rock changed by heat or pressure introduced by Charles Lyell in the third volume of his *Principles of Geology* (1833). The narrator's outstretched left arm underscores the remains of Ovid's name, reinforcing the idea of metamorphosis as a principle of ancient landscape formation (mythic as well as lithic) and pointing to a connection between geological and literary history. In blurring the line between nature and culture, the illustration suggests that human history—as exemplified, made, and marked by its famous men—is subject to erosion and reshaping, like the earth itself. To understand what is at stake in Burne-Jones's visualization of vanishing, his evocative conflation of the geological

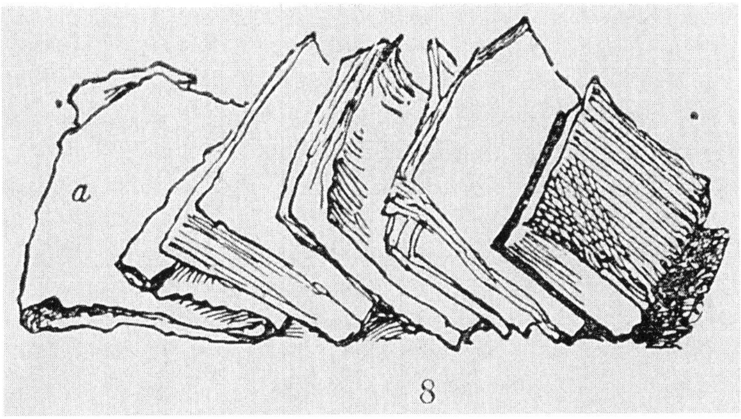

Fig. 2.2. Ruskin likened mountain structure to a row of slanting books. John Ruskin, illustration in *Of Mountain Beauty*, vol. 4 of *Modern Painters* (London: George Routledge and Sons, 1856), between 154 and 155. Courtesy of University of Toronto Libraries, Toronto, Canada.

and literary/historical records must be considered in terms of not only Chaucer's poem but also nineteenth-century writing on and imaging of geology and glaciers by Ruskin and others, and images of melting and modernity in Marx. Through such contextualization, the intersections and contests between the insights of the arts and sciences traced by the illustration can gradually be brought into view.

Chaucer's poem has been read not only as a meditation on the transience and frequent arbitrariness of earthly fame but also as a metatextual commentary. Kathy Cawsey has argued that the figure of the narrator peering at names in the ice is like "that of a reader trying to decipher an old manuscript." The idea of the sun gradually destroying the names in the ice is a paradox, as traditionally fame and being in the sun or the light were linked. Cawsey suggests, however, that the endurance of the names protected, if somewhat obscured, in the shade makes sense if one "envision[s] the mountain as a manuscript, its words fading from over-exposure and over-use." The names in the ice, then, offer an image of the vicissitudes of manuscript transmission, including the phenomenon whereby "some not-so-famous texts survive and other, more famous texts are lost."[7] Which of the mountain's inscriptions wear away depends on chance as much as merit.

For Burne-Jones, the link between manuscripts and geological forms would have been familiar from other sources, especially Ruskin's numerous writings on mountains and ice. Ruskin, whose works strongly influenced Burne-Jones even after he was long past his hero-worshipping stage, also compared mountains to books: in *Of Mountain Beauty* (1856), volume four of *Modern Painters*, he described the book as "a tolerably accurate representation of the mountain in substance, as well as in external aspect; nearly all these tabular masses of rock being composed of a multitude of thinner beds or layers, as the thickness of the book is made up of its leaves." The "mountain volumes" of the Alps are raised in sloping positions, "generally several resting one upon another, like a row of books fallen down," as he illustrated with a drawing (fig. 2.2). The metaphor was not, for Ruskin, purely morphological, as "every one of the mountain leaves is usually written over, though in dim characters, like those of a faded manuscript, with history of departed ages."[8]

The history inscribed in "mountain volumes" is written not only in stone but also in ice, and the connection between rock and water in Ruskin's geology is both literal and figurative. He frequently describes stone as exhibiting traces of an elemental fluidity, with great mountains "tossing" their "boundless crests" in "silent wave[s]," and mountain stone (such as "slaty crystallines") characterized by "undulation" and "quivering" in its frozen substance.[9] That same mountain rock is sculpted by its watery complement, the "fluent" glacier, in the dirt bands of which "record[s]" are also "pencilled."[10] Many of the writers on glaciers with whom Ruskin engaged also used script or print metaphors to describe them. James Forbes, for instance, characterized the glacier as "an endless scroll . . . upon whose stainless ground is engraven the succession of events."[11]

If the idea of history inscribed in glacial ice infused the lettered forms in Burne-Jones's illustration, Fame's palace at the pinnacle of a mountain of ice may also have suggested representations and descriptions of Alpine glaciers to the artist. The Alps themselves were deemed palatial by Lord Byron and others, but the Mer de Glace above Chamonix in particular, as described, painted, sketched, and photographed by eighteenth- and nineteenth-century artists and explorers, offered a perfect model for the House of Fame's icy foundations. Marc-Théodore Bourrit, for example, described the Mer de Glace in his 1776 collection of travel writings as "a mountain all of ice, with steps ascending to the top, which seemed the throne of some divinity . . . the sun which shone upon it, gave a sparkling brilliance to the whole: it was as a glass."[12] A rendering by Samuel Birmann from 1823, now in the Kunstmuseum Basel, shows the Mer de Glace rising in jagged, rounded, and glittering forms into the peaks above, and a Ruskin watercolor from 1860 held at the Whitworth Art Gallery likewise suggests an icy stairway climbing to ethereal reaches. A midcentury daguerreotype by Ruskin or his valet captures the glacier's landscape of ridged, rounded, and undulating lithic shapes

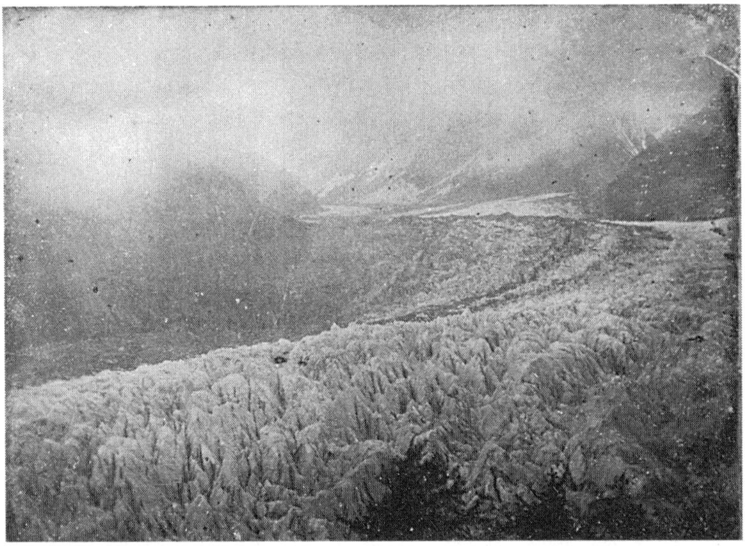

Fig. 2.3. Ruskin photographed and painted the glaciers of the Alps. John Ruskin and John Hobbs or Frederick Crawley, *Chamonix. Mer de Glace, Mont Blanc Massif,* 1854. Daguerreotype, 6 × 8.1 in (15.2 × 20.6 cm). © The Ruskin—Library, Museum and Research Centre, University of Lancaster.

less diaphanously, in a way that resembles Burne-Jones's illustration of circa forty years later (fig. 2.3).

While Burne-Jones did not, as far as we know, ever visit the Mer de Glace, he may have seen glaciers during his travels on the continent. On his first trip to Italy, in 1859, he witnessed the power of ice when an avalanche destroyed the road in the Mont Cenis Pass. On his second trip, in 1862, funded and accompanied by Ruskin, while boating on Lake Lucerne before crossing to Italy via the Saint Gotthard Pass, the pair "became locked in a discussion on current scientific discoveries about the formation of the earth"—in which, given Ruskin's sustained interest in the topic, glaciers surely featured; in the 1870s, he would enter the controversy over glacier movement that mountaineering scientists, contesting authority in the popular press as well as more rarefied venues, had made a matter of widespread interest.[13] Burne-Jones would have encountered representations of the Mer de Glace and other glaciers in Ruskin's writings and collections as well as many other sources. He may have seen St John Tyrwhitt's mural depicting the Mer de Glace in the Oxford Museum (ca. 1859), and he would have been familiar with Viollet-le-Duc's glacier paintings, some of which were exhibited alongside his own cartoons for stained glass at the Grosvenor Gallery's Winter Exhibition in 1881.[14]

Within Pre-Raphaelite circles, John Brett's mesmeric *Glacier of Rosenlaui*, exhibited at the Royal Academy in 1857, set a key precedent (plate 3). Brett's painstaking, documentary detail and evident awareness of Ruskin's ideas in *Of Mountain Beauty* make the painting a declaration of Pre-Raphaelitism.[15] It has also been described as "a résumé of mid-century ideas about the function and impact of glaciers," and "a bridge between science and art."[16] Brett had read Forbes's *Travels through the Alps of Savoy* and Louis Agassiz's *Études sur les glaciers*, a particularly important source for the painting.[17] For all its exactitude and icy splendor, *Glacier at Rosenlaui* is a disquieting work. As Keith Bendiner has pointed out, it offers a view of "uninhabited nature," a desolate landscape that confounds our sense of scale and of time. Because it contains no figure or "object of standard size," the viewer cannot be sure of the size of the boulders in the foreground; the scene might be "microscopic [or] gargantuan." The disconcerting scalar oscillation of the painting is echoed by its temporal uncertainty. The site has been said to reveal "the appearance of the prehistoric world"—the Ice Age Agassiz theorized. It simultaneously suggests, through the implied motion of the ice's "wave-like crests" and undulant rocks, the present moment and "the impermanence of the past."[18] If the artist's signature, inscribed on a foreground rock, is a mark of his possession and conquest of the landscape, the painting anticipates the name's inevitable effacement by the seemingly inexorable advance of the glacier.[19] The landscape's "chilling loneliness" is foreboding, and the painting might also be read as a future scene, the signature simply the ancient graffito of a tourist soon to vanish along with the rest of humanity.[20]

Burne-Jones's illustration is not an exercise in fidelity to observation like Brett's figureless work, and by virtue of its medium, it also lacks the chromatic splendor that drew many nineteenth-century artists to ice. Although echoes of Brett's swooping central ice form and foreground boulders can be found in the Kelmscott illustration, differences in framing and viewpoint attest to the artists' distinct preoccupations. Brett's elevated perspective and rocky foreground reveal the vast extent of the glacier and intimate its forward motion and geological impact. Burne-Jones's closeup view creates, at least at first, a sense of stasis. Nevertheless, the painting's sense of impermanence and intimation of human erasure through temporal and optical confusions are equally foundational to Burne-Jones's work, as I will demonstrate.

William Morris's accounts of his voyages to Iceland in 1871 and 1873 offered one further glacier source for Burne-Jones. Morris undertook his first journey as a Chaucer-like "pilgrim" to Iceland's "holy places," the saga-steads. He encountered numerous glaciers there, as the journal he

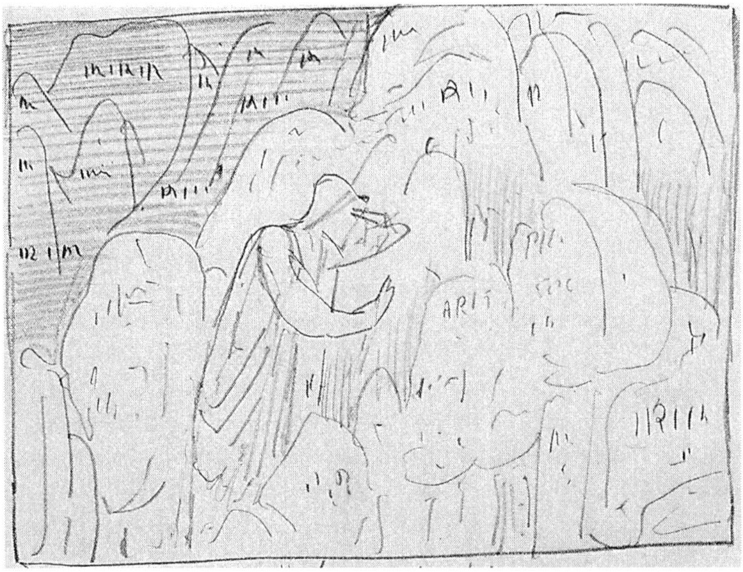

Fig. 2.4. In this preliminary drawing for the illustration, the mountainous surrounds at the upper left are also inscribed. Edward Burne-Jones, study for *The House of Fame: Names of the Famous Inscribed in Ice*, graphite on paper, 4.6 × 6.3 in (11.8 × 16.1 cm). Fitzwilliam Museum (1079/117). © Fitzwilliam Museum, Cambridge, UK.

gave to Georgiana Burne-Jones attests. Early in his first trip he wrote of the sight of distant "ice mountains," and the rest of the journal details his gradual discovery of the country's desolate, fantastic, dreamlike landscape. He sees "bright white *jökulls*" (glaciers) of varying forms, some "tumbling," one with "spiky white waves" and "great blocks," another that takes the form of a "boundless waste of ice all gleaming" and "looks as if it ended the world." The island, with its "half ruined" mountains and inhuman ice, however, is also everywhere inscribed with the history of the sagas. As Morris wrote in his journal, "every place and name marks the death of its short-lived eagerness and glory," like Fame's mountain.[21]

Preliminary studies for the *House of Fame* illustrations, held in the Fitzwilliam Museum, show that Burne-Jones began with, then quickly abandoned, the idea of representing the Chaucer figure in the act of climbing, choosing instead to foreground the act of interpreting textual, visual, and material data. Burne-Jones also experimented with slightly different landscapes before determining the final, glacial form. One sketch that comes close to the final design shows a figure among a heap of lumpy rocks, some inscribed with fragments of names: we can make out ARIS[TOTLE?] and SOC[RATES?] to his right, while the rest are shorthand scribbles of lines and curves (fig. 2.4). The distant rock forms in the upper left are also marked with inscriptions, which disappear in a more finished drawing. The distinction in the final illustration between the fractured forms of the inscribed ice and the unlettered rock at the upper left makes the ice appear more glacierlike, framed against a mountainside (in this respect also echoing Brett's painting). And the figure's newly wrapped right hand, bare in earlier sketches, suggests glacial cold.

In the finished illustration, the figure reaches out his left hand to determine the glassy mountain's material, but his gesture may also be read as an index of uncertainty as to its solidity and stability—the poem's narrator judges the House of Fame's foundations a "feble foundement / To bilden on a place hye."[22] Burne-Jones's understanding of the mutability of ice was indebted to Ruskin, who perceived not only the seeming fixity and actual fluidity of glacier forms but also their state of dissolution. He speculated that "the glaciers of the Alps had once triple or quadruple the extent that they have now," and in 1873 noted a loss of "one-third, at least, in the depth of all the ice of the Alps . . . in the last twenty years."[23] He recognized the significance of the attrition, which led him to declare that "the change of climate thus indicated is without any parallel in authentic history. In its bearings on the water supply and atmospheric conditions of central Europe, it is the most important phenomenon, by far, of all that offer themselves to the study of living men of science."[24] Tyndall and others acknowledged the loss of glacial ice, but while most scientists and the public came to regard "glaciers as a powerful agent in the production of modern landforms" by the end of the century, Ruskin's warning about the consequences of their vanishing went unheeded.[25]

Ruskin's prescient interest in melting glaciers resonates not only with present-day concerns—since he wrote *Of Mountain Beauty*, Alpine glaciers have lost about half their total area, according to glaciologists—but also with Chaucer's own focus on the instability of ice.[26] For Burne-Jones, Chaucer's melting mountain architecture and its Ruskinian resonances seem to have conjured up a constellation of associations with loss and change specific to nineteenth-century modernity, one in which the disappearance of glacial ice becomes a sign of cultural, and perhaps even human, evanescence.

In Chaucer's poem, the names preserved in the shade are those of "folk that hadden grete fames / Of olde tyme."[27] Readers and writers are imagined as engaging with a long tradition, even if the ancients' preeminence was not beyond question (numerous scholars have argued that the poem challenges canonical authority, which "melt[s] away as texts protected by the veneration of tradition are subjected to the magnifying lens of the critic's eye" and "fixed" interpretations vanish[28]). Burne-Jones's late Victorian illustration, however, was created when the cultural primacy of the classics was waning and debates over education, engaged in by science advocates such as Tyndall, challenged their place in curricula.[29] In Burne-Jones's rendering, even the very names of the ancients are fading, and implicitly the knowledge and history they stand for—and there is no visible route to a better-preserved shady side. The possibility of substantial engagement, critical or veneratory, vanishes as names and oeuvres are lost to view.

In the poem, Chaucer's dreaming narrator does not read out or record any of the icy text. Burne-Jones, by contrast, invites the viewer/reader to try to decipher the letters in the vanishing ice, the erasure of cultural and natural heritage becoming the occasion for a kind of melancholy game. Some of the partially legible names in Burne-Jones's illustration seem to correspond to historical and mythological figures mentioned in *The House of Fame*. Scipio Africanus, for example, a Roman general renowned for his prophetic dreams, is invoked twice by Chaucer, inviting cross-referencing between the text and image. Others rely on (and thus fail in the absence of) the viewer/reader's knowledge. All the fragments, though, are puzzles for the viewer, and visually, the array of letters, combined with lines of fissuring and stratification visible on the icy blocks of text, evoke the anagram, acrostic, diamond, word square, metagram, and other word puzzles that became so popular in the Victorian period.[30] Published in periodicals and riddle books, such games asked readers to supply words or missing letters; some even took the form of inscriptions in stone to be deciphered.

The letters in Burne-Jones's illustration, whether darker and thus still seemingly deeply etched in the ice or lighter, partial, and in the process of vanishing, are largely parallel to the picture plane and frequently do not seem entirely moored to their blocks. The N, for example, of what might be ROLAND to the far right of and just above the narrator's left hand, seems not so much to wrap around two differently angled surfaces as to float somewhere in front of them. Other letters, too, seem almost projected across shifting surfaces and even separate blocks (see, for example, HOR IVS, near the bottom right, stretching across three differently angled surfaces of two separate blocks). The slipperiness of the letters detached from a strong foundation, their potential mobility, is key to the kinds of literal and figurative translation games the picture seems to be playing. O ID V / N S , for example, just above the Chaucer figure's left forearm, presumably once spelled OVIDIVS NASO. Since *U* and *V* are indicated by the same letter in the Roman script, though, the *V* also looks as though it has slid across from its place between *O* and *I* in OVID,

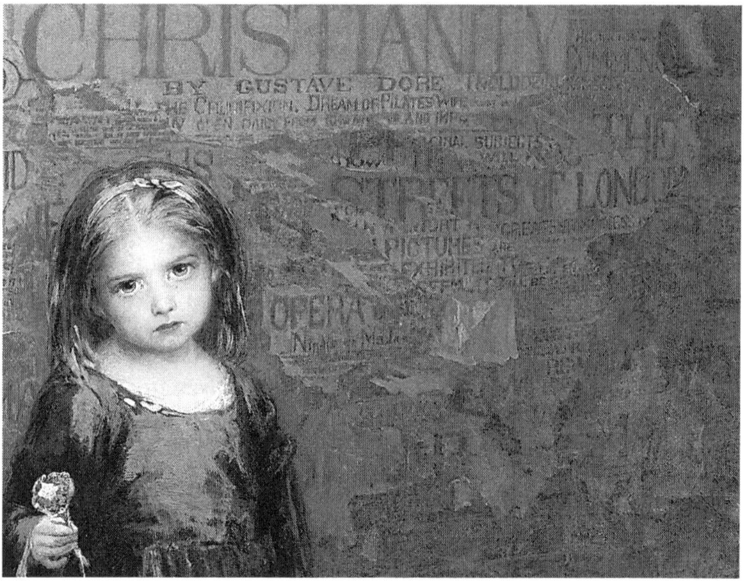

Fig. 2.5 Victorian London was filled with ephemeral texts. Augustus Mulready, detail of *Uncared for*, 1871. Oil on canvas, 39.8 × 29.9 in (101 × 76 cm). Bonhams.

the anglicized version of the Roman poet's name. Thus, the reader, in recognizing O ID V as OVID, literally truncates as s/he translates the Classical tradition. V RG / LI / M R at the top right encourages a similar rearrangement of letters to form VIRGIL. This example invites further puzzling within Chaucer's classical-medieval canon: while the *Aeneid* is invoked in *The House of Fame*, Burne-Jones could be referring to Publius Vergilius Maro, the Roman poet; Virgilius Maro Grammaticus, the medieval author; or deliberately invoking both names. We are presented not with a reliable "rock record" here but with fragments subject to projection and misinterpretation even as they are in the process of disappearing from sight and mind. Many are already too lost to decipher. The *QV* to the left of the dreamer's right arm on the tombstone-shaped iceform could be the beginning of QVINTILIAN, but the remains of the name come closest to asking the question that all the letters beg, QVI (who)? Who or what, given that the ice is melting, will endure? Cultural foundations seem to be vanishing into thin air as the narrator looks on.

Lack of textual fixity could have a positive valence. One scholar contends that the "shambles" Chaucer's poem "makes of medieval textual authority" enables the emergence of "the literary potential of the English language."[31] And aesthetically, melting was not always mournful or unsettling. Ruskin, for instance, described the work of "all the great *painters*" as having a "melting" quality sometimes called "freedom."[32] His preference for Joseph Mallord William Turner's "poetics of dissolve" has been linked to the viewer's "liv[ing] in the moment of contradiction and change."[33] Burne-Jones's image, in attesting to changes in contemporary reading practices, does suggest something of the visual excitement of modernity. Surrounded by walls of text, the figure in the illustration is like a pedestrian in Victorian London struggling to read a confusing array of texts like posters vying for the viewer's attention, which could account for the repetition of certain fragments in the illustration.[34] The letters of posters, subject to the elements like Chaucer's names in ice, were similarly prone to disappearing, as numerous paintings by Augustus Mulready attest (fig. 2.5).

Mulready's works link the transient visibility of posters to human forms of precarity and ephemerality, and indicate that, despite associations with stimulating change, the various vanishings associated with modernity also had more negative connotations. Morris, for instance, lamented the ongoing loss of traditions of popular art: "It is a grievous result of the sickness of civilisation that this art is fast disappearing before the advance of western conquest and commerce—fast, and everyday faster."[35] One of the most relevant images of evaporating culture in the nineteenth century, and one that strongly resonates with the glacial allusions of the Kelmscott illustration, comes from Karl Marx (for whose ideas Morris, if not Burne-Jones, was prepared by Ruskin's writings[36]). In the *Manifesto of the Communist Party*, Marx and Friedrich Engels describe "all fixed, fast-frozen relations" being "swept away" by capitalism. "All that is solid melts into air" and "man is at last compelled to face . . . his real conditions of life, and his relations with his kind" when "the icy water of egotistical calculation" has dissolved ancient bonds.[37] Without claiming a direct connection between the Kelmscott illustration and this passage (although Burne-Jones may have encountered it), I want to highlight the way both picture the old *and* the new melting away under capitalism—in Burne-Jones's case through the conflation of the literary tradition and the more ephemeral textual artefacts of mass communications.

In so doing, Burne-Jones's illustration echoes the way some of his contemporaries viewed modern print culture. The mass production and circulation of print by steam- and then electric-powered presses and railroads, lamented by Morris as "the present age of superabundance of books," could be imaged in terms of water and evaporation in a way that also evokes Marx's formulation.[38] The jurist and historian Frederic Harrison, for instance, warned against the "remorseless cataract of daily literature which thunders over the remnants of the past," with the result that people "neglect all the abiding parts of knowledge for the sake of the evanescent parts."[39] Morris similarly used the image of liquidation in his denunciation of "the utilitarian production" of cheap books, which "has swept away the book-producer in its current."[40] His response to evaporating quality in this inundation of print was to seek to create something "solid." The Roman type he designed for the Kelmscott Press, for example, was to be "solid, without the thickening and thinning of the line, which is the essential fault of the ordinary modern type."[41] The vanishing and difficult-to-decipher Roman letters on their unsolid ground in the *House of Fame* illustration may thus evoke poor-quality contemporary printing as much as the fading of ancient manuscripts.[42]

A lack of solidity, though, affects the illustration as a whole, and not just its lettering: it is more delicate than most of the others in the volume, which, with their thicker lines and deeper blacks, are more of a piece with their robust borders. Here, a slightly faded appearance suggests that the illustration itself is in the process of vanishing as one looks on at the forms of melting ice. All of Burne-Jones's illustrations for the Chaucer, like the ancient names this one pictures, were subject to a number of modern translations and erasures in the process of making the book. The artist's drawings were photographed by Emery Walker and Walter Boutall; reworked by Robert Catterson-Smith in Chinese white (to eliminate "everything but the essential lines"), pencil, and then ink; photographed onto wood blocks; and then cut by William Harcourt Hooper.[43] But in this metatextual illustration, the process takes on particular significance. Wood engraving, an art in the process of becoming largely obsolete, entails excising wood from the block around the lines to be inked. Here the evanescent ice, or empty white ground of the page, is made visible through dematerialization of parts of the block that, like the labor expended in their carving, are rendered invisible in the finished product.[44] Attentiveness to the image's play between presence and absence on the levels of subject, material, and technique thus allows various vanishings and erasures to become visible.

If, for Morris and others, contemporary print culture emblematized the melting or liquefaction of tradition and substance associated with capitalism, competition, and commodity culture, so too did the contemporary landscape of glass. Glass was, for the Victorians, a "congealed liquid" and key material of nineteenth-century consumer culture and modernism, as Isobel Armstrong has shown.[45] Ice and glass, as Chaucer's medieval poem makes clear, are visually similar, and later writers including Ruskin identified other shared physical features.[46] Chaucer's text reflects medieval glassmaking traditions, with which the poet had a degree of familiarity: it highlights the material transformation of sand into glass (imaged in the poem by the glass temple's location in a desert) and glass's role as a "medium for narrative" through stained-glass windows in the Temple of Venus as well as the inscriptions in Fame's glassy foundations.[47] Burne-Jones's oeuvre, which includes many designs for stained-glass windows, partook at once of a medievalizing aesthetic and the Victorians' "new glass consciousness" that evolved as plate glass remade the urban sphere into a realm of commodity spectacle through shopfronts and conservatories.[48] Glass in the Victorian period both materialized and dematerialized contradictions: it was simultaneously ancient and modern, "ethereal" (made with and of breath) and thoroughly material—indeed, a form of "crystallized" labor and, for Marx, "congealed labour time," associated by both Ruskin and Marx with capitalist abuse of workers. It could thus be simultaneously transparent, idealizing, and distorting.[49]

Returning to Burne-Jones's illustration now, we may start to appreciate the mountain's glassiness and even a kind of shimmer. Places where parts of letters have disappeared, leaving only whiteness—for example, to the right of the figure's left hand—suggest glare off a reflective surface, subjecting the names to an optical as well as a material erasure, a loss of sight as well as substance. This impression is heightened in the vellum copies of the Kelmscott Chaucer, as glare on the shiny, slightly pearly vellum can efface parts of the image, depending on the angle of viewing.[50] The figure of the dreaming narrator peering at the script that surrounds and even inundates him, like nineteenth-century advertising, starts to resemble not only an urban pedestrian surrounded by posters but also a Victorian window shopper, peering through one of the "glazed shopfront[s] that burnished commodities," in Armstrong's words.[51] He may even seem poised, with his cloth-covered right hand, to reach out to polish its perhaps sooty surface.[52] And some of the letters, hovering slightly in front of the ice forms, appear as though they are etched on a pane of glass through which we are viewing the entire scene. The erosion of tradition suggested by the melting names becomes conflated with commodity phantasmagoria, with the erasure of the value of labor in commodity culture, the glacier disappearing in favor of the glazier.[53]

In Chaucer's allegory, Fame's palace atop the mountain of ice is the site of all sorts of distortions and condensations as well. The walls are made of beryl, a crystal out of which glasses were made in the medieval period. Like the commodity-burnishing glass of shopfronts, the beryl walls "shoon ful lighter than a glas" and "made wel more than hit was / To semen, every thing." If there is a Victorian architectural counterpart to this fantastic, scale-confounding construction, as full of "windowes, / As flakes falle in grete snows" and thronged with those competing for, or having won, Fame's favor from around the world ("Auffrike, Europe, and Asye"), it is the Crystal Palace, "the Palace made o' windows," which Burne-Jones visited in 1854 at Sydenham.[54] An edifice to which both glaciers and icebergs were later likened, and which even contained "glaciers of frosted glass," the giant conservatory mobilized precisely the slippage between glass, ice, melting, and evaporation into air that I have been tracing.[55] As one visitor to Hyde Park during the Great Exhibition wrote, the Crystal Palace's "airy unsubstantial character" disconcerted the eye used to "the solid heavy details of stone," appearing like "a splendid phantasm, which the heat of the noonday sun would dissolve."[56] Another described it as appearing, "when the sun shines through the acres of glass," as "an exhalation," "the colour of the evening mist exactly."[57] At Sydenham, Elizabeth Eastlake saw the whole "airy structure," "resting like a cloud on the horizon," seem to vanish as "the scales of its transparent armour melt[ed] into a far more liquid atmosphere."[58] The ephemerality suggested by the enormous glass edifice wrote large the loss of substance associated with contemporaries' "acute awareness of a reflective world."[59]

The original Hyde Park structure encompassed an empire's wealth of material goods and, in bringing "multiple heterogeneous times and spaces together," was meant to embody an "allegory of progress."[60] Yet contemporary and later readings recognized the fragility of the exhibition's organizational principles as well as its edifice, which bespoke "a counternarrative of anxious competition," the "fears unleashed by the spread of the free market and the project of capitalist modernization."[61] At Sydenham, a sense of fragility and precarity likewise pervaded the experience of the vast "commercial speculation," as one review put it.[62] The exhibits (which encompassed cast courts, natural history displays, commercial booths, and more) were marred by a disconcerting "absence and confusion of scale."[63] Even the historical exhibits suggested impermanence. The casts of ancient sculpture, for example, left Eastlake with "a mournful sense of the perishableness of all things."[64] The building's quantities of glassy surface may have suggested the conservatory, but there was little conservation to be hoped for. Burne-Jones lamented its "painfully dazzling reflection," "transparency," and the "powerlessness" of its materials to "effect an Architecture"; he considered it an apartment fit for ephemera such as perfumes and "strawberry ices."[65]

The sense of evanescence associated with the Crystal Palace must be considered in terms of the "nascent planetary consciousness" that Jesse Oak Taylor and others have argued it brought into being. Taylor considers its role within Victorian London, a key "geohistorical location at which to trace the emergence of anthropogenic climate change," as "at once metonym and metaphor for the world-as-glasshouse." Its creation coincided with a "sharp spike" in atmospheric carbon content, and through it we may also trace a nascent understanding of the "greenhouse effect," a metaphor introduced in the nineteenth century. At a moment in which "climate and human society were beginning to appear as *mutually* influential," the vast array of people, materials, products, and networks the Crystal Palace brought together—and the perceptual challenges its dizzying shifts in scale created—suggested the difficulty of comprehending the forces changing the globe.[66]

The figure in Burne-Jones's Victorian illustration of Fame's mountain, pondering its glassy extent, seems to linger nostalgically on the cusp of change, on a moment of awakening consciousness of the material forces at work transforming both human and natural history and futures. The image suggests that the nature of history—and of nature—as Burne-Jones knew it is already partially lost to view, that the past and its traditionally valued actors will inevitably be, like those of Shakespeare's *Tempest*, "melted into air, into thin air," and "shall dissolve" like the "cloud-capp'd towers, the gorgeous palaces, / the solemn temples, the great globe itself."[67] The state-skipping Shakespearean image of melting into air recurred across literary, aesthetic, and scientific realms in the Victorian period, with variable connotations. In *Modern Painters* Ruskin lauded the way Turner would let a tower at a distance "melt into air."[68] Tyndall, who usually used the term *melting* to refer to a change from solid to liquid, pictured a universe bound by the law of conservation of energy in his 1862 lectures on heat, in which "suns may resolve themselves into floræ and faunæ, and floræ and faunæ may melt in air" without loss, for "to Nature nothing can be added; from Nature nothing can be taken away."[69] Samuel Moore, who translated the *Manifesto of the Communist Party* in 1888, also deliberately invoked Shakespeare with his phrase "all that is solid melts into air." The German original, "Alles Ständische und Stehende verdampft," could be translated more literally as "all the social estates and unquestioned things evaporate." Moore's Shakespearean formulation, as Tobias Meneley has argued, takes the stress off "specific sociological condition[s]," to foreground the idea that "elemental transmutation stands for the total experience of the modern age."[70] Both original and translation, though, invoke state-skipping metamorphoses. While Tyndall's image of solid beings melting into air condenses cosmic stretches of time for the modern reader, Marx's focuses on the revolutionary human time of modernity.

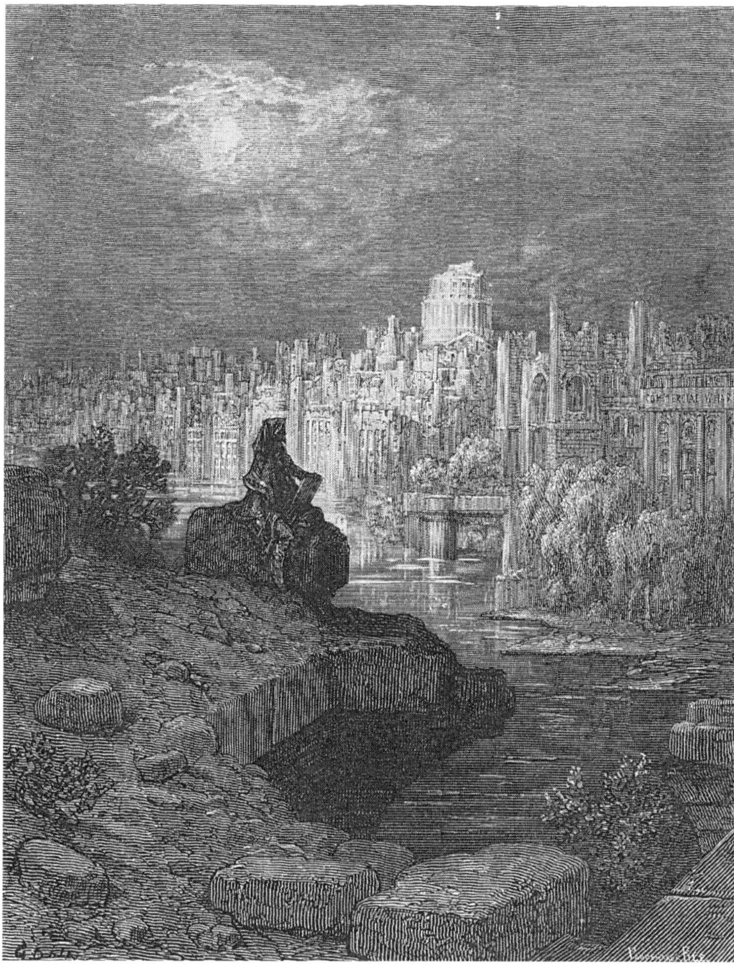

Fig. 2.6. A number of Victorian works imagined the future ruin of London and the British empire. Gustave Doré, *The New Zealander*, from Blanchard Jerrold and Gustave Doré, *London: A Pilgrimage* (London: Grant, 1872), between 188 and 189. Courtesy of the Thomas Fisher Rare Book Library, University of Toronto, Toronto, Canada.

Created in the last decade of the nineteenth century, Burne-Jones's melancholy illustration lacks any consolatory valence, and suggests that such elemental transmutation is inevitable and seemingly irreversible. If, in Chaucer's poem, those names preserved by shade "do signify that Fame, though arbitrary, can promise the afterlife of a shielded tombstone," Burne-Jones's entropic image offers no such certainty, despite the gravestonelike shapes of some of the ice formations.[71] Their cracked and melting forms literalize the lines of John Keats's epitaph, "Here lies one whose name was writ in water," and imply their future dissolution. His letters also make present the threat of future erasure implied for Brett's signature in *Glacier at Rosenlaui.* Burne-Jones pictures the transience of all human words and deeds, the material substrate and visual history of which seemed to be fading at an accelerating rate in his commodified Victorian modernity. The medieval does not offer a way out of the modern here: on the contrary, it anticipates it. For Burne-Jones, Chaucer's imagery of melting mountain ice prefigures Ruskin's and, in the artist's fin-de-siècle context, suggests that the eroding capacity to visualize and inscribe the past—in the print the Chaucer figure's writing hand is shrouded and perhaps even amputated—raises real concerns about our futures.

In this respect, Burne-Jones's illustration can be read in light of the future ruins trope, a motif linked to the rise of capitalism and speculative economies.[72] Specifically, we can compare it to the rendering of "The New Zealander" by Gustave Doré, whom Burne-Jones admired, in Doré and Blanchard Jerrold's 1872 *London: A Pilgrimage* (fig. 2.6).[73] The New Zealander was a figure invented by the historian Thomas Babington Macaulay in 1840, in a review of a book about the history of the papacy: the Catholic Church, he predicted, "may still exist in undiminished vigour when some traveller from New Zealand shall, in the midst of a vast solitude, take his stand on a broken arch of London Bridge to sketch the ruin of St. Paul's."[74] Macaulay's traveler, which drew on the eighteenth-century literary figure of the future tourist, quickly lost any religious significance and was invoked as a way to contemplate the future fall of empire—and the British empire specifically. Robert Dingley reads the figure of the New Zealander and fantasies of future ruins as responses to anxieties produced by "the breathtaking speed with which the results of the Industrial Revolution appeared to be transforming the structure of British life."[75]

Set in the Victorian present day, *London: A Pilgrimage* describes a city transformed by capitalism, in which all "classes and ranks of Englishmen . . . have the air of men seriously engaged in the sordid cares of commercial life." In Doré's last full-page illustration, the future tourist, whose clothing seems more medieval than antipodean, contemplates "something matching— / 'The glory that was Greece— / The grandeur that was Rome" while sketching the craggy remains of a city on the banks of the much-diminished Thames, its once thriving commercial artery.[76] The figure is silhouetted against the crumbling white ruins, ossified yet also icy in the moonlight, of warehouses lining the river beneath the domeless husk of Saint Paul's. Rather than sketching the cathedral, however, he seems to be drawing one of the most complete buildings in the picture, with the only legible inscription: COMMERCIAL WHARF. London, as pictured here, was for Victorian viewers "transparently contemporary," and thus suggestive of imminent ruin.

As David Skilton has shown, from about 1770 to 1870, the New Zealander and other future tourists visiting the ruins of London associated "the fall of the metropolis and the loss of empire with the decline and fall of Rome, establishing by its classical associations the transfer of learning (*translatio studii*) as well as the transfer of empire (*translatio imperii*) from ancient Rome to modern London, and predicting in the same device the eventual *translatio imperii* from Britain to the next great imperial power."[77] Burne-Jones's illustration, which, like Doré's, could stand as an "emblem for British cultural anxieties in an age of unprecedented transition," invokes yet suggests the end of this tradition.[78] Classical learning is, as we look on and as we have seen, being lost in translation, and no future imperial power is in sight.

Burne-Jones's vision of disappearance rather than transfer may in part account for what Velma Bourgeois Richmond identifies as the "nightmarish quality" of the *House of Fame* illustrations featuring the Chaucer figure, which, she contends, reveal Burne-Jones's "volatile state of mind, with moods of despondency and fears of isolation."[79] Burne-Jones, who identified with the medieval poet to some degree, has modified the portrait of Chaucer found in Harley MS 4866 in the British Library: etiolating and stretching the penless poet, who looks rather haggard; secularizing him by removing the prayer beads; and reversing the pose to offer a pale Victorian reflection of the more robust and colorful medieval artist.[80] For Richmond these portraits—as much self-portraits of Burne-Jones as images of Chaucer—betray "the tensions of late Victorian England."[81] While she focuses on the "fearful uncertainty" of the artist and the role of art in the fin de siècle, these questions are indissociable from the larger Victorian "tensions" I have suggested thus far, over loss of tradition and learning, the commercialization of life and art, and, as importantly, environmental changes.

The connection between capitalism and environmental ruin was not lost on Burne-Jones and his circle. Morris loathed the effects of coal on the landscape and supported various preservation and conservation societies, but he saw little recourse against "the present gospel of Capital" responsible for the environmental degradation he railed against. "There is no hope of bettering it," he wrote, so "let us eat and drink, for tomorrow we die—choked by filth."[82] As noted, Ruskin was aware of significant climate change in Europe, through his studies of both glaciers and clouds. In 1869 he raged that the "air which once inlaid the clefts of all [the Alps'] golden crags with azure is now defiled with languid coils of smoke, belched from worse than volcanic fires; their very glacier waves are ebbing, and their snows fading, as if Hell had breathed on them. . . . The light, the air, the waters, all defiled!"[83] Burne-Jones gave expression to similar anxieties in his works.

The *House of Fame* illustrations show or suggest inhospitable, even quasi-apocalyptic environments. The glass Temple of Venus, which Chaucer describes as filled with inscriptions and portraits, including one of Venus rising from the sea, appears, in Burne-Jones's rendering, awash in flood waters. The narrator leaves the temple to find himself in a vast field of sand, an image Richmond finds particularly "disturbing," with its "barren ground" and "blinding light" conveying extreme heat even as Fame's mountain suggests icy cold.[84] Next we see the poet, in a kind of rocky pit, caught in the claws of a gigantic eagle, which can be read both as an emblem of natural forces and as a messenger of the gods. After Fame's precarious, melting, and possibly smog-besmirched mountain of ice comes an illustration not of Aeolus trumpeting the names of the worthily famous with his gold clarion, but rather blowing his black horn, at Fame's unjust bidding, to unleash slander and bad air in the form of foul-smelling, multicolored smoke, like that from a smelter, which grows and grows.[85] The narrator's last stop in Chaucer's unfinished narrative is the wicker House of Rumour, which whirls through the air of a barren world like an aimless ark, its thoughtless whispers catching like sparks that blaze into city-destroying conflagrations, according to the narrator.

Various commentators have noted that Burne-Jones chose to render "passive rather than active scenes" in the Kelmscott Chaucer, giving the book "a still, timeless quality."[86] This is largely true, as far as the figures are concerned. They are often frozen, in a state of narrative as well as literal suspense, seemingly incapable of action. The same cannot be said, however, of his depictions of nature. If we, like the narrator, gradually learn to see the ice as melting in *The House of Fame*, many other illustrations give a more immediate impression of the environment as fluid, even volatile. In Burne-Jones's often barren, almost lunar landscapes, pools ripple, rivers churn and froth, waves chop and swell, rising seas bringing not only death and shipwreck but imaging the destruction of the landscape itself, with rock formations collapsing and the very page roiling, as in "The Frankeleyns Tale." Burne-Jones's fragile human figures are often powerless bystanders, seemingly doomed to vanish while the globe transforms into a landscape of empty deserts, sterile canyons, melting ice, and rising waters.[87]

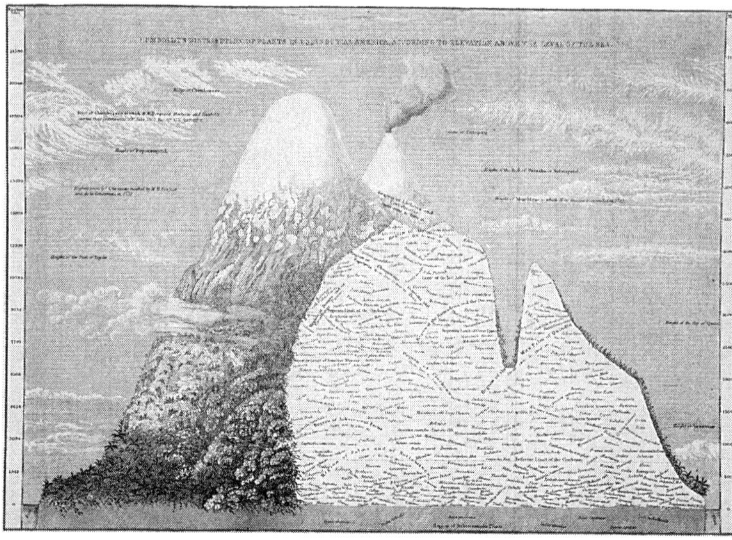

Fig. 2.7. Humboldt's legible mountain emphasizes human agency and knowledge. *Humboldt's Distribution of Plants in Equinoctial America, According to Elevation Above the Level of the Sea*, in *General Atlas of the World* (Edinburgh: A & C Black, 1854). Courtesy of Thomas Fisher Rare Book Library, University of Toronto, Toronto, Canada.

In 1872, Tyndall acknowledged that if conditions did not change, the Swiss glaciers would eventually be reduced to "mere spectres of their former selves." Ruskin berated him for offering "no evidence, nor even suggestion, as to the causes of the change itself."[88] Although Tyndall is generally credited with discovering the absorption of heat by water vapor and carbon dioxide, and thus a key factor in the greenhouse effect, he "drew no explicit connection between climate change and human activity."[89] The language of Tyndall's full sentence, however, does suggest a cause, if unwittingly—he described the depleted form of the Mer de Glace as exhibiting, "in a striking degree . . . excess of consumption over supply."[90] While referring specifically to water, his assertion that excessive consumption drives the disappearance of glaciers, made in the language of the market, is, in retrospect, dramatic irony. Burne-Jones's illustration of the names in ice, by contrast, draws prescient connections between disparate phenomena, in effect making visible the global effects of capitalism and the prospects onto which they open through a form of visual condensation.[91]

In this respect, the illustration merits one final comparison, to a different mountain engraved with Greek and Latin names, and one that also brings together disparate phenomena: Alexander von Humboldt's

Tableau Physique.[92] Burne-Jones, who had read Humboldt's *Views of Nature* and *Cosmos* by 1851 and "followed with imaginative interest the vast schemes with which they deal," may have encountered this illustration in one of its German or French printings, or in reduced form in a British atlas (fig. 2.7).[93] He may also have come across similar representations by Humboldt, such as the *Journey Towards the Summit of Chimborazo*. Like the "vast schemes" outlined in Humboldt's books, the *Tableau physique* draws connections between vegetation, elevation, air pressure, rainfall, fauna, and human accomplishments. John Tresch calls it a "cosmogram" aiming "to raise awareness of the interconnections of nature and to place human endeavors within a wider frame of meaning."[94] It invites viewers to actively read and cross-reference. It also, as Michael Dettelbach has argued, asks them "to introspect, to *see* 'the various climates [vertically] succeed one another, layered one on top of the next like strata.'"[95] Burne-Jones's illustration likewise asks the viewer to "introspect" and to envisage changes in climate, but over time rather than elevation, and in a global context.

Created in a more optimistic moment, the *Tableau physique* differs from Burne-Jones's in other ways, not least in terms of function but perhaps most tellingly in terms of legibility. The carefully engraved rock face of Humboldt's text-heavy representation suggests the enduring power of human knowledge on its stony substrate. Burne-Jones's illustration, which, on the contrary, spells nothing out on its vanishing ground, foregrounds the perceptual difficulty of seeing change, or loss, in action. As such, it anticipates more recent attempts to represent the Anthropocene, which, as Nicholas Mirzoeff notes, can be visualized but not seen.[96] The illustration demands a slowing down of perception, one capable of registering gradual changes of state as well as the relative insignificance of human time, which may be better represented by ice than stone. Perhaps as importantly, while Humboldt's *Tableau* foregrounds human agency by noting heights scaled, geography explored, and data recorded, Burne-Jones's vanishing names suggest what Mirzoeff describes as one of the most disturbing consequences of the Anthropocene: "the potential realization that no one . . . has authority."[97]

Elsewhere in his oeuvre, Burne-Jones "was producing work that stood as a melancholy meditation on the dying of history painting."[98] In the *House of Fame* illustrations, his theme is more sweeping, a meditation on much else that was dying or becoming lost in his modernity. His illustration of the names inscribed in ice renders all that is solid melting into air literally and figuratively. The image's conflation of medieval dream vision, Victorian print culture, glazed London shop fronts, melting glaciers, and future ruins suggests the artist's attentiveness to the complex relationships linking industrial capitalism and its forms of commodity fetishism and entertainment to climate change elsewhere in the world. Patience and slow looking allow us to perceive and contemplate the accretion of literary, scientific, visual cultural, and historical allusions on Fame's mountain, and their intimation of invisible but ongoing change. Taken together, they offer a glimpse of the very material forces that have shaped our natural and cultural histories in a way that resonantly suggests our evaporating futures, adding a critical extra dimension to the preservation project that the Kelmscott Chaucer represents.

CHAPTER 3

GRASPING THE ELUSIVE

VICTORIAN WEATHER FORECASTING AND ARTHUR HUGHES'S ILLUSTRATIONS FOR GEORGE MACDONALD'S *AT THE BACK OF THE NORTH WIND*

Carey Gibbons

George MacDonald's *At the Back of the North Wind* ran in the children's periodical *Good Words for the Young* from 1868 to 1870 and was published in book form in 1871.[1] The story follows the personified North Wind and Diamond, the angelic young son of a London cabman, as they undertake a series of journeys together. Arthur Hughes's illustrations, engraved by the Dalziel Brothers, suggest Hughes's understanding of the emerging fields of both meteorology and psychology. The waves of the North Wind's hair become symbolic of her ability to concurrently possess multiple aspects, resonating with the undulating lines in weather maps and charts that expressed weather conditions across different regions simultaneously. Although Hughes's representations imply an ability to achieve comprehension and mastery of the North Wind's multifaceted identity through their similarities to scientific imagery, the North Wind's association with the mysterious, unpredictable unconscious mind challenges any fixed understanding. This lack of clarity does not distance Hughes's project from the work of meteorologists, however; indeed, the inability to completely understand and accurately record the weather resonates with the mobility and indeterminacy of the North Wind. Challenges and difficulties continually interfered with meteorologists' forecasting efforts, hiding behind the apparently stable, empirical forms on Victorian weather maps and charts. Grasping the identity of the North Wind and predicting the weather become similarly elusive processes, compellingly signified by the form of the wave.

Hughes's wave forms, which initially convey the sense that the North Wind's various aspects can be grasped simultaneously, relate visually and functionally to the wave forms that appear in Victorian weather maps and charts, which similarly seek to capture and communicate an array of information at a glance. Beginning with discussions of the work of Victorian meteorologists and Hughes's efforts to convey the multiple aspects of the North Wind, I will then explore the North Wind as a personification of the unconscious mind, ultimately connecting this aspect of her being to the struggle to secure meaning in Victorian weather maps and charts. Like meteorological wave forms, the North Wind's wavy strands of hair ultimately challenge efforts to find stable, comprehensive meaning, instead conveying the complex and elusive nature of identity. I locate the wave form within a broad field of knowledge and cultural discourse, emphasizing its status as an image with internal activity and complexity, a form that could encourage open-ended, imaginative thinking.

Wind and Waves

Arthur Hughes was a follower of the Pre-Raphaelite Brotherhood, becoming intrigued by the movement after reading a copy of the short-lived Pre-Raphaelite periodical the *Germ* while training at the Royal Academy schools. Although he was a painter, known for a few masterpieces of Pre-Raphaelite technique such as *The Long Engagement* of 1853–1859 (held at the Birmingham Museums and Art Gallery), Hughes achieved more success as an illustrator, designing around eight hundred total published illustrations, most of which were for children's literature. Much of the current scholarship on Hughes originates in literary studies and overlooks the significance of his career-long interest in searching for new ways of envisaging and representing the body through the medium of illustration.[2]

Hughes completed hundreds of illustrations for the Scottish author, poet, and preacher George MacDonald, including seventy-six for *At the Back of the North Wind*, more than double the number of illustrations he produced for the other stories he illustrated in *Good Words for the Young*. Viewed as an early example of modern fantasy writing, the work features, as noted, a captivating female personification of the wind with long, flowing hair, who acts as a liminal being between life and the afterlife. With this character, MacDonald developed a unique Christian theology, emphasizing the ideas of spiritual growth and unity with the divine. Although no correspondence between Hughes and MacDonald survives, their families were close and saw each other regularly. It seems likely that they discussed the illustrations for *At the Back of the North Wind*, and Hughes appears to have worked intimately and extensively with MacDonald's text.[3] MacDonald's son and biographer, Greville MacDonald, discusses the story in relation to his "father's surest gift

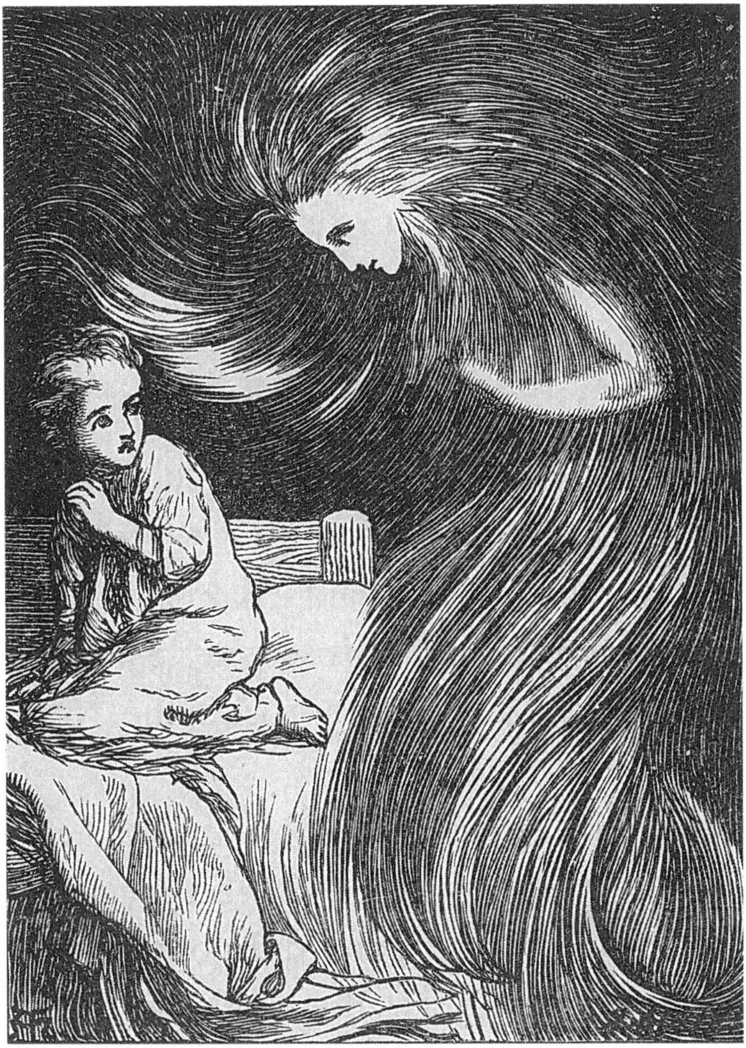

Fig. 3.1. When the North Wind makes her first appearance to Diamond, Diamond looks at her with hesitation and awe. Arthur Hughes, illustration for George MacDonald's *At the Back of the North Wind*, engraved by the Dalziel Brothers (London: Strahan, 1871), 11. By permission of the National Library of Scotland.

of faerie-allegory" and acknowledges the "like-inspired illustrations" by Hughes.[4] Similarly grouping text and illustrations together in a review of the book, *Art Journal* wrote: "The book is lavishly illustrated; the engravings are 'peculiar,' as the story is . . . the artist, as well as the author, may be tried by a high standard and maintain unquestioned right to public approval."[5] Additionally, the publisher Alexander Strahan was extremely pleased with the illustrations, proposing that the original drawings should be bound in volumes of ivory and gold for MacDonald's children.[6]

The character the North Wind becomes lifelike through MacDonald's vivid, descriptive language, particularly the repeated references to the energy of her wavy hair, which is compared to the movements of air and water. In a collapse of the boundaries between the female body and the natural world, tresses "lift and rise and spread and stream and flow and flutter," while "great billows of mist-muddy wind" are "woven out of . . . North Wind's infinite hair."[7] As the North Wind and Diamond fly through the "ocean of the air" together, they take "the shapes of the wind, eddying and wreathing and whirling and shooting and dashing about like gray and black water."[8] Hughes gives visual form to MacDonald's language, evoking the freedom and fluidity associated with the North Wind's hair repeatedly through his illustrations. Like MacDonald, Hughes confuses the boundary between hair and air; his illustration of the North Wind's first appearance (fig. 3.1) conveys the energy of her streaming black tresses that makes the "darkness in the hayloft" appear to be "made of her hair."[9] Her body, seemingly composed of wave-like strands, fills the picture, while Diamond raises his arm in a gesture of surprise and looks at the magical woman with awe and hesitation, taking in her appearance and attempting to make sense of her form.

MacDonald's descriptions and Hughes's illustrations of the North Wind were created during a period when scientific subjects were omnipresent. Popular periodicals played an enormous role in shaping the public's understanding of scientific theories and discoveries.[10] Articles, stories, illustrations, and even poetry on a range of new subjects provided Victorian readers with a variety of opportunities in which to think about their role as observers and participants in scientific discussions and developments. *Good Words for the Young* and its companion periodical for adults, *Good Words*, were characterized by a broad Christian perspective that often intersected with scientific content. For example, *At the Back of the North Wind* appeared in *Good Words for the Young* in 1869 alongside Charles Kingsley's "Madam How and Lady Why," a story that teaches the reader to see the design in nature through a series of lessons on geology from two powerful fairies. Kingsley's piece has numerous parallels with MacDonald's story; in both, supernatural female figures take on aspects of nature and a series of imaginative digressions convey nature's vastness and awe-inspiring power.

Given their substantial contributions to the field of children's literature and the fact that they both had children of their own, MacDonald and Hughes were likely aware of contemporary scientific literature for children and understood the possibilities for encouraging an imaginative contemplation of nature among readers—both young and old, since parents purchased literature for their children and read to them. When it was published in book form, *At the Back of the North Wind* appeared alongside numerous books for children centered around the wonders of the natural world. As with much of the content of *Good Words for the Young*, natural phenomena were often given religious or supernatural significance; for example, Margaret Gatty's popular book *Parables from Nature*, first appearing in 1855 and reaching its eighteenth edition by 1882, presents interaction with nature as a means of understanding God.[11] Similarly, John George Wood's best-seller *The Common Objects of the Country* (1858) compares the act of learning to read the Bible to learning to read nature.[12] The popularity of Wood's book was largely due to its lavish engraved illustrations of animals and scenes from nature, fascinating a public with a newfound appetite for images.[13] Imagery of all forms became more widely available during the nineteenth century and contributed to the development of a reader receptive to the artistic, scientific, and cultural associations of Hughes's illustrations.

At the Back of the North Wind was also published during a period of growing enthusiasm for observing and analyzing the weather.[14] Newspapers and periodicals published meteorological facts and opinions relating to changes in the atmosphere, giving rise to an increased interest in tracking the types of winds and their variable nature. In 1848, London's *Daily News* began publishing "telegraphic daily weather reports," and the first same-day weather maps were sold at London's Great Exhibition in 1851. The first *Daily Weather Report* prepared by the Victorian meteorologist Robert Fitzroy was issued to several newspapers in 1860.[15] Fitzroy, who founded the Meteorological Board of Trade in 1854 (renamed the British Meteorological Office in 1867), adopted the term *forecast* for his daily weather predictions, which were inaugurated in 1861. Additionally, the Royal Society approved the *Quarterly Weather Report* featuring extensive maps and charts that began publication in 1867. By 1872 the Meteorological Office had introduced a series of four same-day weather maps that tracked pressure, temperature, wind and sea, and clouds and rain. These maps, consisting of observations from about sixty stations taken at eight a.m., appeared in both the *Times* and the *Hour* by 1875.[16] In 1876, the *Times* started publishing additional daily weather charts based on data compiled the previous evening, and by 1879 the *Standard* and the *Daily News* had started including daily weather maps. Popular books containing maps and charts were also published, including Alexander Buchan's *Handy Book of Meteorology* (1867), which remarked on the

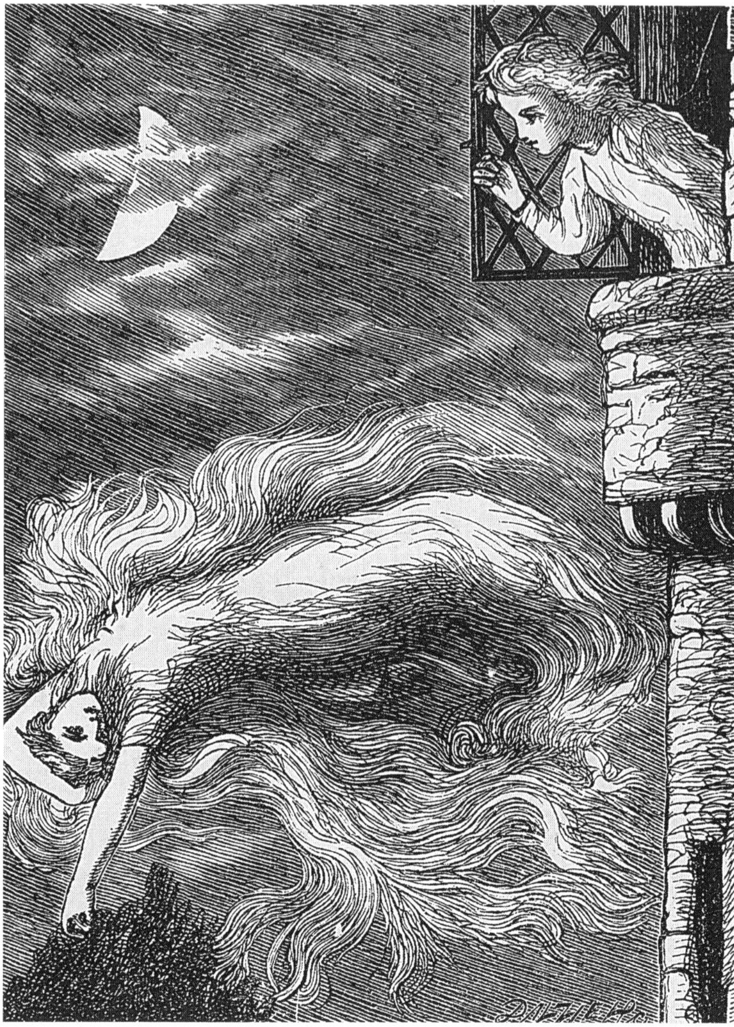

Fig. 3.2. Diamond watches the North Wind, who shows off the curving, waving forms of her body and conveys a sense of fluidity and elasticity. Arthur Hughes, illustration for George MacDonald's *At the Back of the North Wind*, engraved by the Dalziel Brothers (London: Strahan, 1871, 368). By permission of the National Library of Scotland.

rapid development of the field, as evidenced by the growing number of independent observers and increased purchase of meteorological instruments.[17] An anthropomorphized wind in the pages of *Good Words for the Young* thus would have been received by an audience that was likely intrigued by the imagery, given the increased popular interest in the field of meteorology.

Victorian meteorologists' discussions of the wind were noticeable for their incorporation of wave form imagery, which cannot be dissociated from the relevance of waves in scientific culture generally at the time. The idea of the wave as a periodic disturbance in space was crucial to nineteenth-century conceptions of light, heat, and sound, which were all understood in terms of waves rippling within the transparent, universal medium known as the ether.[18] Gillian Beer discusses the wave in relation to the need for a single explanation to account for all phenomena, drawing attention to the universalizing impulse of wave theory: "Nineteenth-century scientists from Helmholtz to Thomson, Clausius to Clerk Maxwell, were pursuing a single explanation of cosmic processes that would include light, heat, and sound and that would construe them all as motion, passing irreversibly beyond the reach of the senses and dissipating irregularly through the ether (that crucial explanatory substance that ebbed quietly out of the universe early in the twentieth century . . .)."[19] According to Beer, the wave was prominent not just within science but also within the broader Victorian imagination due to a "manifest social complex of referents which could overlap, incompletely but persuasively, with fresh scientific theory."[20] For example, educated people read the works of writers such as Heraclitus, Lucretius, and Ovid, who "emphasized the wavelike flow of energies." As the inhabitants of an island nation, Britons were also very conscious of the waves of the sea; crossing the Channel in a boat could give a memorable physical experience of wave activity, and sea journeys were crucial for emigration, imperialism, and trade.

The same imagery that dominates Hughes's illustrations of the North Wind (fig. 3.2) in the wavy strands of her hair, as well as the fluid curves of her elastic, seemingly boneless bodily structure, is found in the language of Victorian meteorologists, including the writings of John F. W. Herschel, a polymath also known for his work as an astronomer, mathematician, chemist, and pioneer of early photography. In his 1864 article "The Weather and Weather Prophets," in *Good Words*, Herschel observes that the atmosphere is "thrown into vast ripples or waves," "when driven in the state of wind over mountain ranges," producing barometric fluctuations.[21] In *Meteorology from the Encyclopædia Britannica* (1861), Herschel compares winds to the currents and waves of the ocean: "We live under what may not improperly be called an ocean of air, which covers the sea and land, and extends far beyond the summits of the highest mountains. Like the sea, this ocean has its currents, which are winds, its waves of vast extent and magnitude, not visible indeed to the eye, but capable of being made so to the intellect by means of the barometer."[22]

William Radcliffe Birt, an assistant to Herschel, studied barometric fluctuations and looked for data that would support the presence of

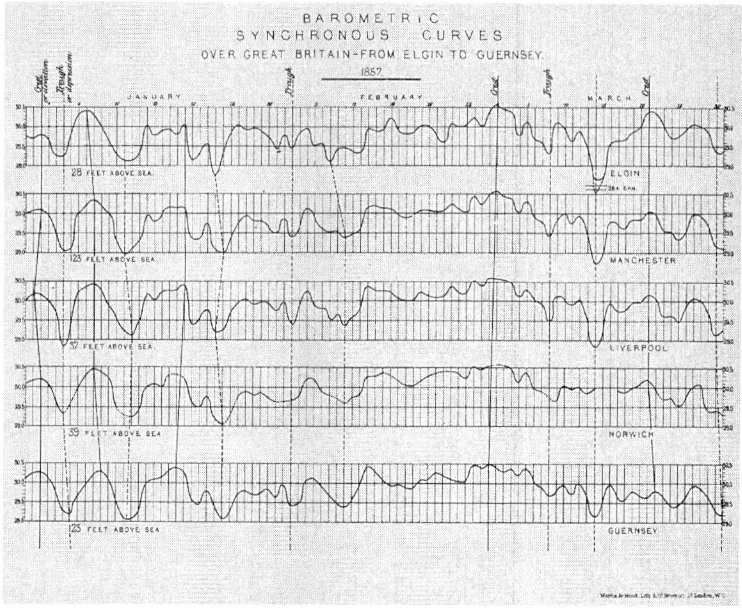

Fig. 3.3. Designed to be simple but comprehensive, Fitzroy's chart measures changes in barometric pressure over time in different locations. "Barometric Synchronous Curves Over Great Britain—From Elgin to Guernsey," 1857. Robert Fitzroy, *The Weather Book: A Manual of Practical Meteorology* (London: Longman, Green, Longman, Roberts & Green, 1863). By permission of the National Library of Scotland.

atmospheric waves.[23] Herschel originally explained his view of waves in 1843 to Birt: "My view of the matter to which I have alluded is that every barometric movement is essentially and intimately connected with an appropriate and corresponding wind—as . . . every wavelike movement in a fluid consists of two distinct things—an advancing form and a molecular movement— . . . the advancing form is indicated by the barometer—the molecular movement by the wind, and between these two phenomena there exists of necessity a close and purely dynamical connection."[24] Although Birt became progressively aware that the term *wave* merely described the curve traced on a graph in response to fluctuations of the barometer rather than referring to specific motions in the atmosphere, he continued to discuss the existence of waves while acknowledging that a clearer conception of them was needed.[25] Herschel firmly held onto his original belief in the correspondence between barometric movements and the movement of the wind.

The prominent meteorologist Robert Fitzroy refers to the presence of an "aerial ocean" in *The Weather Book* (1863), declaring, "That there

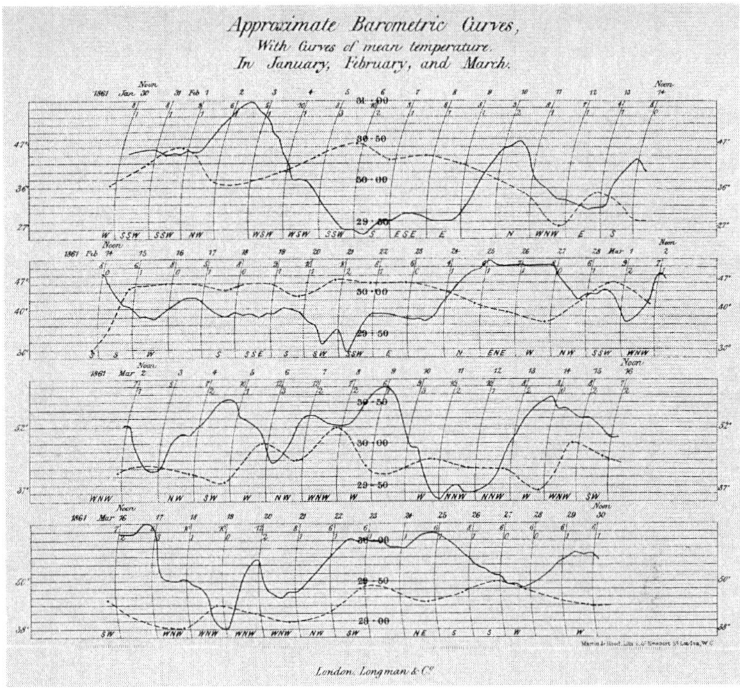

Fig. 3.4. Fitzroy seeks to convey atmospheric changes over time at a glance, combining barometric curves with curves of mean temperature in this chart. "Approximate Barometric Curves, with Curves of Mean Temperature in January, February, and March." Robert Fitzroy, *The Weather Book: A Manual of Practical Meteorology* (London: Longman, Green, Longman, Roberts & Green, 1863). By permission of the National Library of Scotland.

are waves of air, atmospheric undulations, or pulsations, we have authorities for accepting."[26] He also describes the appearance of waves (as well as "gyrations") when currents circulate: "These circulations of polar and tropical currents . . . raising or lowering the barometer, as they pass over any country, have caused the appearances, often noticed, of 'atmospheric waves,' corresponding to barometrical oscillations; as well as to the 'gyrations' of wind, so well elucidated by that eminent meteorologist [Heinrich Dove]."[27] Fitzroy, however, struggles with the "wave theory." He presents a view contrasting with Herschel's insistence that atmospheric waves are "rendered sensible by the increase or diminution of barometric pressure"; he acknowledges that a line on paper corresponding to the rise and fall of a barometer has a form "like the outline of a wave of water," "in which the crests will correspond to one current and the hollows (or troughs) to the opposite," but does not demonstrate "a mechanical, watery wavelike undulation of the body of atmosphere."[28]

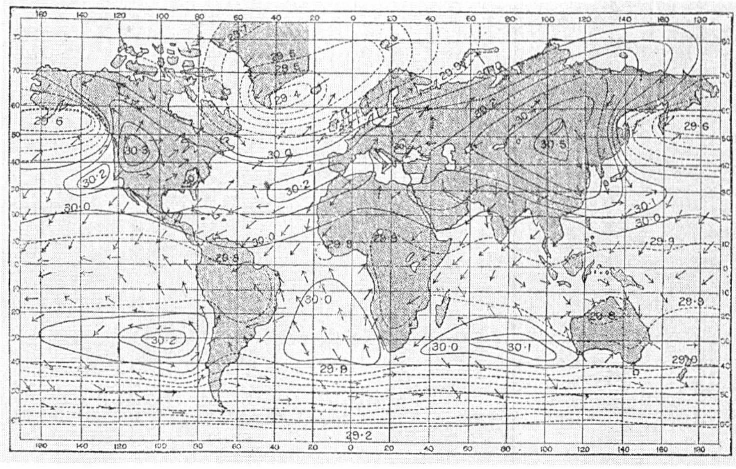

Fig. 3.5. Isobars, or lines formed by connecting places with the same barometric readings, produce wave forms, curves, and circles on this map. "January Isobars of the Globe and Prevailing Winds, 1870–80." Fig. 14 in B.S. [Balfour Stewart], *Encyclopaedia Britannica* 9th ed., vol. 16 (1883), s.v. "meteorology." © British Library Board 12214.a.1.

Still, the wave remained a pervasive idea within Victorian meteorology, despite disaffections with the idea of atmospheric waves among Fitzroy and others. Fitzroy acknowledges that his charts of storm observations "*seemed* to decidedly show consecutive 'waves' of the atmosphere, by the curves indicating altitudes of the barometric column," and observes that the works of the American meteorologist James Pollard Espy, with their "multitudinous curves closely covering many large sheets," initially appeared to "*prove* the case indisputably."[29] Furthermore, a reviewer of *The Weather Book* remarks on Fitzroy's ability to "give us the most perfect idea of the rise and progress of the atmospheric waves that we anywhere possess," despite the inconsistencies in his text.[30] The mental images evoked by Fitzroy of an "aerial ocean" and the meteorologist drawing "the outline of a wave of water" while tracking the wind are powerful and enduring.

In addition to using wave imagery in their writings discussing the wind, meteorologists incorporated linear wave forms into their maps and charts. Fitzroy was particularly interested in transforming his meteorological research into effective visual representations, and he and his assistant produced hundreds of charts of observations of the north Atlantic, designed to convey atmospheric changes "at a glance" and express the "consecutive simultaneous states of the atmosphere—as if an eye in space looked down on the whole North Atlantic" at regular intervals.[31]

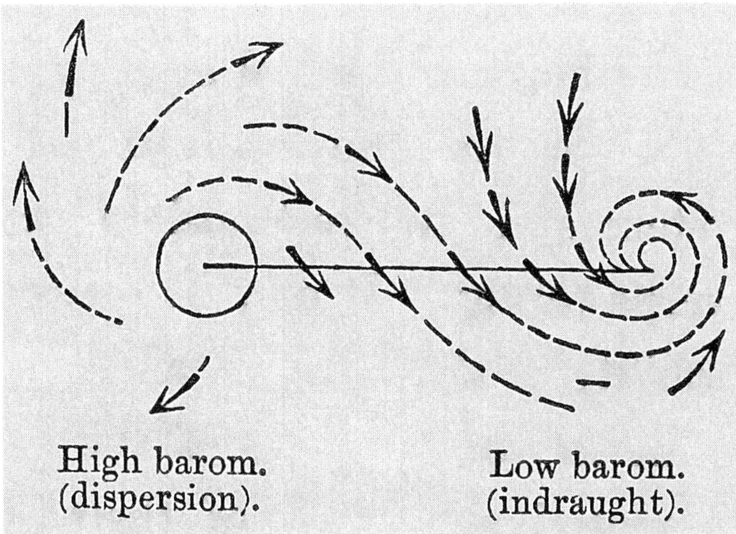

Fig. 3.6. Visualizing the "anticyclone" pattern of high-pressure areas surrounded by clockwise winds, and the "cyclone" pattern of low-pressure areas surrounded by high-velocity counterclockwise winds, this diagram joins the two wind conditions through a wave-like formation. Fig. 2 in Francis Galton, "A Development of the Theory of Cyclones," *Proceedings of the Royal Society of London* 12, no. 54 (January 1863): 386.
© British Library Board Ac.3025/21.

The wave forms in his charts of "Barometric Synchronous Curves" over Great Britain (fig. 3.3) and "Approximate Barometric Curves" with curves of mean temperature (fig. 3.4) from 1857 provide a means of summarizing and grasping changes in barometric pressure across a wide area over time. The crests and troughs of each line are emphasized, while the rows of lines join together in the mind to convey the sense of a simultaneous occurrence. The inclusion of a gridlike structure behind the undulating forms conveys regularity and predictability, adding to the sense of mental mastery over complex information.[32] In addition to appearing in charts of barometric curves, wave forms can be seen in the maps of meteorologists in the form of isobars, or lines joining places or points on the map with the same barometrical readings. Meteorologists like Robert Scott asserted that isobars are "the most important lines we can give on a weather chart," since they show "what the winds and weather are likely to be."[33] Isobars can be seen in the map featured in the "meteorology" entry in *The Encyclopaedia Britannica* (1883), which summarizes the atmospheric pressure of the globe and the prevailing winds from 1870 to 1880 (fig. 3.5).[34] Maps and charts such as these were intended to be simple but comprehensive, enabling any mind to quickly grasp atmospheric conditions across time and space.[35]

Curving, waving forms are also found in the meteorological visualizations of Francis Galton, who served on the oversight committee of the Meteorological Office and encouraged meteorological work within the Royal Society. In his *Meteorographica* (1863), Galton conveys the direction of the wind in maps through an arrangement of squares containing curves that face different directions, encouraging the mental aggregation of these curves into wavelike formations. Galton wrote that "a few judicious sweeps and shadings of a draughtman's pen may embody the simultaneous observations of hundreds of meteorologists," and encouraged "the ultimate condensation of bulky reports into small intelligible maps."[36] Additionally, Galton's research on the "cyclone" pattern of low-pressure areas surrounded by high-velocity counterclockwise winds, and "anticyclone" conditions in high-pressure areas that were surrounded by clockwise winds, led to the production of diagrams incorporating curving and wavelike lines that helped explain the movements of the wind (fig. 3.6).[37]

With the development of weather forecasts, charts, and maps that affected everyone, and of meteorology as a science "within the grasp of every mind," forecasting the weather became a daily practice applicable and accessible to all, while its visualizations became familiar and recognizable.[38] The telegraphic network made modern weather prediction possible, connecting the work of independent meteorologists with public knowledge and offering what Fitzroy described as a "means of *feeling*—indeed one may say *mentally seeing*—successive simultaneous states of the atmosphere."[39] As observers followed the streams of air flowing and forming winds from all directions and of various sizes and speeds, the linear form of the wave united the experiences of many, expressing the weather conditions across different regions simultaneously, while yielding information that affected a range of individuals and communities.

The North Wind

Placed in this context, Hughes's illustrations adapt the meaning of the wave form in weather maps and charts in order to enable a summarization or instinctive grasp of the North Wind's multivalency. Hughes repositions the wave form, translating it from a simple, recognizable form that helped forecast the weather and unified many conditions, places, and individuals at once to a representational device that indicates the simultaneous presence of the North Wind's varied aspects. In order to demonstrate the wave form's ability to function in this way, it is necessary to move beyond the North Wind's identity as a personification of the wind

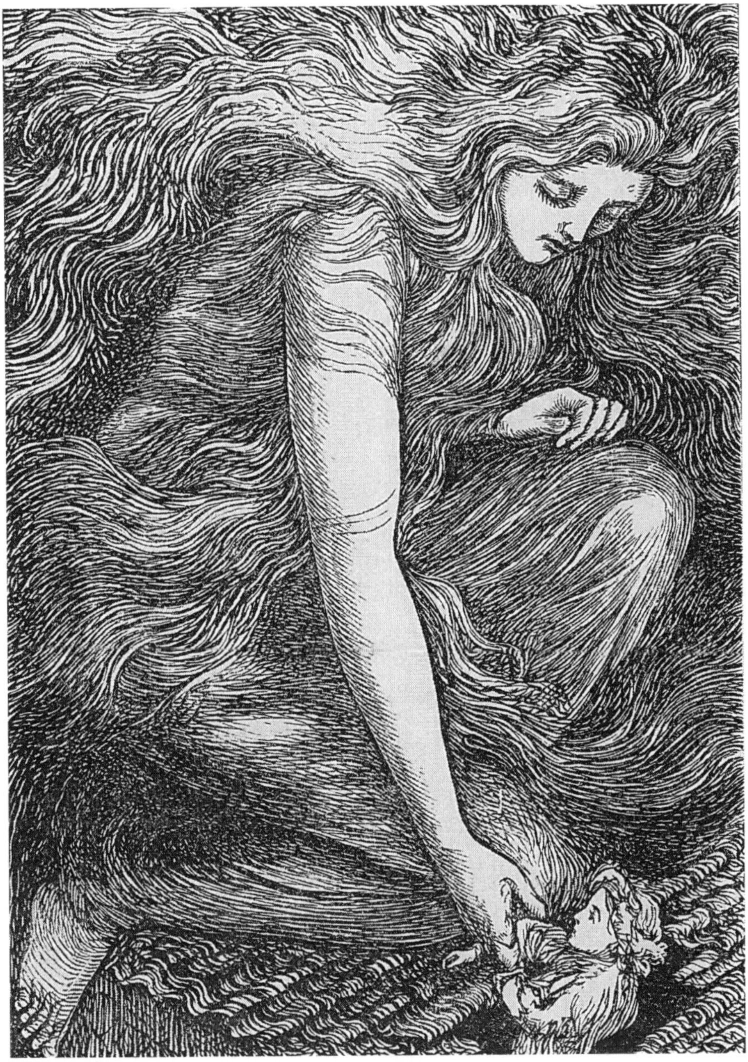

Fig. 3.7. Hughes conveys the North Wind's maternal and divine qualities, as she appears both nurturing and awe-inspiring. Arthur Hughes, illustration for George MacDonald's *At the Back of the North Wind*, engraved by the Dalziel Brothers (London: Strahan, 1871), 67. By permission of the National Library of Scotland.

and consider her maternal, divine, and erotic qualities. MacDonald relates how the North Wind's voice sounds more like Diamond's "mother's voice than anything else in the world," and after placing Diamond in her lap while positioned in a nest at the top of a great beach-tree, she hushes him "as if he were her own baby."[40] The North Wind emerges as a Mother

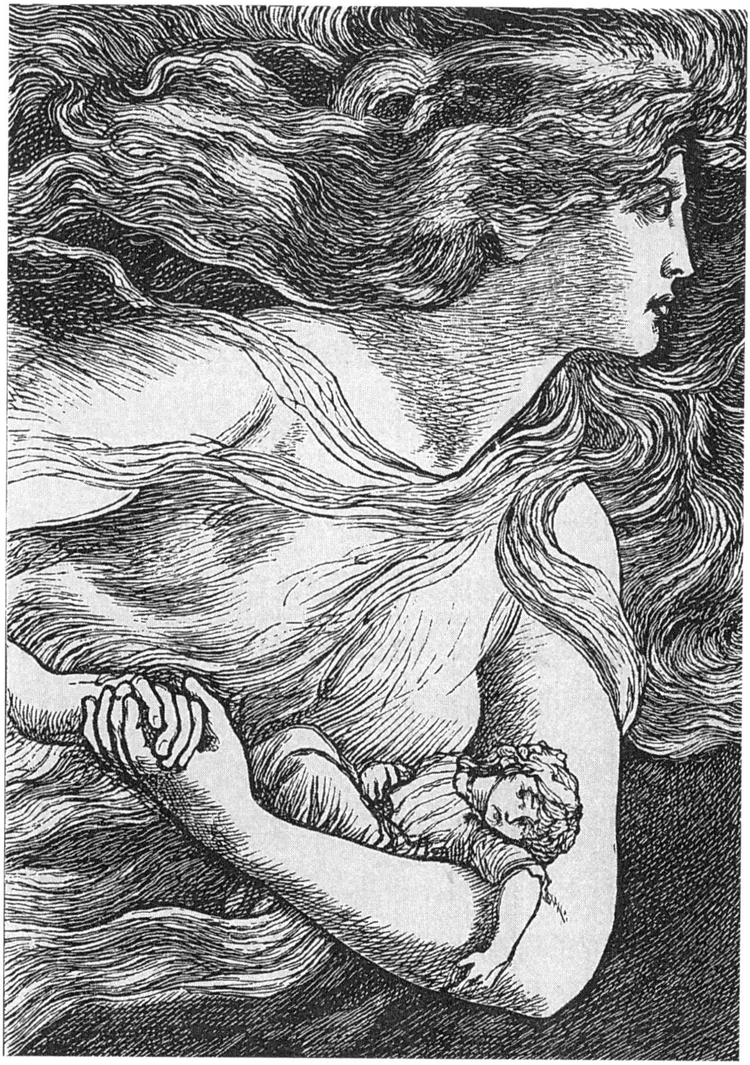

Fig. 3.8. The North Wind, cradling Diamond, prepares to carry out her divine mission of sinking a ship, as her luxurious waves of hair suggest her erotic potential. Arthur Hughes, illustration for George MacDonald's *At the Back of the North Wind*, engraved by the Dalziel Brothers (London: Strahan, 1871), 72. By permission of the National Library of Scotland.

Nature–like figure within the context of this genre of children's literature, but she also continues the development of the modern nature-goddess that occurred during the nineteenth century, with Hertha, the German earth-goddess, in Algernon Swinburne's poem by that name from 1867, being the most powerful of such figures to emerge in the period.[41]

Bonnie Gaarden has considered MacDonald's literature in relation to this development and suggested that he Christianized his goddesslike figures as a way of dissenting from widespread contemporary Christian doctrines and articulating his unique theology through his open-ended, mythlike associations and insistence on spiritual and psychological growth.[42]

The North Wind's maternal and divine functions are fused in an illustration by Hughes that draws attention to her large size and awe-inspiring presence (fig. 3.7). Appearing supernatural while also taking on a maternal function, she looks down on Diamond from above, watching and protecting as her body towers over him. The massive waves of hair crossing her limbs and garment signify her supernatural power, while conveying the movement and fluidity associated with the wind. Her hair impressively streams outward in multiple directions, pushing against the picture frame and even crossing the bordering line of the picture. Hughes's representation resonates with MacDonald's description of the North Wind's massive size and transformative properties: "Her head went up and toward the stars; and as she grew, still trembling through all her body, her hair also grew—longer and longer, and lifted itself from her head, and went out in black waves."[43] Hughes constructs a sense of mobility and power that does not obey boundaries.

Although she becomes a sublime, mystical figure through her impressive abilities and larger-than-life, imposing presence, the North Wind tells Diamond that she actually takes orders from an authority higher than herself, acting as an instrument of God for governing the world. She explains to Diamond that her tasks are managed by a baby, reflecting MacDonald's belief in "the divine idea of childhood, which moved in the heart of God when he made that child after his own image," or the idea of the "child as the visible likeness of the Lord himself."[44] The North Wind states that sometimes she is instructed to perform "disagreeable work" and announces that she must sink a ship. When Diamond expresses confusion, she explains, "I can do nothing cruel, although I often do what looks like cruel to those who do not know what I really am doing. The people they say I drown, I only carry to—to—to—well, the back of the North Wind—that is what they used to call it long ago, only I never saw the place."[45] Her actions suggest the following of God's orders, and her divine persona is further emphasized through her ability to take individuals to the back of the North Wind, a place understood to be the afterlife. In Hughes's illustration that shows her preparing to sink the ship (fig. 3.8), the divine, maternal, and natural aspects of the North Wind are joined; her wavy strands blend with the air as she holds Diamond in a maternal embrace, moving forward with purpose and conviction to carry out her divine mission.

The erotic nature of the North Wind is another aspect explored by both MacDonald and Hughes. When the North Wind first visits Diamond, she strikes his chest and sweeps off his clothes. Later in the novel, MacDonald draws attention to the dangerous aspect of her character and Diamond's attraction to it: "He looked out of the bed, and saw a gigantic, powerful, but most lovely arm—with a hand whose fingers were nothing the less ladylike that they could have strangled a boa-constrictor, or choked a tigress off its prey—stretched down through a big hole in the roof. Without a moment's hesitation he reached out his tiny one, and laid it in the grand palm before him."[46] As Nina Auerbach suggests, the arm takes on the totemistic aura that parts of the female body can acquire when severed from the whole.[47] Becoming both threatening and mesmerizing, the North Wind's luxurious waves of hair take on a similar fetishistic power.[48] Additionally, MacDonald's descriptions and Hughes's illustrations of the North Wind's hair can be interpreted within the context of Victorian physiology and phrenology, which contributed to the prevalence of ideas about the sexuality of women's hair, associating dark, abundant hair with strong, energetic, passionate, and even dangerous women.[49] Suggesting the North Wind's erotic potential and the strength of her resistance to conform to notions of virtuous womanhood, Hughes depicts her hair as massive and full of vitality, continually pushing up against the frame.[50]

In the illustration portraying the sinking of the ship, the mesmerizing hair takes on the destructive power associated with the femme fatale, a female category interpreted by some scholars as a nineteenth-century response to anxiety about women's increasing demands for equality and independence.[51] Expanding upon MacDonald's description of the winds writhing "like a storm of serpents" on the facing page, the strands of hair form thicker waves than in the other illustrations, recalling the mythological figure of Medusa with snakes in place of hair.[52] Closely associated with a fear of female sexuality, the figure of Medusa works to eroticize the threatening aspect of the North Wind, whose hair flies through the air, suggestively falling against her bosom and pushing against the picture's borders to suggest fervent, unrestrained, deadly sexuality.[53]

When the North Wind takes Diamond on her journey, she says, "I shall only want one arm to take care of you; the other will be quite enough to sink the ship," implying her ability to take on nurturing and destructive qualities simultaneously.[54] Diamond expresses confusion at this act of violence, and in response she explains that the North Wind he knows and understands to be good is the same as the North Wind he does not yet know. The dialogue between the two suggests the presence of more than one persona within the same "me":

> DIA:"'Here you are taking care of a poor little boy with one arm, and there you are sinking a ship with another. It can't be like you.'
> "'Ah! But which is me? I can't be two mes, you know.'
> "'No. Nobody can be two mes.'"[55]

The North Wind ultimately explains, "Then the me you don't know must be the same as the me you do know,—else there would be two mes."[56] Diamond agrees that the two different aspects must exist within the same "me." Hughes's illustration of the sinking of the ship convincingly suggests the coexisting aspects of the North Wind—as her sensuous form flies through the air, she cradles the frightened Diamond and obeys the orders of a higher divine authority to carry out her destructive mission.[57] Hughes avoids depicting the North Wind in terms of objective ideals or restrictive stereotypes, acknowledging her ability to concurrently possess multiple qualities, and engaging with what Susan Casteras refers to as the "largely untouched dimension of femininity" between the poles of "aberrant female" and "mainstream" during the Victorian period.[58] The North Wind's waves of hair assert not only her departure from expectations of virtue but also her unwillingness to be categorized or bounded in terms of a singular identity.

Like the meteorologists who used wave imagery in maps and charts that were reproduced and distributed in order to convey meaning "at a glance," enabling a large number of individuals to grasp a range of weather conditions and observations, Hughes similarly seeks to subsume the multiple or complex within an elemental form. Although certain illustrations may emphasize a particular aspect more than others, he continually suggests the coexistence of personas or states within the North Wind through his use of the wave form. Additionally, Hughes's consistent wave imagery unites the illustrations and emphasizes the all-inclusive power of the North Wind, seeking to convey her qualities simultaneously rather than isolating discrete aspects of her personality. Tracing the pattern of the blocks of illustrations in the layout for the *Good Words for the Young* version of the story actually produces a larger wave across the two-page spread, reinforcing Hughes's wave forms within the reader's imagination. Hughes carefully delineates each undulating strand of the North Wind's hair in each illustration, yet repeats a basic linear form, reflecting the view held by the illustrator Laurence Housman in 1908 that Hughes's work possesses a "really extraordinary power—and by power I mean economy—of design," and an ability to "get to his point straight, with no extraneous detail, no thoughtless elaboration."[59] Like meteorologists, who were interested in the idea of a whole vision that could address a range of interests and communities yet be understood at a glance, Hughes creates wave forms that are both accessible and all-encompassing.

The North Wind as the Unconscious Mind

The wave imagery found in the work of Hughes does not eliminate the presence of elusiveness and mystery, despite the wave form's empirical, summarizing potential. Hughes suggests the ability for the form to equally signal unpredictability or uncertainty, elaborating on MacDonald's connection between the North Wind and the unconscious mind through the act of dreaming. Diamond at first believes he is dreaming when the North Wind visits him and mysteriously vanishes, stating that time seems to pass "as if he were in a dream," and, after not seeing her for a week, he begins "to feel as if North Wind were a dream of some far-off year."[60] When Diamond hears horses talking in the stable in the middle of the night, he again begins "to feel as if he had been dreaming," and after glancing "to see if North Wind was anywhere visible," he returns to bed.[61] The presence of the North Wind during the act of dreaming is also suggested by her ability to bring Diamond both into and out of the dream state; as Diamond tells his friend Nanny: "I believe North Wind can get into our dreams—yes, and blow in them. Sometimes she has blown me out of a dream altogether."[62] At times the dream world and reality are fused, and MacDonald does not explicitly state that Diamond is transferring from one to the other.

The Romantic idealization of the child's imagination was challenged in the nineteenth century as it became increasingly linked with a lack of reason.[63] As Stephanie Schatz observes in her discussion of midcentury medical theories by thinkers like Henry Holland, Forbes Winslow, and Henry Maudsley, children's imaginative pursuits were often pathologized.[64] Daydreams or reveries were thought to be dangerous, since they merged unconscious thoughts with the conscious mind and could lead to a conflating of reality with fantasy. Maudsley claimed that young children believe their imaginative fantasies since they are unable to "correct sense [with] reflection," drawing a parallel between the mental state of children and adults who suffer from insanity.[65] Schatz argues that Lewis Carroll creates a "dream-child persona" that deviates from contemporary medical characterizations, instead making a case for "the creative and intellectual potential of childhood abstraction, fancy, and daydreams."[66] Carroll (Charles Lutwidge Dodgson) was a close friend of the MacDonald family for decades, and MacDonald similarly embraced the imaginative possibilities that resulted from the mingling of unconscious and conscious thoughts.[67]

Although the nineteenth century saw a particular fascination with the mind of the child, the idea of the dream world as a space of unrestrained invention was a subject of fascination and discussion that extended beyond childhood. *At the Back of the North Wind* was published at the time that a range of scientific thinkers, including William B. Carpenter, Thomas Laycock, Alexander Bain, Herbert Spencer, and George Henry Lewes, were exploring and debating the role of the unconscious mind.[68] Theories were popularized in books and periodicals, such as the *Westminster Review, Fraser's Magazine,* and *Macmillan's Magazine,* that aimed to explain new developments in mental science to a growing middle-class readership in an accessible manner. As Nancy Rose Marshall notes in her contribution to this volume, the nascent field of psychology was fascinated by the learning potential of dreams and hallucinations. There was an increasing realization that the cognitive and behavioral abilities of human beings were largely dependent on the unconscious portion of their nature, over which they had little voluntary or volitional control.

Carpenter's theory of "unconscious cerebration" and his analysis of memory, discussed in his 1855 edition of *Principles of Human Physiology,* was particularly influential.[69] Carpenter notes how the cerebrum retains or stores latent memories or impressions that are "forgotten in the waking state" but "present themselves to the mind of the dreamer," and observes that "dreams are most remembered in the waking state," when they have passed through the "transitional phase" or "passage between the states of dreaming and waking."[70] He developed the idea of unconscious cerebration more fully in his *Principles of Mental Physiology* (1874), positioning the unconscious mind as a productive and an integral part of perception, memory, and behavior and emphasizing the intertwining of unconscious and conscious experience. He argues that ideas outside of consciousness are stored in memory and can be "revived again in full vividness under certain special conditions," comparing the emergence of a forgotten word or phrase to a sudden "flashing" that can occur when "thinking of something altogether different, or on awaking out of profound sleep."[71]

Although MacDonald may have not been directly familiar with the work of Carpenter or other theorists of mental science, he references their ideas by alluding to an assertion by his close friend John Ruskin about the unconscious mind in 1893:

> As to this matter of *creation,* is there . . . any genuine sense in which a man may be said to create his own thought-forms? Allowing that a new combination of forms already existing might be called creation, is the man, after all,

> the author of this new combination? Did he, with his will and his knowledge, proceed wittingly, consciously, to construct a form which should embody his thought? Or did this form arise within him without will or effort of his—vivid if not clear—certain if not outlined? Ruskin (and better authority we do not know) will assert the latter, and we think he is right. . . . Such embodiments are not the result of the man's intention, or of the operation of his conscious nature. His feeling is that they are given to him; that from the vast unknown, where time and space are not, they suddenly appear in luminous writing upon the wall of his consciousness. Can it be correct, then, to say that he created them? Nothing less so, as it seems to us. But can we not say that they are the creation of the unconscious portion of his nature?[72]

MacDonald explores these issues of agency and creation in *At the Back of the North Wind*, linking them with theology. The North Wind functions as a kind of intermediary figure connecting the unconscious mind to God, and her presence signals that Diamond is dreaming or engaged in unconscious mental activity, where God's presence and operation are most immediate. The story reflects MacDonald's belief that "God sits in that chamber of our being in which the candle of our consciousness goes out in darkness, and sends forth from thence wonderful gifts into the light of that understanding which is His candle."[73]

While Diamond and his mother are sitting on a beach, a book on the sand blows and flutters about. The book contains nonsensical nursery rhymes that Diamond's mother dismisses. She tries to find something better to read, but the wind keeps blowing the pages back to the same verses:

> it's all in the wind
> that blows from behind
> and all in the river
> that flows for ever
> and all in the grasses
> and the white daisies
> and the merry sheep
> awake or asleep
> and the happy swallows
> skimming the shallows
> and it's all in the wind
> that blows from behind[74]

The lines go on for pages, with variations of the same or similar words and phrases, which Diamond says must have been written by someone who visited the back of the North Wind. The narrator relates: "Now I do not exactly know what the mother read, but this is what Diamond heard, or thought afterwards that he had heard. He was, however . . . very sleepy, and when he thought he understood the verses he may have been only dreaming better ones."[75]

Diamond cannot find the poem when he looks for it later, but it appears to come back to him in his dreams in the form of a song. MacDonald describes the interaction between the dream world and reality as Diamond awakens from sleep:

> When he knew he was coming awake, he would sometimes try hard to keep hold of the words of what seemed a new song . . . a song in which the words and the music somehow appeared to be all one; but even when he thought he had got them well fixed in his mind . . . one line faded away out of it, and then another . . . till at last there was nothing left but some lovely picture of water or grass or daisies, or something else very common, but with all the commonness polished off it, and the lovely soul of it, which people so seldom see, and, alas! yet seldomer believe in, shining out. But after that he would sing the oddest, loveliest little songs to the baby—of his own making, his mother said; but Diamond said he did not make them; they were made somewhere inside him, and he knew nothing about them till they were coming out.[76]

When Diamond sings the songs, he is not able to determine where exactly they come from; he is "only trying to remember what he heard the river sing at the back of the North Wind," and later explains that the baby "pulls them out of me."[77] The verses on the beach, which Diamond may have experienced in a dream state, appear to stay with him in his unconscious mind, becoming revived when he awakens from sleep and sings to the baby.

Carpenter's description of latent or unconscious thoughts suddenly "flashing" forth, and the likelihood for this to happen while emerging from sleep, corresponds with the sudden appearance of the verses in Diamond's consciousness as he awakens and the baby "pulls them out." MacDonald emphasizes the role of the baby, who reflects his belief in the divine aspect of childhood, in formulating and bringing forth ideas from the unconscious mind. Through the baby, and the North Wind as a divine representative connected to the act of dreaming, MacDonald implies that God acts on the unconscious, playing an important role in the creation of human thoughts. Complete access to God is achieved with Diamond's death at the end of the novel, when he enters the place where the verses originated, or the back of the North Wind, and the unpredictable waves of the North Wind give way to a rayless light appearing to emanate from flowers.[78] The North Wind, then, is a redemptive force, taking her place among the many supernatural female characters in MacDonald's work who often take on the role of leading the protagonist away from mystery toward fundamental truth and understanding.[79]

Hughes's illustrations move beyond MacDonald's characterization of the North Wind as an intermediary figure leading Diamond toward an ever-increasing experience of unity with the divine, peaceful equilibrium, and the attainment of higher truths, instead portraying her as an embodiment of the unconscious mind. An examination of the Scottish journalist and author Eneas Sweetland (E. S.) Dallas's two-volume publication *The Gay Science* (1866), in which he applies mid-Victorian psychology to an analysis of the poetic imagination and aesthetic pleasure, reveals the importance of metaphor and analogy in communicating conceptions of the mind in nineteenth-century psychological discourse and contains imagery that particularly resonates with Hughes's illustrations. Dallas describes the unconscious as a "mental existence within us," and "secret flow of thought," which "haunts us like a ghost or dream and is an essential part of our lives."[80] He continues, "how vast is its extent, how potent and how constant is its influence, how strange are its effects." The unconscious action of the mind is conceived of as an anthropomorphized supernatural figure, not unlike the North Wind; in addition to "ghost," Dallas describes the presence of a "fairy in our thoughts, a willing but unknown and tricksy worker which commonly bears the name of Imagination."[81] Dallas also conceives of the human soul as a double, "or at least as leading a double life," containing a conscious portion, as well as an unconscious portion known as the "Hidden Soul," a transformative, creative force that shapes mental activity.[82]

Dallas was not alone in conceptualizing the mind in this way; a range of metaphors and models were used to explain mental activity. The sense that the unconscious, dreaming state and the conscious, waking state might correspond to two separate but interactive worlds or beings was remarked on by other writers who sought to explain the new developments in Victorian psychology to a general audience. For example, Francis Power Cobbe, who wrote for popular periodicals and contributed to important psychological debates relating to memory and the limits of individual autonomy, produced what Carpenter admiringly referred to as "graphic sketches" of everyday instances of unconscious activity.[83] She vividly describes the process of forgetting and then remembering something, after "we have ceased consciously to seek for it," observing that we have no "sense of labor in the task, any more than if it were performed by somebody else." Additionally, when a forgotten word or line of poetry is suddenly remembered, she describes the process in terms of "an obedient secretary or librarian . . . hunt[ing] up a missing document or

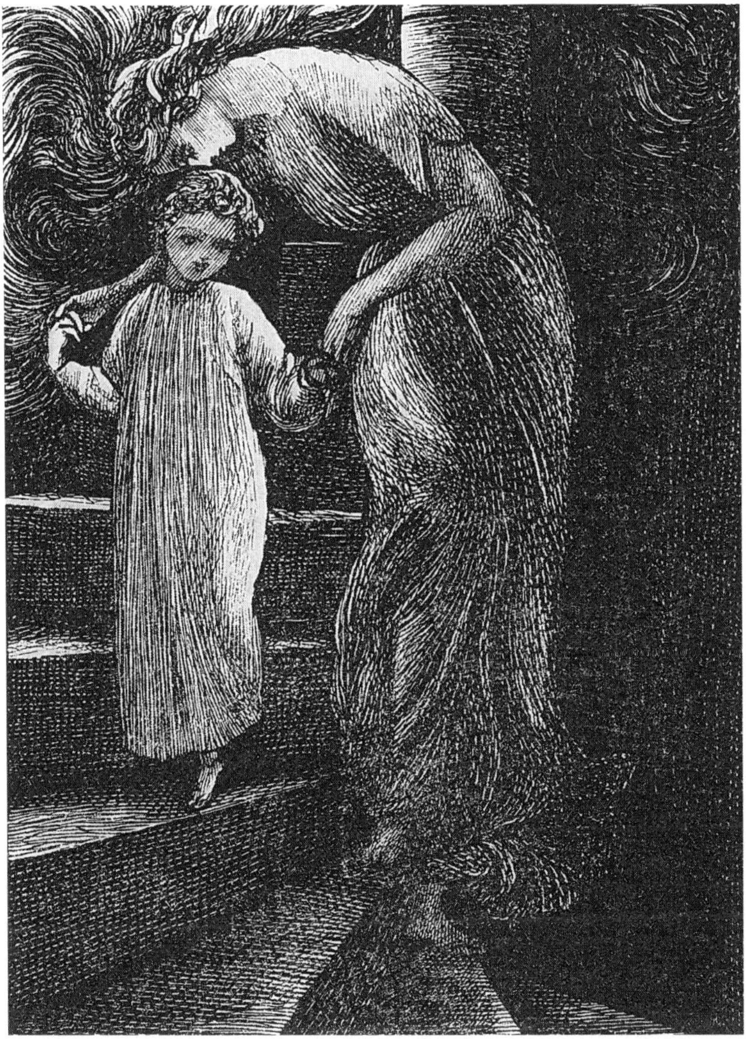

Fig. 3.9. The North Wind, associated with the world of dreaming, becomes an embodiment of the unconscious mind as she walks down a spiral staircase with Diamond. Arthur Hughes, illustration for George MacDonald's *At the Back of the North Wind*, engraved by the Dalziel Brothers (London: Strahan, 1871), 80. By permission of the National Library of Scotland.

turn[ing] out a word in a dictionary, while we amused ourselves with something else."[84]

Therefore, it would not have been unusual for readers to interpret Hughes's illustration featuring the North Wind and Diamond walking down a spiral staircase (fig. 3.9) as a kind of metaphor for mental activity that presents a composite notion of the self, with the North Wind functioning as the unconscious mind and Diamond as the conscious self. The right portion of the North Wind's body fades into the shadows of the stairwell, corresponding with Dallas's allusions to the secrecy and darkness of the unconscious mind, while her energetic, wavy strands of hair give visual form to his vivid language comparing unconscious thought to the sea: "When one is most struck with the grandeur of the tides and currents of thought that belong to each of us, and yet roll beyond our consciousness, only on occasions breaking into view, one is apt to conceive of it as a vast outer sea or space that belts our conscious existence something like the Oceanos of Homer."[85]

Additionally, Dallas's explanation of the conscious and unconscious realms in terms of concentric rings, and the ongoing interactions between them, presents a compelling comparison with Hughes's imagery: "The thing to be firmly seized is, that we live in two concentric worlds of thought,—an inner ring, of which we are conscious, and which may be described as illuminated; an outer one, of which we are unconscious, and which may be described as in the dark. Between the outer and the inner ring, between our unconscious and our conscious existence, there is a free and a constant but unobserved traffic for ever carried on."[86] The hands of the North Wind and Diamond are clasped so that they form a ring, around which an outer ring is formed by the circular structure of the staircase. Outside of the inner ring she forms with Diamond, the North Wind's hair rises and flows in multiple directions, suggestive of constant streams of unconscious activity. By depicting the North Wind leading Diamond into a dark space, with her body partially obscured in darkness, Hughes implies, like Dallas, that unconscious thoughts occur secretly or without one's knowledge, and that cognition and behavior do not require complete awareness. The wavy strands gather out of the darkness only to become obscured again, and move both toward and away from Diamond's body. The illustration suggests the multiple relations and combinations among ideas that can occur during the unconscious state of dreaming, as well as the potential for unconscious impressions or thoughts to spontaneously enter or leave consciousness.[87]

A sense of the interactive, collaborative relationship between the conscious and unconscious minds is also revealed through Hughes's depiction of a scene in which the North Wind fears she will drop Diamond and thus fashions a place for him in her hair by weaving and knotting together strands to create a pocketlike nest (fig. 3.10). The woven portion of hair conforms to the shape of Diamond's body, possibly implying that the North Wind, or the unconscious mind, actively weaves or generates processes that are integral to his perception, memory, and behavior.

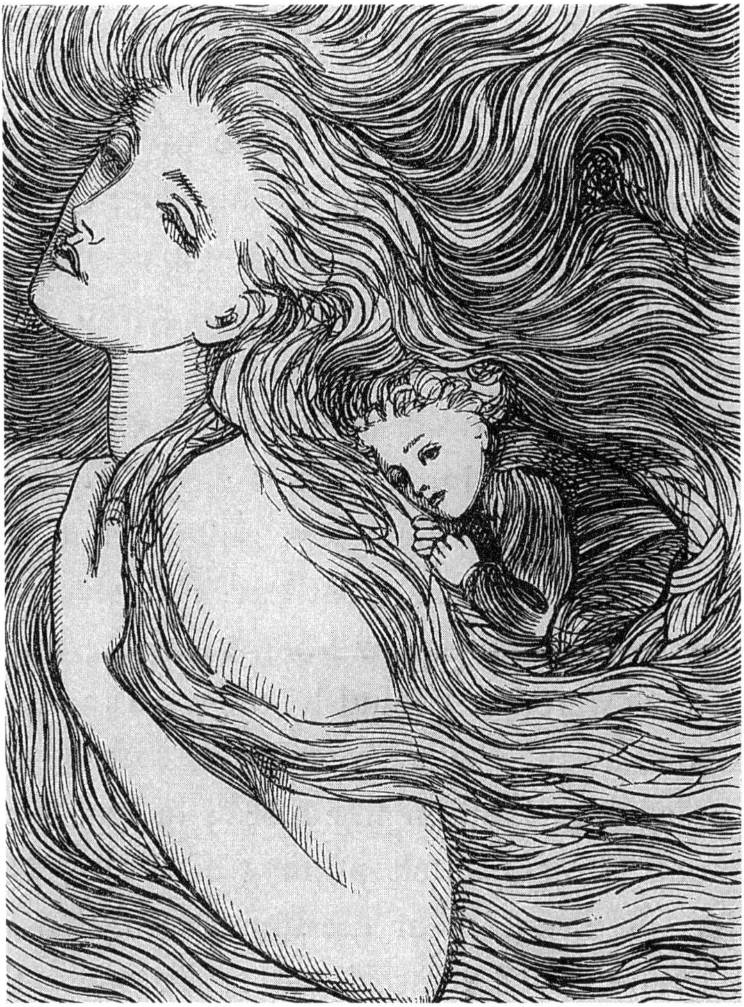

Fig. 3.10. This illustration, which shows the North Wind taking Diamond on a journey after creating a woven, pocketlike nest for him in her hair, can be interpreted within the context of the interactive, collaborative relationship between the conscious and unconscious minds. Arthur Hughes, illustration for George MacDonald's *At the Back of the North Wind*, engraved by the Dalziel Brothers (London: Strahan, 1871), 39. By permission of the National Library of Scotland.

The illustration strikingly corresponds with another passage from Dallas's text: "Outside consciousness there rolls a vast tide of life, which is, perhaps, even more important to us than the little isle of our thoughts which lies within our ken."[88] Diamond's body is curled up, forming what resembles an island surrounded by the waves of the sea. The expansive, undulating strands, occupying most of the picture and pushing against the borders, provide an enduring visual companion to Dallas's words: "In the dark recesses of memory, in unbidden suggestions, in trains of thought unwittingly pursued, in multiplied waves and currents all at once flashing and rushing, in dreams that cannot be laid . . . we have glimpses of a great tide of life ebbing and flowing, rippling and rolling and beating about where we cannot see it."[89] Both elemental and dynamic, the defined, linear wave forms compellingly align with Dallas's conception of the unpredictable, active, and mysterious world of the unconscious mind, despite that world's apparent resistance to notions of delineation.

The wave forms in Hughes's illustrations continually convey the potential for mobility and variation. Unable to escape their relationship to the unconscious mind, the undulations of the North Wind struggle against the idea of conveying information that is secure, measurable, or comprehensive. The review of the second edition of *At the Back of the North Wind* in the *Aberdeen Journal* observes the challenge of grasping the North Wind, referring to her spirit as "a force that throws around the young bright life something of that air of mystery with which all life is felt to be encircled."[90] Hughes uses wave forms to present a multifaceted, mysterious notion of female identity, as the undulating lines ultimately indicate the lack of closed or complete meanings. Like the North Wind, the self can possess proliferating, shifting, and even contradicting categories and definitions. Furthermore, individuals are not abstracted or dissociated but are connected to others, subject to influences, and part of a system of transformative interactions.

Unruly Meteorological Wave Forms

Although Hughes's exploration of the wave form in relation to the unconscious mind might initially seem to challenge the parallel established between his work and the work of meteorologists, a closer consideration of the challenges faced by meteorologists actually strengthens the correlation. Victorian meteorologists continually searched for an array of results that could address a wide audience and be summarized "at a glance," only to realize the difficulties of observing vast, changeable, and unreliable forces. They could not ignore the fact that the weather was often fleeting and inexplicable. The wind could be influenced by a number of factors, including "local causes, such as the nature of the ground . . . the physical configuration of the surface, whether level or mountainous; and the vicinity of seas or lakes."[91] They discovered that the infinite permutations and combinations of the wind and other aspects of the weather made grasping the structural energies of nature difficult. Since the meteorologist was forced to weigh conflicting observations and confront omissions and irregularities, the interpretive work of the forecasting process produced an absence of definite, transparent meaning. Buchan expressed the challenges meteorologists faced, declaring that "meteorology is the most difficult and involved of the sciences," due to "the complexity and intricacy of the phenomena, and the manifold influences by which they are modified and determined," as well as the inability to "give a rational and perfectly satisfactory explanation of them."[92] A complicated assembly of forces affected the motion of the air, and results could not always be accounted for, leading Herschel to similarly conclude that meteorology was a peculiar science in which "theory affords us no assistance."[93]

Galton described the difficulty of forecasting the weather in detail in his *Meteorographica*: "No means exist of obtaining access to any considerable portion of these observations, without great cost, delay, and uncertainty; much less are they to be obtained in a 'reduced' and never a meteorographic form."[94] He acknowledges the incompleteness of his research due to the absence of reports from many stations and reflects that the observations he received were all liable to error. Recognizing the difficulty of the forecasting process, he asserts that "patience and some training of the eye" is needed "to interpret the wind currents satisfactorily."[95] He also states that there is "wide room for fancy, especially when the air is much disturbed and thrown into eddies." The results at one station can contradict an adjacent station, making it seem impossible that they could be "indications of the same great system." Additionally, he comments on the influence of preconceived ideas on observations, remarking that if forecasters "wish to see cyclones, they will find them everywhere." Commentators on Galton's work similarly acknowledged "the difficulties to be encountered in translating figures into lines and curves, without producing seeming anomalies, especially in reducing wind-observations," and stressed the need for clear delineation.[96]

Subjective impressions often competed with a statistical or numerical approach as meteorologists wrestled with the attempt to reduce complex information into logical symbolic form. Popular, local knowledge of the weather, or "weather wisdom," based on factors such as the observation of the skies or the behaviors of animals, competed with the work of official meteorological observatories. Fitzroy's method was criticized for not being based on standardized rules that could be clearly "communicated in the shape of instructions," and he advised his readers to closely observe the appearances of the sky in addition to monitoring the barometer.[97] Additionally, barometric measurements could not be divorced from the barometer's connection to subjective experience; as Katharine

Anderson notes in her recent study of Victorian meteorology, mercury, also known as quicksilver, was associated with the "animation, vital fluids, and the circulation system of the body."[98] The telegraphic network, used in Britain for weather forecasting since 1859, linked the data of dispersed meteorologists but was similarly associated with the human body; in *Principles of Mental Physiology* (1874), Carpenter drew comparisons between the telegraph and the human nervous system.[99] The sense of interplay between subjective and objective observations of the weather destabilized confidence in the meteorological results expressed in maps and charts. Rather than providing a reassuring, coherent image, the wave forms used by Victorian meteorologists ultimately assert the instability associated with not only the weather but also the human mind and body.

An 1865 article in the popular periodical *Cornhill Magazine* captures the difficulty of forecasting the weather, specifically focusing on the mystery and irregularity associated with atmospheric waves: "Meteorology is a very young science—if it deserves the title of science at all—and until observations for a long series of years shall have been made at many stations, we shall not be in the possession of trustworthy facts on which to ground our reasoning. It is merely shoving the difficulty a step farther off to assign these irregular variations to atmospheric waves. It will be time enough to reason accurately about the weather and its changes, when we ascertain what these atmospheric waves are, and what causes them."[100] Although the writer expresses hope that ultimately atmospheric waves will be understood, he underscores the ambiguity and uncertainty of the work of meteorologists and links it to the wave.

As Birt increasingly realized that the linear wave form did not imply an actual molecular movement that could produce wind, he wrote, "the term atmospheric wave has been used to designate *ideal individuality* which the mind attributes to the process which it observes of the successive change of place which the barometric maxima and minima undergo."[101] Barometric waves thus functioned on a more idealistic or conceptual level in Victorian weather maps and charts, enabling the mind to combine a collection of instances into a visual whole, and attempting to give graphic structure and shape to the invisible and complex forces of the atmosphere. In his *Manual of Meteorology* (1926), Napier Shaw suggests the tension surrounding the reliance on the wave form in his observation that the introduction of the weather map created "a curious alienation" of the "theoretical or experimental physicists," interested in minute mathematical and physical details, from the observers and mapmakers, who were more focused on the broad features, outlines, and visual suggestions of forces.[102] The reliability of the visualizations of

Victorian meteorologists is continually challenged by their connection to conceptual concerns and the questionable relationship between graphic forms and reality.

Hughes's illustrations similarly function in a conceptual or suggestive manner; Hughes uses expansive curves and exploits outlines rather than rendering minute details. Reproduced through the linear medium of wood-engraving, his undulating lines provide graphic structure and seek to convey a clear, inclusive viewpoint, mirroring the graphic forms of meteorologists with similar aspirations, before ultimately suggesting the tenuousness of this vision. Like Victorian meteorologists, Hughes is presented with the challenge of representing ongoing motion and fluctuation within a static image. The wave forms of Hughes and meteorologists are not simply in dialogue with each other, however; the suggestive, expansive approach employed by both encourages an intersection and overlapping of meanings, embedding the wave within the broader culture of the nineteenth century. Waves of hair, atmospheric waves, the wave forms of meteorological maps and charts, and the undulating lines indicating unconscious mental activity, which also evoke the waves of the sea, come together and reinforce one another, conveying the potential for unruly movement and modification. Additionally, this mental plasticity encouraged by both Hughes and meteorologists even opens possibilities for a rethinking of other forms; for example, the undulating lines in Hughes's spiral staircase illustration appear in conjunction with "rings" of conscious and unconscious activity (fig. 3.9), mirroring meteorologists' use of isobars streaming in a wavelike formation alongside curvilinear forms resembling concentric rings (fig. 3.5).

Although the image of the wave repeats throughout MacDonald's story in Hughes's illustrations, functioning on an abstract, emblematic level, the illustrations also create a tension between consistency and variation that inspires an exploratory, participatory form of thinking. There are variable factors in Hughes's depictions of the North Wind—the size of the wavy strands, the degree of separation between them, the exact placement of curves, their ability to move in different directions, their proximity to Diamond, and their tendency to move beyond the picture's border—that resonate with the crests, troughs, and slopes in meteorological maps and charts that seek to make sense of a dynamic array of interactive forces. Hughes builds up a summarizing, all-compassing form, only to show that the form represents a living system marked by continual mobility and transformation. Every curvilinear variation employed by Hughes and meteorologists, ranging from slight to unmistakable wave forms, ultimately becomes suggestive of the strange and secret waves of the sea described by Dallas—rushing, ebbing, flowing, and breaking—that signal unconscious mental activity, and thus challenge attempts to obtain an all-inclusive, stable form of vision. Still, like the meteorologists who searched for patterns but experienced ongoing irregularities, recognizing the difficulty of achieving definite knowledge but nevertheless pursuing and valuing the observational and interpretive process, Hughes emphasizes the possibility for subjective, imaginative participation opened up by the spontaneous vacillations and fluctuations of his wave forms. Although her identity is not stable, transparent, or easily summarized, the reader engages fully, repeatedly trying to discover and comprehend the elusive and complex North Wind.

Hughes's illustrations for MacDonald's *At the Back of the North Wind* reflect his conception of the wave form as a symbol of both instability and dynamism linked to the weather, the human mind and body, and female identity. The idea of the wave form as a comprehensive symbol of the North Wind's multiple aspects corresponds with Victorian meteorologists' visualizations that attempt to succinctly convey the weather over time and across a wide region, but, as we have seen, the form in both contexts resists coherence and order. Although Hughes's wave forms continually mirror the forms used in the charts and maps of Victorian meteorologists, which reflect both the difficulty of securing the mobile, unpredictable, changeable forces of the weather and the ongoing challenge of capturing and condensing complex information into readily intelligible symbolic form, they also suggest the presence of a wider field of culture, discourse, and knowledge. The forms correspond with the descriptive imagery used to describe the unconscious mind, which references the waves of the sea, and engage with Victorian understandings of unrestrained, abundant waves of hair. For Hughes, femininity extends beyond the erotic associations of wavy hair, however, as the turnings, windings, and whirlings of the North Wind suggest the continual attempt to understand a multifaceted and mysterious female presence, engaging with the broader challenges of securing identity, with its dynamic dimensions and continual transformations.

More than this, however, the wave itself as a form comes into focus as a generative trope that disrupts the boundaries between science and art in this period. The North Wind contains intermingling, interrelating meanings that reveal how vivid the wave form was within the Victorian imagination. Despite its consistent, firm presence in the mind, and ability to function as a delineated emblem, the form's power lies not in its ability to outline, clarify, or harmonize knowledge but in its link to nature and the human body, and the resulting sense of internal activity and mobility. Hughes's illustrations embrace the wave form's capacity to signify ongoing, irregular development and transformation, and recognize its potential for generating expansive, open-ended modes of conceptualization and imaginative forms of engagement. The wave form encourages mental elasticity while expressing the presence of ambiguities and complexities, powerfully marking the process of grasping for that which continually eludes.

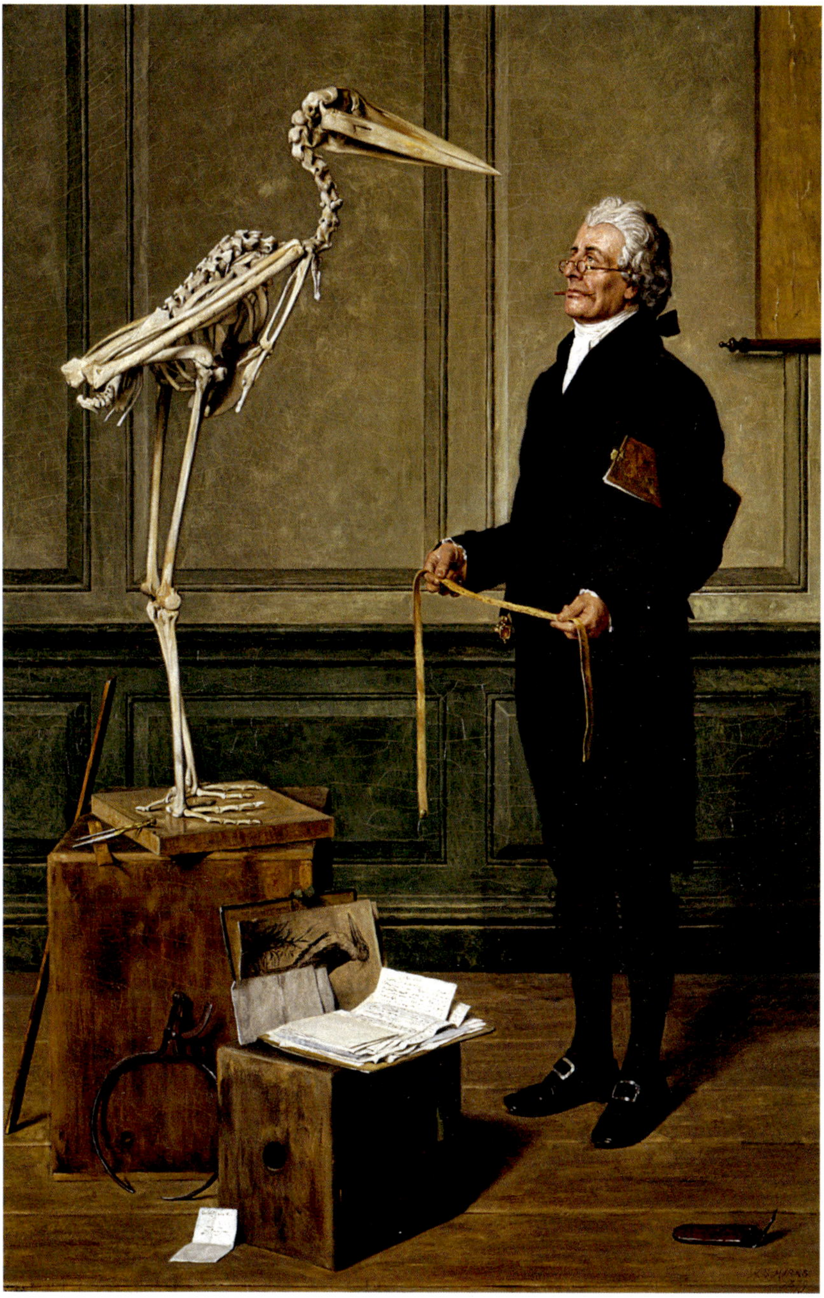

Plate 1. Reviewers saw in Mark's painting "a doctrine of our more recent times" for its apparent support of the idea that through empirical observation and record-keeping, scientific laws "could be discovered and made useful to man." Henry Stacy Marks, *Science is Measurement*, 1879. Oil on Canvas. 36 × 24 in (91.4 × 60.9 cm). Royal Academy Collection, London. Credit: Royal Academy.

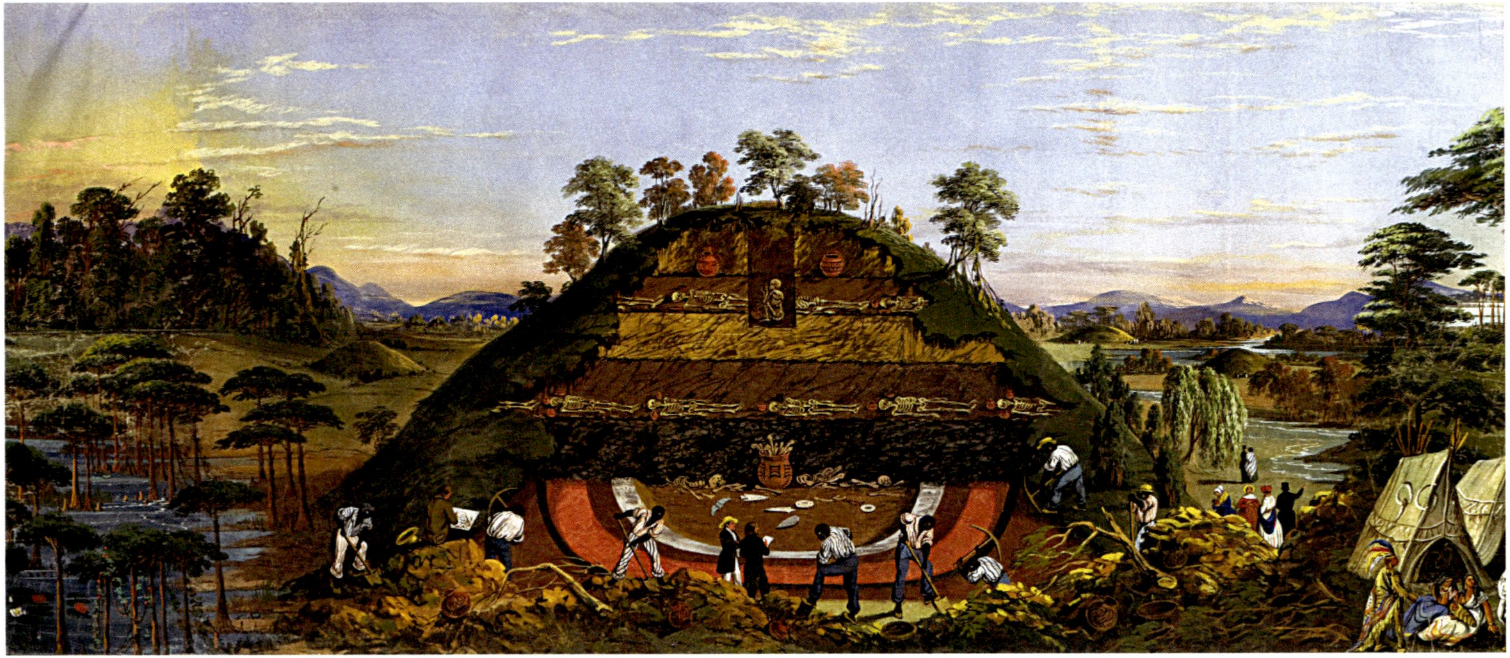

Plate 2. In the panorama, Egan depicted Dickeson's excavation of a Native American burial mound on the Louisiana plantation of William Feriday. John J. Egan, "Huge Mound and the manner of opening them," *Panorama of the Monumental Grandeur of the Mississippi Valley*, ca. 1850, tempera on lightweight fabric, 90 in × 348 ft (228.6 × 10607 cm). Eliza McMillan Trust, 34:1953, Saint Louis Art Museum, Saint Louis, Missouri. Courtesy of Saint Louis Art Museum.

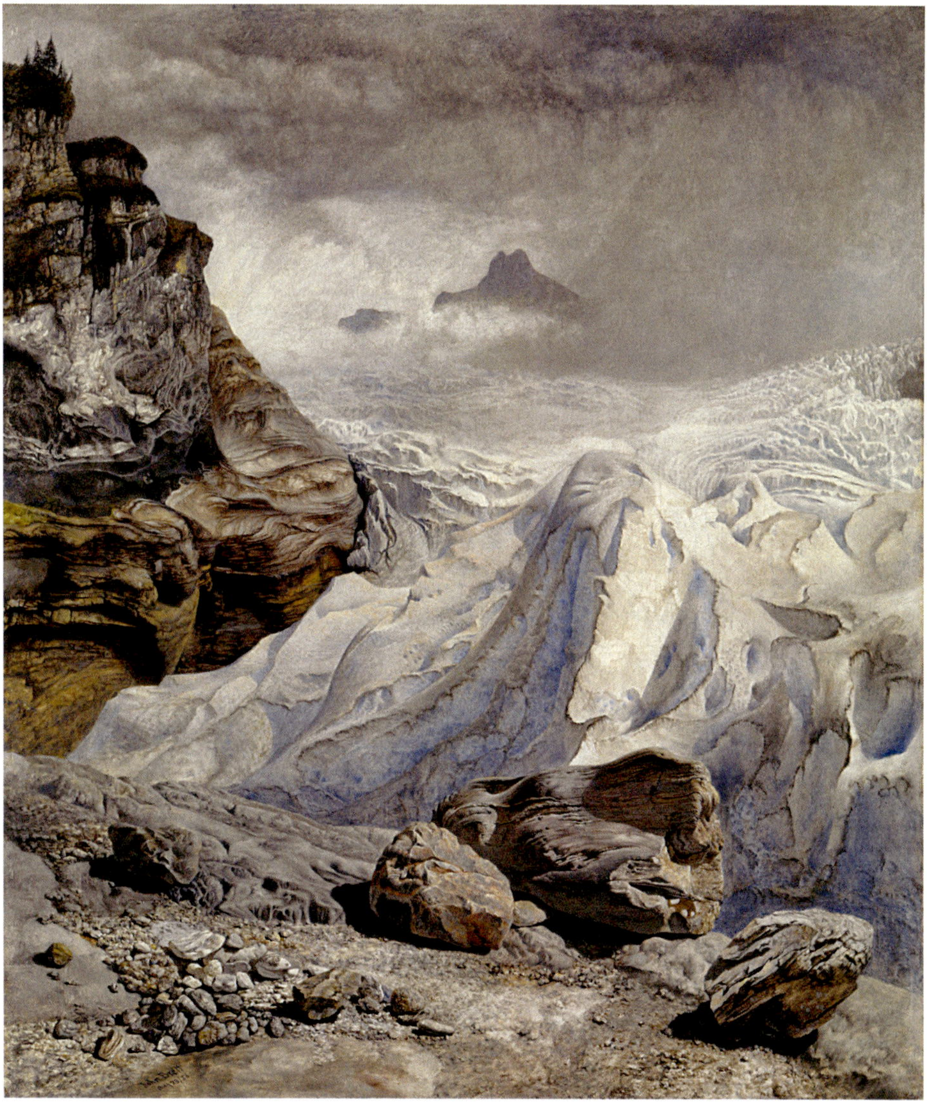

Plate 3. Brett's painting reflects his reading on glaciers as well as his adherence to Ruskin's aesthetic principles. John Brett, *Glacier of Rosenlaui*, 1856. Oil on canvas, 17 1/2 × 16 1/2 in (44.5 × 41.9 cm). Tate Britain © Tate, London 2019.

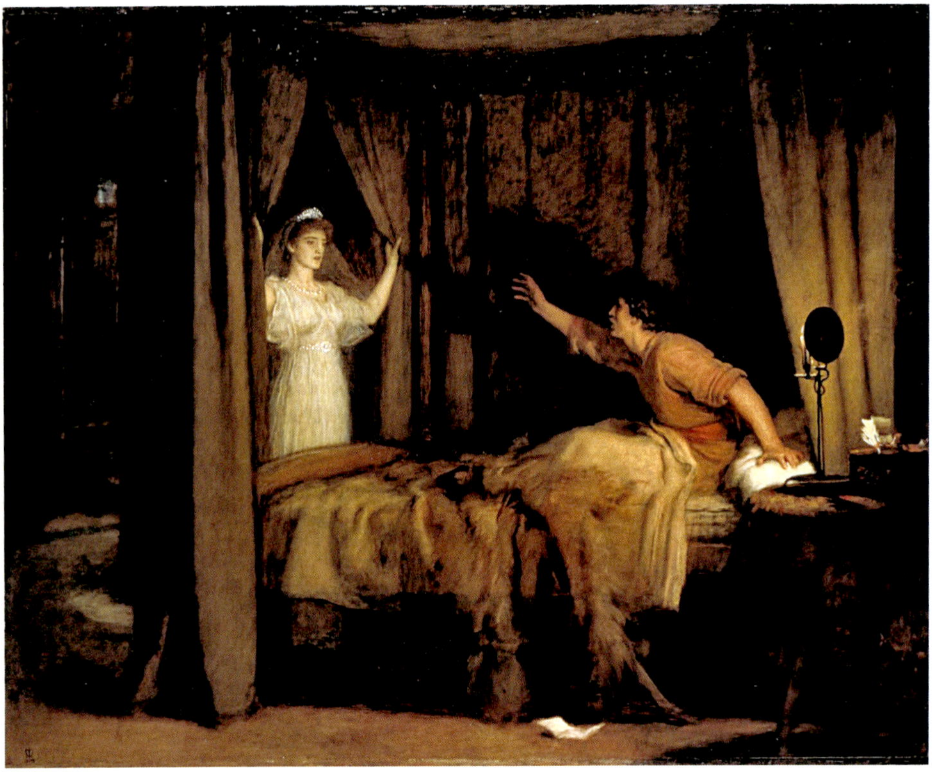

Plate 4. Milllais's paintng is ambiguous about whether the woman in white is an actual woman, a hallucination, or a "real" ghost. John Everett Millais, *Speak! Speak!*, 1895. Oil on canvas. 65 × 81 in (167.6 × 210.8 cm). Tate Britain, London. Photo@Tate.

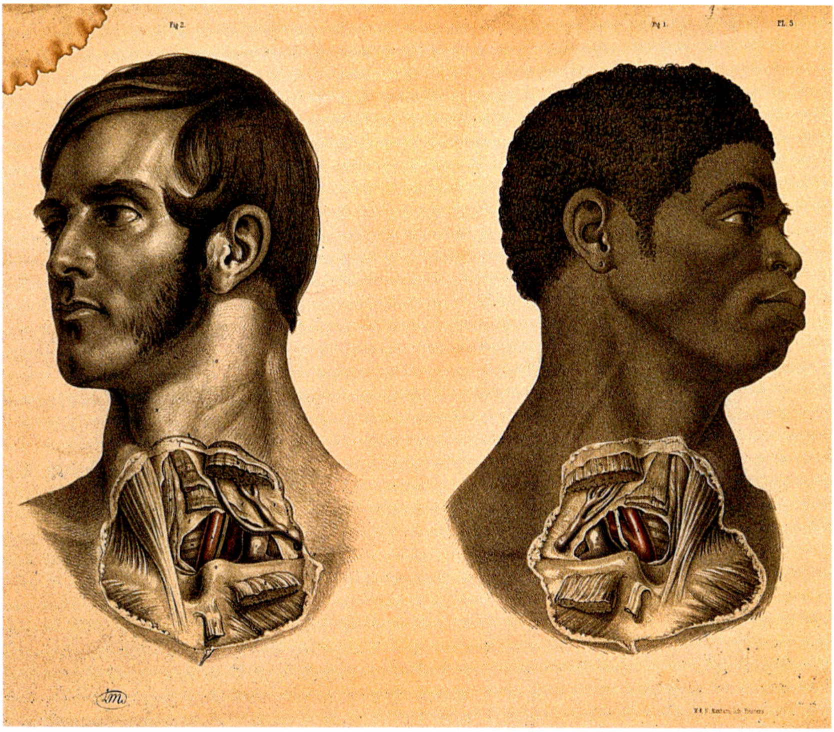

Plate 5. Joseph Maclise highlighted the similarities rather than the differences between the anatomical structures of men of different races. Joseph Maclise, "Two heads of men (one white, one black African), showing dissection of muscles and blood-vessels of the subclavian region of the chest," lithograph with watercolor, sheet 21 1/2 × 14 3/4 in. (54.5 × 37.7 cm), 1851. Courtesy of Wellcome Library, London.

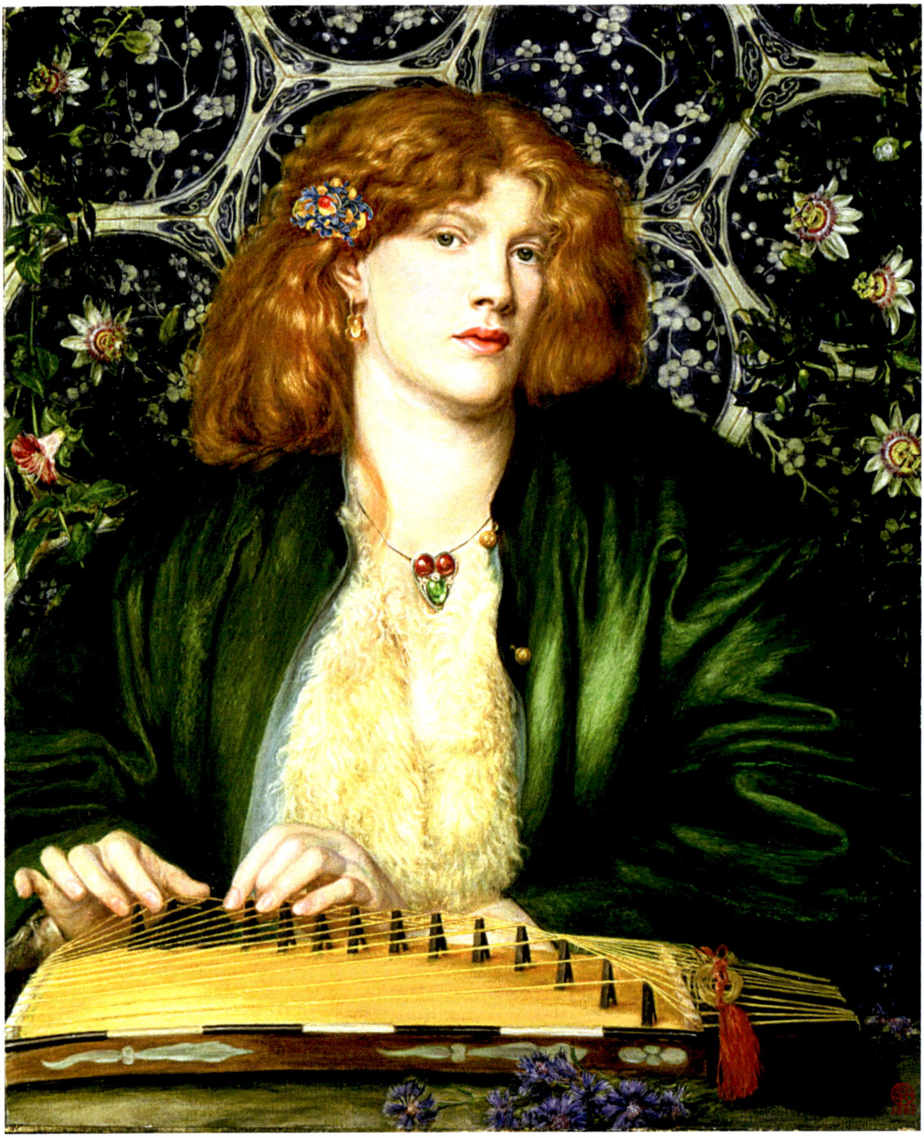

Plate 6. Rossetti emphasizes the reproductive organs of passionflowers, and behind the sensual head of the woman pictured, prunus blossoms are displayed in what appears to be a cellular pattern infused with protoplasm. Dante Gabriel Rossetti, *Blue Bower,* 1865. Oil on canvas 33 × 28 in (84 × 70.9 cm). Credit: The Henry Barber Trust, The Barber Institute of Fine Arts. University of Birmingham/Bridgeman Images.

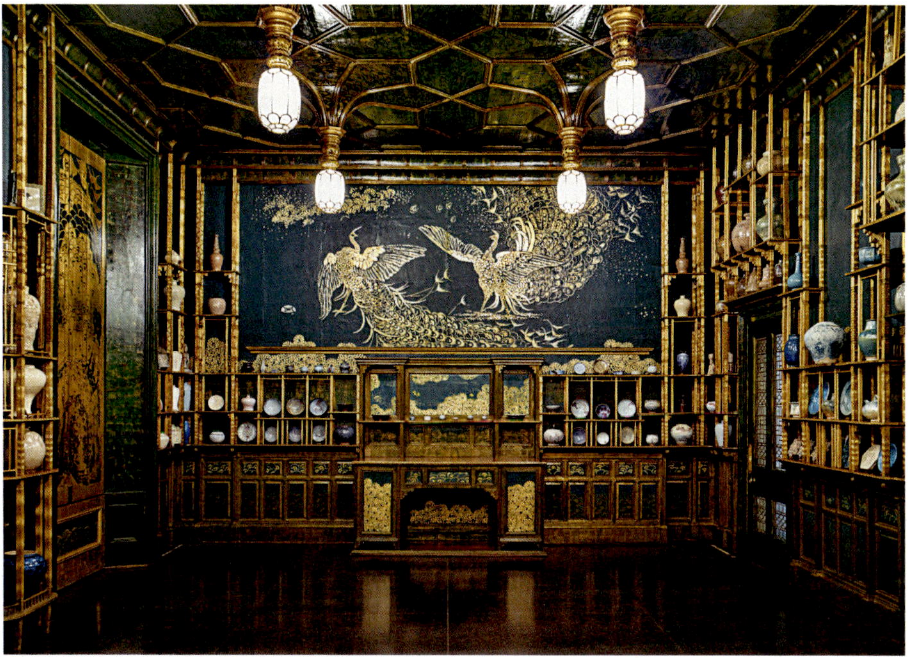

Plate 7. James McNeill Whistler transformed his patron Frederick Leyland's dining room into the space now known as the Peacock Room. James McNeill Whistler, *Harmony in Blue and Gold: The Peacock Room*, 1876–1877. View facing south. Oil paint and gold leaf on canvas, leather, mosaic tile, and wood. 166 × 241 1/2 × 404 in. (421.6 × 613.4 × 1026.2 cm). Freer Gallery of Art, Smithsonian Institution, Washington, DC: Gift of Charles Lang Freer, F1904.61. Courtesy of Smithsonian Institution.

CHAPTER 4

A HAUNTING PICTURE, IN LIGHT OF VICTORIAN SCIENCE

JOHN EVERETT MILLAIS'S *SPEAK! SPEAK!*

Nancy Rose Marshall

So long as men can breathe or eyes can see,
so long lives this and this gives life to thee [me?]

—*William Shakespeare, Sonnet 18, 13–14*

This chapter considers English artist John Everett Millais's compellingly strange picture, *Speak! Speak!* (1895, plate 4), through the lens of Victorian sciences. Rather than identifying specific aspects of Millais's scientific reading or knowledge, I reflect on both the subject matter and the formal qualities of the work in light of ideas generated from the fields of chemistry and psychology. A candlelit painting of an apparition was uniquely positioned to evoke a range of nineteenth-century scientific conversations about the urgent and related issues of the nature of material reality. It also allowed Millais, at the end of his life, to meditate both on the meaning of death and on his entire career in order to generate some reparative, reassuring ideas regarding the nature and function of life and of art.

This chapter serves as a methodological inquiry into what happens when we look at a painting in terms of the ways its depicted subject intersects dominant scientific discourses of the period in which it was produced. Considering how empirical research and creative works of art might share conditions of epistemological possibility offers an interpretive strategy rich with potential. I argue that candles and ghost stories both generated new meanings in an era of the rapidly coalescing fields of psychology and chemistry. Arguably, the rise of spiritualism in the Victorian period was a byproduct of advances in understandings of the material world, which provided rational encouragement for a belief in the persistence and metamorphosis of things—while in turn, a culture saturated with faith that dead loved ones remained present just beyond the veil of human perception may well have prodded scientists to investigate the invisible.

Techniques and tools of art history can be uniquely generative in revealing resonances between the forms of artworks and of systems of thought. Crucial to the ways in which Millais represents vision and doubt, and to *Speak! Speak!* generally, are the kind and quality of the lighting, taken up here as the formal element most relevant to an analysis of the picture. Light relates to the concept of perception—how we know what we know—which constituted a primary field of inquiry for the new science of psychology. The candle also was a notoriously mysterious light source, the secrets of which had partially been unlocked by chemists in the nineteenth century who discovered that, while seemingly consumed in process of burning, the candle in fact merely morphed into a different form. Millais's painting, then, resonated with the uncertainty of the visible, of making the invisible visible, and of transformation and of translation; it suggested moreover that the human body comprehended these things from a perspective that was, to paraphrase Hamlet, simultaneously bounded in a nutshell but also potentially king of infinite space.

Portraying a man interrupted in the reading of old letters by the sudden appearance at the foot of his bed of a female figure, *Speak! Speak!* implies through its Shakespearean title that the pale woman is no more substantial than Hamlet's father's ghost, to whom Horatio addresses the titular imperative. Yet the glittering diamonds and apparent solidity of the woman in white provoked the same confusion in Royal Academy viewers that philosophical and psychological researchers experienced in analyzing the nature of hallucinations, that of determining any precise boundary between the objectively and the subjectively real. That the status of the woman's "reality" was a crucial issue is clear from the abundance of critical commentary to this effect.[1] Complained one reviewer, "I do not know . . . what I am intended to think or feel. Is the visitor a living intruder or a welcome ghost?"[2] The painting cleverly sets in play thoughts about how we might ever know with certainty of the existence of any observed phenomenon. Significantly, Millais saw doubt as the appropriate response to the painting; biographer M. H. Spielmann recounted the artist's rejoinder to Spielman's complaint that he "could not tell whether the luminous apparition were a spirit or a woman": "That's just what I want . . . I don't know either, nor," he added, pointing to the painting, "does he."[3]

The fact that the critical debate in reviews of *Speak! Speak!* centered on a need for certainty as to the nature of the phantom points, on the one hand, to the presence of the emerging discourses of psychology, which, as Anne Helmreich argues, "stimulated new theories of perception and aesthetic judgment that became critically important within artistic circles."[4] The function and reliability of vision, newly urgent concerns in the period, are underlying themes in the painting as well, flagged by the kind and quality of the lighting, candlelight. The candle was a notoriously mysterious light source, the secrets of which had partially been unlocked by chemists in the nineteenth century who discovered that, while seemingly consumed by the process of illumination, it in fact merely morphed into a different form.[5] Millais's painting, then, spoke of the uncertainty of the visible, of making the invisible visible, and of transformation and of translation. The picture, then, was ultimately also about art itself, in essence an artistic manifesto.

The Artist and the Picture

Millais clearly envisioned *Speak! Speak!* as a substantial statement when he submitted it to the annual exhibition of the Royal Academy in 1895; in interviews, he claimed that the subject had preoccupied him for as many as forty years, underscoring its close relationship with his whole career.[6] At the academy, the picture commanded academygoers' attention in its dramatic, even hokey, staginess. Framed by the verticals and horizontals of the bed and its curtains, the tableau resembles a theatrical production, as indeed is acknowledged by the Shakespearean title. "Speak! Speak!" is a repeated imperative, the second word added eagerly after frustrating silence in response to the first one (as in fact the line is sometimes delivered). Deployed perhaps in observance of the folk belief that ghosts could not speak unless spoken to, the line expresses an intense and painful desire.[7] Against the rectangular "stage," the man's full-body gesture of yearning thrusts in a powerful diagonal toward the woman. The dramatic visual rhetoric points up the idea of performance, with the histrionic elements conveying the old-fashioned quality of earlier nineteenth-century stagecraft—or art—as opposed to the naturalism in vogue at the fin de siècle. The viewer is thereby positioned as a spectator, separate from the depicted world and meant to absorb and respond to what is clearly a work of art, with lessons to impart. But what are we supposed to learn?

In part, *Speak! Speak!* was intended to restore Millais's reputation. As a precocious Royal Academy (RA) school student at age eleven and, not long after that, as one of the original Pre-Raphaelite Brotherhood—the midcentury group of young artists, writers, and critics who advocated a reform of what they viewed as British painting's tired, formulaic conventions through direct observation of the subject—Millais had long ranked as one of the most influential English painters. Earning as much as 10,000 pounds per year, he had become a baronet in 1885, but by the 1890s his work had fallen off in the public's admiration.[8] Aged sixty-six, the artist had suffered recent critical misfortune, and he must have hoped that *Speak! Speak!* would address this setback.

Appearing at the moment Millais had been chosen to fill the ailing Frederic Leighton's spot as president of the academy (PRA), *Speak! Speak!* was also subject to intense scrutiny regarding the new PRA's relative place in the field. Indeed, his choice of model, Mary Lloyd, boldly demanded a direct rivalry with Leighton, who also favored her distinctively expressive features and whose *'Twixt Hope and Fear* (held in private collection), for which Lloyd modeled, was hung opposite *Speak! Speak!* in the academy's honored spot, the Grosvenor Room.[9] Millais seems to have met his own expectations; as his son commented, "Never before . . . had I seen him so well pleased with any work of his own."[10] Many critics, despite confusion over the subject, were willing to confirm the stature of the picture, rejoicing that "this is the Millais we used to know and whom we welcome back after the comparative failure of the past few years."[11] As a mark of its perceived cultural significance, *Speak! Speak!* was purchased for the nation by the Chantrey Bequest.[12] The fact that this painting, which may not seem particularly substantial today, took on such significance for both artist and public suggests that it bore special relevance and topicality for late Victorian Britain.

The question of Millais's own connection with scientific knowledges or networks is difficult to establish in any detail. Although he was not particularly intellectual and preferred hunting and fishing in Scotland to participation in scholarly communities, these outdoor interests actually produced a measure of observant engagement with his environment. Indeed, Julie Codell notes that Millais's son, John Guille Millais—himself a writer for the magazine *Nature*—made just such claims for his father's interest in natural sciences.[13] As, too, a gentlemanly member of the art world who regularly painted portraits of politicians, cultural figures, doctors, and others, Millais was certainly well placed in the flow of conversation regarding the topics of the day. Acquainted with many Victorian scientific thinkers, such as the philosopher and physiologist George Henry Lewes, he was also well placed to absorb general ideas about, if not detailed understandings of, the latest advances.[14] Several scholars have also argued for ways in which aspects of Millais's work might be productively situated in relation to discourses of science. In the early days of the Pre-Raphaelites, Millais followed Ruskin's directive to observe nature directly and render what he saw, taking up the challenge of representing character and psychological states through physiognomy and body language in works such as *Isabella* (1849, held at the Walker Art Gallery, Liverpool), which portrayed a range of nuanced responses to the depicted relationship between lovers.[15] As discussed below, Helmreich relates Millais's stylistic shifts to changing notions of mechanisms of perception.[16] Moreover, Millais's close friend F. G. Stephens and the other Pre-Raphaelites celebrated chemistry and geology as sciences engaged in the same precision and observation as their artistic methods. Stephens demanded that artists see "glory" in chemistry, "the modern science, possessing thousands of powers as great as any used yet," and new advances produced the pigments that enabled the bright colors of the early PRB.[17] Millais, then, was certainly well informed enough to know precisely what play of ideas might be set in motion by a painting of a ghost.

The Painting, Light, and Chemistry

Speak! Speak! at first seems simplistically built around a number of conceptual and formal binaries, yet ultimately it complicates all of them. Most obviously it sets up male and female sides of a charged pictorial space. In contrast to the man's energetic desperation, the woman stands stark still, framed in the emphatic triangle produced by the parted curtains. In this way the picture neatly illustrates film theorist Laura Mulvey's famous thesis regarding how iconic female beauty stops narrative, in contrast to active male figures, who function in three dimensions.[18] Yet the woman is unearthly, eerily alien, while the man's action is disproportionately vigorous and clearly futile, as if the two modes in the same frame become discordant and jarring.

The naturalization of the light effects into female and male segments of the work also at first appears to uphold gender norms of the day: the painting tidily cleaves into the still, silent, female half of moonlight and the active, vocal, male half of lamplight, with the woman merely a passive figure summoned by the man's desire. Such a division helps support conventional Victorian associations of man with reason, and woman with madness, especially as the color white has a long association with mental imbalance.[19] The confrontation of the couple in a bedroom sets up notions of a sexual encounter, the white gown evoking wedding nights and, from a Victorian perspective, an anticipation of the deflowering of the innocent white figure.[20] Yet this immobile, zombielike being appears unafflicted by maidenly jitters and in fact is herself doing the

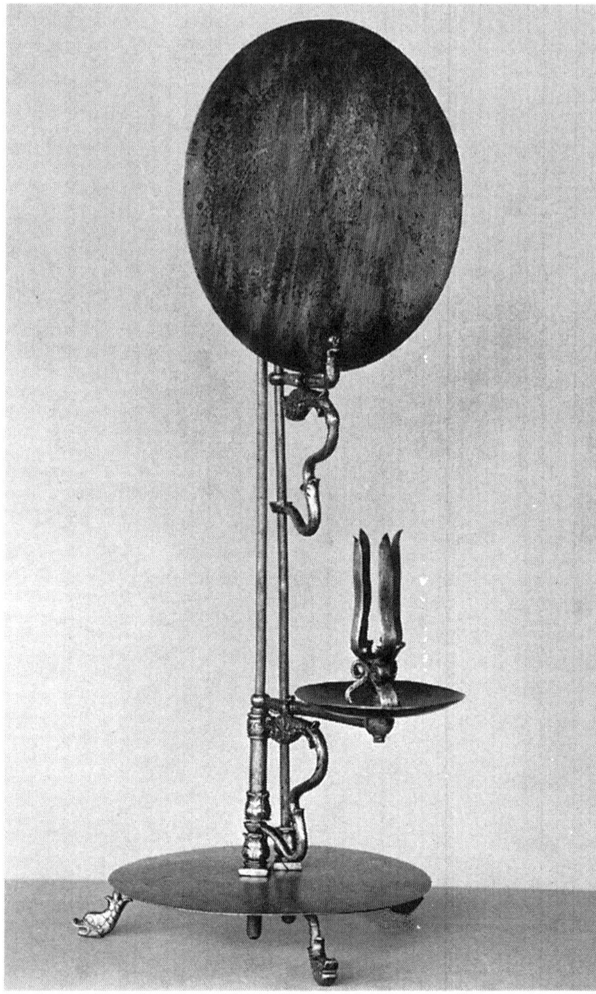

Fig. 4.1. Millais copied the lamp in *Speak! Speak!* from this one in the V&A museum collection. Candle-lamp, early eighteenth-century Dutch. Victoria and Albert Museum. © Victoria and Albert Museum, London.

penetrating by parting the man's bed hangings. In light of the fact that only married women wore diamonds at this time, she seems less virginal bride and more sexually knowledgeable, desiring female.[21] The resemblance of Millais's picture to Henry Fuseli's *The Nightmare* of 1781 (held at the Detroit Institute of Arts), well known in the Victorian period in engraving form, enhances the view of the man as passively in thrall to the terrifying demon peering through the curtains.[22] Moreover, white is the color of death as well as of purity, and the woman might have come to

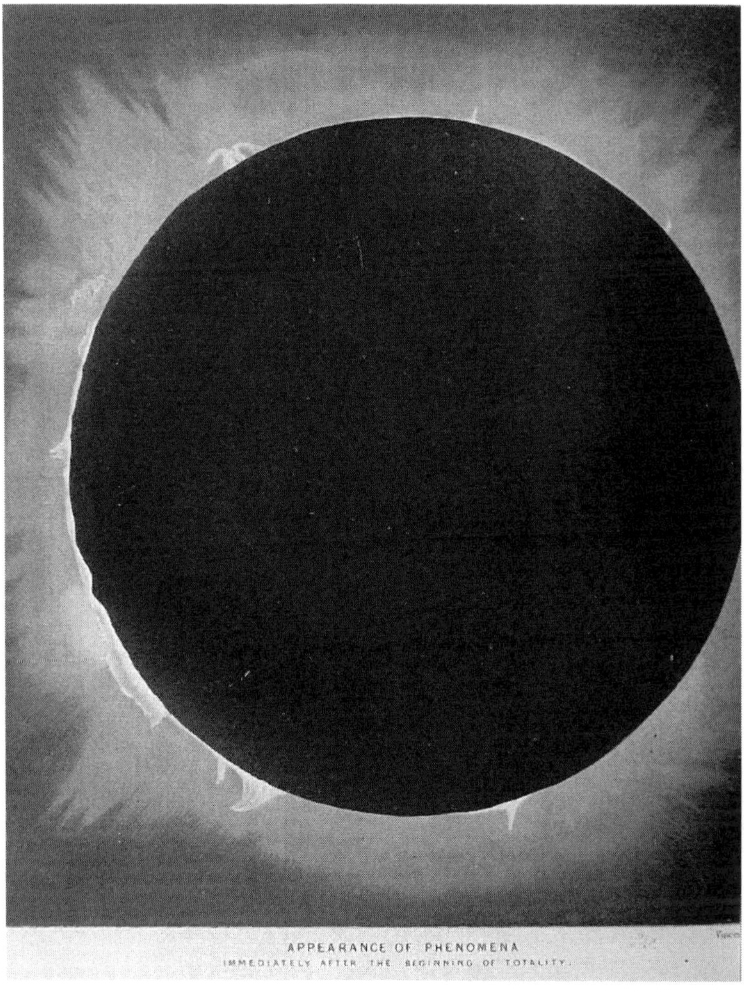

Fig. 4.2. Millais renders his lamp in such a way as to echo illustrations of eclipses, invoking another phenomenon that, like ghosts, was both uncanny and under scientific study. Warren de la Rue, *Eclipse of the Sun,* July 1860, Rivabellosa, Observed at Near Miranda de Ebro, Spain. Reprinted from *Philosophical Transactions of the Royal Society of London* 152 (1862): plate 7.

claim the man's life, as was the unfortunate predilection of many folkloric specters.[23] Scotland, where Millais painted *Speak! Speak!*, was particularly associated with tales and folk traditions of fatal spirits, and Millais himself was of a somewhat superstitious turn of mind.[24]

However, as the viewer ponders the central question about the nature of the female apparition, they are also forced to entertain the possibility that the man may be hallucinating, and that it is his reason that is in fact disordered. We look, and look again, attempting to define through our gaze the true identity of the female form, which Millais cleverly lights in such a way as to make us contend with the nature of vision itself. The candle lamp shines on her form so brightly that she reflects it back.

Creating striking light qualities in this picture was extremely important to Millais. Artist Frederick Barwell recounted that the artist had visited the Victoria and Albert Museum in London in search of just the right lamp (fig. 4.1), of which he made a replica to in order to paint it "exactly under all the conditions of lighting the effect demanded."[25] The lamp's reflector produces a black, eclipselike shadow that almost but not completely obscures the candle flame just barely appearing to the left edge of the dark circle.[26] Strikingly similar to eclipse illustrations in popular astronomy accounts of the period (fig. 4.2), this detail sets in play the notion that the ostensibly supernatural, while apparently fearsome, possesses logical patterns and predictable rules for those who can understand them; in much the same way British explorers' accounts of eclipses routinely contrasted the behavior of the cowering, superstitious "natives" with their own superior rationality.[27] Through this detail the picture signals an oscillation between scientific explanation and mysterious unknowability.

The resulting canvas toys with the drama of shadow and dueling light sources in a manner rare in Millais's oeuvre. Reviewers acknowledged this unusual aspect, referring to it as "harmonious discord" and citing Millais's only previous work in this vein as *The Rescue* of 1855 (held at National Gallery of Victoria), which had portrayed contrasting firelight and moonlight in a scene of a fireman rescuing children from a burning home.[28]

Light, then, and in particular, candlelight, seems significant here; both were relevant to new scientific investigations. Signaling the past and mystery, as opposed to the modernity and harsh clarity of electricity, candlelight as a phenomenon in the 1890s would have been informed by decades of new discoveries about fire in the field of chemistry, physics, and thermodynamics. To begin with, using or painting candlelight was clearly a declaration in favor of ambiguity and obscurity, of leaving things unseen and uncertain, in the wake of the increasing dominance of brightly illuminating electric light in the period (no doubt the rise in candlelight as a pictorial subject relates to this fact as well as it came to be synonymous with the past). From 1861, electrical illumination began to replace gaslight, which, as Lynda Nead points out, possessed its own poetic elements.[29]

The invention of new incandescent forms of electric lighting had allayed people's initial fears and the technology spread rapidly, aided by

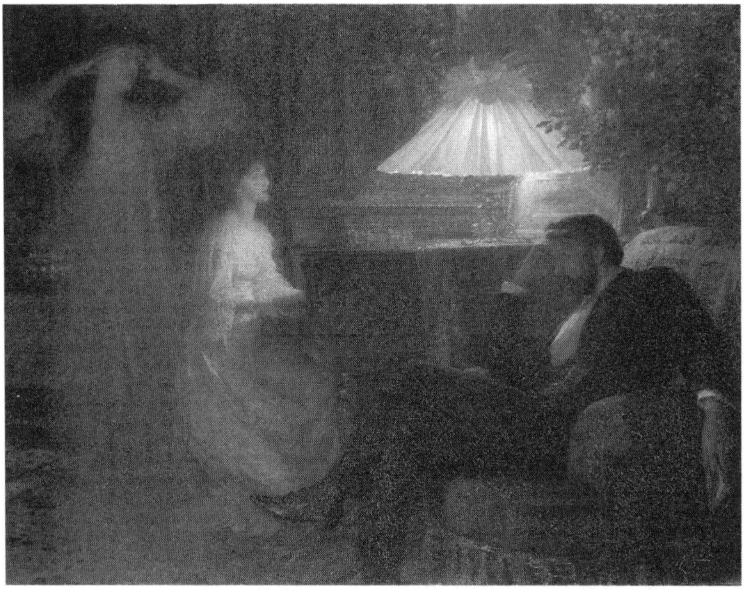

Fig. 4.3.As an indication of the popularity of the ghost picture at this time, Frank Dicksee submitted this painting to the same Royal Academy Exhibition in which Millais's work appeared. Frank Dicksee, *A Reverie*, 1895. Oil on canvas. 41 × 53 in (104 × 134.6 cm). Walker Art Gallery, Liverpool. Courtesy of National Museums Liverpool: Walker Art Gallery.

the 1892 Electrical Exhibition at the Crystal Palace.[30] Millais himself was knowledgeable about the effects of electric lighting on interiors, as he had been drafted—along with fellow artists Walter Crane and Elihu Vedder—to supervise the choice of colors in the design for the Electricity Building in the 1893 World Columbian Exposition in Chicago.[31]

The 1890s were in fact a key decade of transition for artists struggling with how to light their studios, a conversation recounted in the journal *Lightning*.[32] Older painters, such as William Powell Frith, opposed the novelty for its harsh effects.[33] Coincidentally, Frank Dicksee, who had been unusually quick to electrify his painting rooms in 1883, contributed another ghost picture in the same room of the 1895 RA exhibition.[34] *A Reverie* (fig. 4.3) invited comparison with *Speak! Speak!*, in its suggestion of the presence of the supernatural by means of unusual light effects, with the ghostly apparition appearing just out of range of both the firelight and the bright glow of what is most likely an electric lamp.[35] The transparency of this forlorn female, possibly a former lover or wife of the contemplative man to the right, was a more common way of rendering the spectral, pointing up the unusual quality of Millais's bridal ghost.[36]

The candle was also significant at this time given its role in the history of chemistry. Understanding fire was the purview of the chemist, who was in the Victorian period still shadowed by his past incarnation as alchemist and known as the "philosopher by fire."[37] In 1774 the English scientist Joseph Priestley's experiments with candles helped him discover oxygen, or, as he understood it, "dephlogisticated air."[38] Analogies between a life and a swiftly consumed candle moved from the realm of poetry to proven fact as scientists came to comprehend the parallels between combustion and respiration. As influential chemist Michael Faraday suggested, "in every one of us there is a living process of combustion going on very similar to that of a candle. . . . It is not merely true in a poetical sense—the relation of the life of man to a taper."[39] And like human life, candlelight lasted until its fuel source was exhausted—or it was forcibly terminated. In Millais's work the flickering of the painted candle in the breeze apparently caused by the appearance of the ghost would have suggested the potential vulnerability of the man—and of human life in general.

Yet despite their obvious intimations of mortality, candles also betokened something new and more particular for Victorians. Announcing his findings in 1779, French chemist Antoine Lavoisier took Priestley's experiments further to realize that, rather than being a substance from which something (the theoretical component "phlogiston") had been removed, the gas was an element unto itself, which he named oxygen.[40] Such studies in turn established the foundational law of the conservation of mass, defined by Lavoisier in his *Traité élémentaire de chimie* of 1789 and popularized by Faraday in his public demonstrations of 1848, published as "The Chemical History of a Candle," in which he showed that burning a taper makes hydrocarbon molecules react with oxygen, producing heat, light, carbon dioxide, and water.[41] An updated version of the popular "bird in an airpump" performance seen in Joseph Wright of Derby's famous painting (as will be discussed below), Faraday's candle experiments were further popularized by Charles Dickens in an article for *Household Words* in 1850. In a fictional dialogue between a boy who has attended Faraday's lectures and his obligingly clueless mother, Dickens and Percival Leigh convey their lesson: "'But what becomes of the candle,' pursued Harry, 'as it burns away? where does it go?'

"'Nowhere,' said his mamma, 'I should think. It burns to nothing.'

"'Oh, dear, no!' said Harry, 'everything–everybody goes somewhere.'"[42] This idea was in common currency by the end of the century, to the point that leading psychiatrist and philosopher Henry Maudsley could state with confidence in 1883, "Nothing perishes absolutely in the universe. Matter is neither created nor destroyed." Maudsley then took the unusual leap of linking this law to the human belief in the persistence of the soul: "There is no death in the sense of annihilation. Here then may have been a foundation of a vague notion that man will not all die."[43] Maudsley succinctly and reassuringly recasts religious beliefs in scientific terms.

It was now clear for those in the know that fire produces metamorphosis rather than destruction, a reshuffling of atoms that nonetheless remain present. Popular chemistry manuals explained that fire is "a strong transforming force" and that "the mark or note of all chemical occurrences is change. As we look at [a] burning . . . coal, we see a slow process of change; a change from a solid . . . lump . . . to smoke, flame, and ashes."[44] The candle in Millais's painting produces a wisp of black soot as it gutters. In Faraday's terms we are observing the process of "imperfect combustion," in which a lack of oxygen produces unburned particles.[45] In this tangible illustration of metamorphosis—the candle is translated from solid matter into smoke and air—we have a metaphor for a human being's transformation from material form to ethereal spirit.

As Kate Flint avers, "by the closing decades of the nineteenth century, the borders between the visible and the invisible worlds were increasingly hazy."[46] In science as in spiritualism, invisible substances could be made visible, identified and understood to follow rules. Ghosts too were subject to this sort of scientific testing.

Vision and Perception

The picture uses the candlelight to draw attention to concepts of reflection. What we see here, then, is the way in which vision itself occurs, with rays bouncing back from the woman's glittering form to the man's eyes, a process to which Millais draws attention by the dramatic beam of light catching the peculiarly bright surfaces of her garb. Notably, scientists examined the action of reflection both to prove and to disprove the existence of phantoms. German chemist Justus von Liebig, for instance, stated firmly: "We do not believe in ghosts, though thousands are said to have seen them, because the laws of light teach us that . . . an unsubstantial being no longer possesses the property of reflecting light, which is essential to its becoming visible."[47] The president of the Society for Psychical Research, the philosopher Henry Sidgwick, on the other hand, contended that in fact there was no need to verify the phenomenon of a reflective ghost, since spirits did not follow the laws of physics: "There are few intelligent believers in ghosts . . . who really hold that a ghost is actually seen as an ordinary material object is seen, i.e., that it affects the percipient's eyes from the outside by reflecting rays of light onto them."[48]

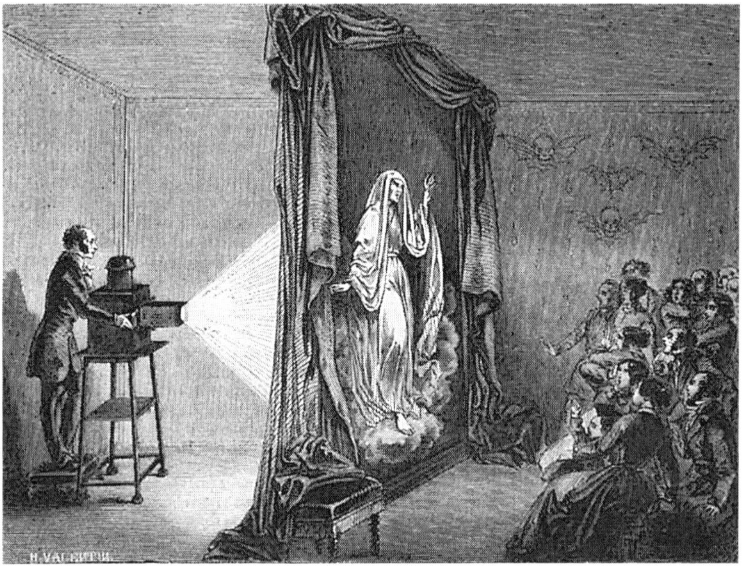

Fig. 4.4. Visual entertainment of the period often deployed ghostly images of women in white. H. Valentin, *Une Scene de Fantasmagorie*. Engraving. Reprinted from *Magasin Pittoresque*, 1848, 55.

For reviewers, it was the woman's reflectivity, particularly her diamonds, that endowed her with substance. Remarked the critic for the *Month*, "The painting of the jewels which she wears, excessive in its brilliant realism, makes it almost impossible to believe that she is meant for a phantom."[49] Diamonds, notably, reflect, and indeed were thought to be best showcased by flame for this reason. As a description of the stone from a book on mining declared, "Candlelight . . . heightens the faculty of luminous reflection."[50] Comparing diamonds and candles, Faraday observed, "The flame shines in darkness, but the light which the diamond has is as nothing, until the flame shine upon it."[51] In this way, the female figure becomes a reflection herself, something that exists only at the pleasure of the primary subject: in this case, the candle—or, of course, the man, who has willed her into being; he is the flame to her diamond. The moonglow around the woman similarly resonates with qualities of secondary illumination. However, in yet another complication of simple binaries, the man himself sheds no actual light, undercutting any power he might have to generate an image. Rather, he projects a shadow across the bedcurtains as an indicator of his own material tangibility. Paradoxically, it is this shadow that proves his reality, blurring the traditional definitions of, and distinction between, substance and shadow (notably the more physically articulated woman does not cast any shadow of her

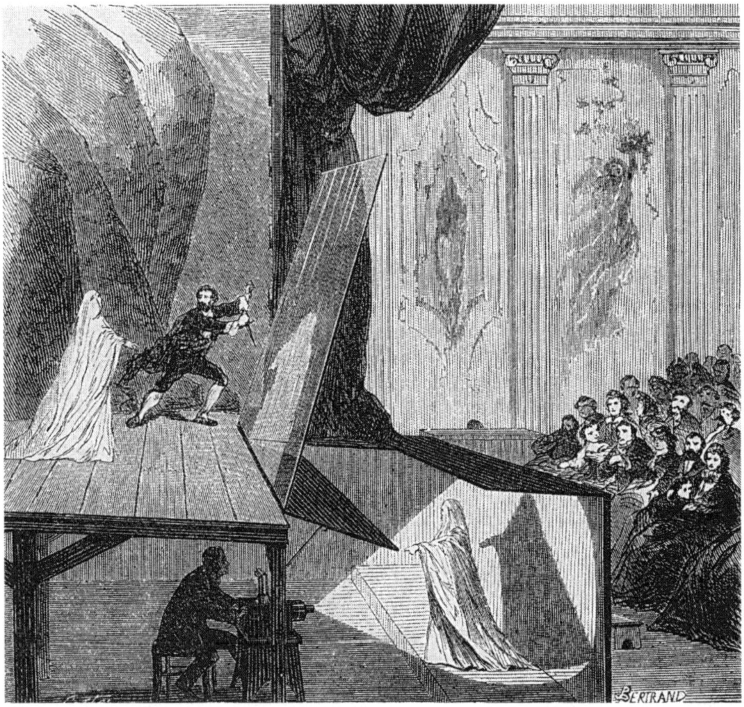

Fig. 4.5. Pepper's Ghost was a classic example of how the apparently supernatural intersected with educational entertainment. "Pepper's Ghost." Engraving. Reprinted from *The World of Wonders: A Record of Things Wonderful in Nature, Science, and Art.* (London: Cassell, Petter, and Galpin, 1873), 201.

own). The impossibility of connection between the man and—we assume—his former lover is signed by the wrenchingly poignant fact that only his shadow reaches her shade. His physical body cannot touch her or hear her because she is not part of his own bounded reality, his limited human perspective from one particular moment in time and space. She seems, by contrast, to function in another world, her unfocused gaze staring blankly past him toward the candle, reinforcing the sense of her dependence on that light source.

I would argue that in its suggestions of the practices of both reflection and projection, which in turn resonate with the ways in which various forms of entertainment and magic shows produced illusions of spirits, Millais's picture demonstrated a highly developed engagement with these technologies. From their invention in the late eighteenth century, magic lanterns and "phantasmagorias" had drawn on the magical macabre. Still popular in the nineteenth century, shows such as those of Belgian physicist Étienne-Gaspard Robertson (fig. 4.4) involved a glass

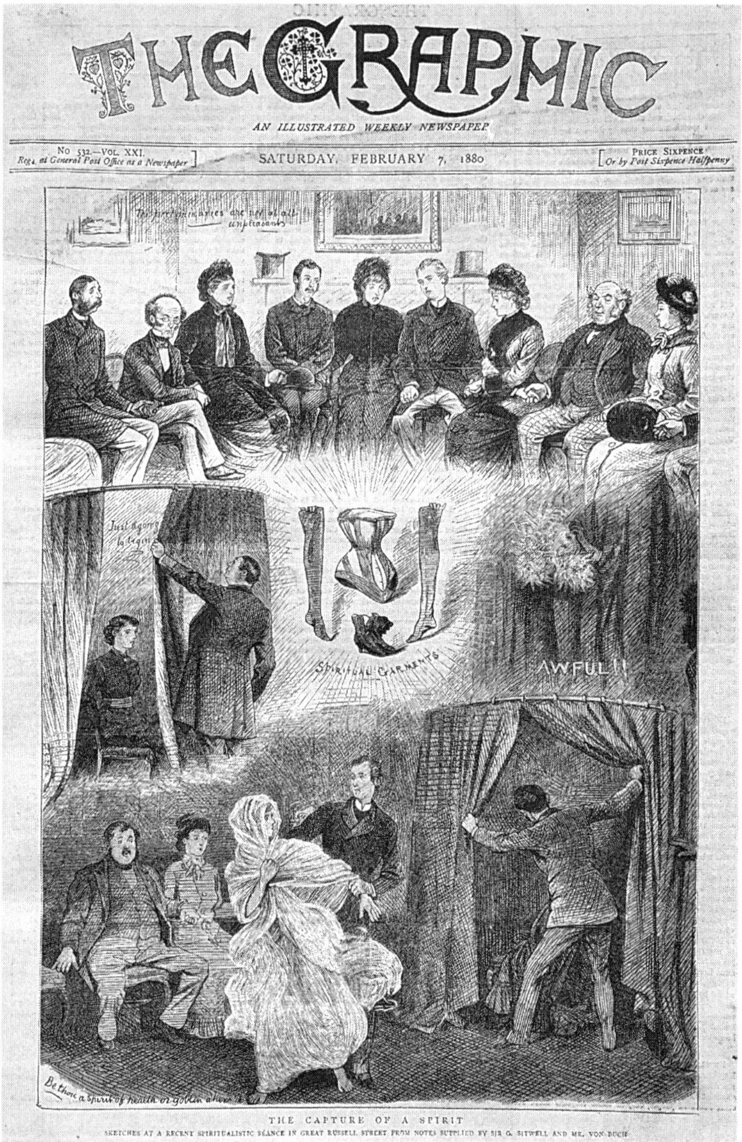

Fig. 4.6. Millais's image echoes the theatrical elements of séances. "The Capture of a Spirit." Engraving. Reprinted from *Graphic*, February 7, 1880, 1.

slide placed in front of a light source and projected on a screen (with the mechanism either hidden or visible).[52] The production of "Pepper's Ghost," perfected by a British chemist, John Henry Pepper, entailed the projection of a bright light (originally a limelight) onto a person in order to capture their mirrored reflection to throw it elsewhere—for instance, on a stage (fig. 4.5).[53] Candles were also used in the production of ghostly phenomena to trick the eye, as in Faraday's entertainment of his niece on a foggy evening.[54] The uncanny qualities of Millais's woman, then, rhyme with those generated in spectacular theatrical illusions of the period.

Yet even these spectacles were concerned with the operation of the senses. Iwan Morus points out that such apparently lightweight entertainments were actually highly sophisticated in the way they foregrounded the operation of the senses in fooling them; such impresarios were presenting "illusions as experiments in the process of making knowledge."[55] The light in Millais's picture, then, calls attention to its own workings, and to the function of vision itself in determining truth.

Psychology

Is the woman in white a ghost or a hallucination? There is clear "evidence" for both, but ultimately what matters less than the answer is the striving for a conclusion, an analytical process that echoed that of several fields of nineteenth-century science brought to bear on the study of perceptual anomalies. In particular, apparitions, dreams, and hallucinations were foundational to the debates, investigations, and experiments driving British psychology in the later nineteenth century.[56]

With her ectoplasmic train trailing behind her, the female figure resembles the manifestations at séances, particularly those that emerged from behind drapes where a medium was stationed; one such event occurred in the same year as Millais's work, in which "several forms in succession . . . appeared at the opening of the curtain . . . [followed by] a little 'filmy' grey figure, clad in drapery."[57] An illustration in the *Graphic* in 1880 portrays another example of this sort of manifestation (fig. 4.6).[58] Like science, spiritualism (the practice of seeking contact with the dead) filled the anxious vacuum left by growing doubts about traditional religious dogma.[59]

And like science, too, spiritualism was understood as a measurable phenomenon. To name just a few recognized individuals engaged in such analysis: chemist William Crookes investigated spiritualism with an interest in proving its possibility, as did physicists William Fletcher Barrett and Oliver Lodge. Skeptical analyses, on the other hand, were offered by sociologist and anthropologist Herbert Spencer; physiologist William Carpenter, Alexander Bain, the founder of the first psychology journal (*Mind*); physicist John Tyndall; and psychiatrist Henry Maudsley, among others.[60]

If she is not an actual spirit, Millais's woman must be generated from the man's—and the viewer's—imagination. The overt reference to

Fuseli's *Nightmare* suggests that she could be a dream figment.[61] Indeed, *Blackwood's* described the man as "a widower, who seems to be starting from his bed in the agonies of a nightmare, rather than by the breath of vision."[62] Such a reading brings the picture into conversation with contemporary psychological investigations of altered mental states. The spectral woman was therefore also linked to the tradition of hallucinations and dreams, particularly the succubus, a female demon who preyed sexually on sleeping men.

Emerging, generally speaking, from strands of philosophy, theology, and physiology, and focused on a study of consciousness and its relationship to perception and the sensorium, psychology as a separate discipline solidified in midcentury England.[63] One of the new psychological organizations engaged in just this course of investigation of both spirit manifestations and hallucinations was the Society for Psychical Research (SPR), started in 1882 by poet and classicist Frederic W. H. Myers.[64] Psychologist Edmund Gurney, philosopher Henry Sidgwick, and Henry James's brother William—trained as a physician, psychologist, and philosopher—were all members of the SPR, as were many psychologists, physicists, and chemists.[65] This group aimed to understand a range of hitherto inexplicable phenomena through scientific observation. From 1889 to 1894, for instance, the SPR implemented the remarkable and ambitious "Census of Hallucinations," the results of which were published in the journal of its proceedings; in its analysis of responses it cited 2,272 accounts of wakeful hallucinations (1,112 visual and 494 auditory).[66]

Myers defined ghosts as "a manifestation of persistent personal energy," taking them as evidence of spiritualism.[67] Several SPR members, including Gurney, who had studied chemistry, physiology, and physics, also pursued proof of the possibility that apparitions represented telepathic projections of one person to another (based on the frequency of such appearances at the time of death of distant loved ones).[68] Others at the time maintained that all hauntings came from within: Dr. William Carpenter, a formulator of physiological psychology, coined the phrase "unconscious cerebration" to explain ghosts.[69] As Carey Gibbons explains in her chapter in this volume, Carpenter argued that ideas are stored in memory to be "revived again in full vividness under certain special conditions" and wrote of the intertwining of unconscious and conscious experiences in light of the fact that dreams and reality often became conflated to the point that people asked, "Did this really happen to me, or did I dream it?"[70] The subject of *Speak! Speak!* was, then, ideally chosen to raise the same questions as a number of contemporary scientific discussions.

Importantly, for the new psychologists, regardless of their convictions regarding the origins or identity of ghosts, the question of whether the object provoking the sensorial response truly existed was moot.[71] Whereas previously the debate had concentrated on identifying the process by which diseased, false, or inappropriate sensory responses generated hallucinations, in order to normalize or correct the malfunction, now scientists focused on the responses themselves. Psychologist Edmund Parish, who opposed the SPR's idea that ghosts were telepathic communications, was yet able to maintain the "reality" of hallucinations, declaring that, "'to believe one sees' and 'to see' are two expressions meaning the same thing," and adding, "in nearly all modern psychological inquiries . . . hallucination is a sense-perception, and . . . the only question of practical importance—viz., whether the object is or is not really there—is psychologically irrelevant." Parish concluded that "all sense-perception is ultimately a psychical phenomenon," citing Gurney to agree that "every psychological phenomenon that takes the character of a sense-impression is a sense-impression."[72] Joining this contemporary conversation, *Speak! Speak!* foregrounds sense perception in unmissable ways; Millais's haunted man demonstrates a need to verify the nature of the observed phenomenon with his senses, requesting auditory confirmation of the specter's existence, gazing at her intensely, and reaching out to touch her (notably, Lewes argued that touch, not sight, was the most important sense).[73] We imagine, too, he can smell the smoke from the candle guttering in the breeze she has produced.

Explaining hallucinations in a way that illuminates some of the larger issues at stake in Millais's painting of the appearance of a spirit, German American physician Charles Goodlove Raue propounded that we can know only images or representations of things, never the things themselves, because "the things outside remain outside and foreign to us. We can never penetrate into their actual being or *be the like of them*, and, for aught we know, they may exist only in our imagination. This has been the view of Idealism for centuries."[74] Raue here refers to the philosophical divide between idealists and realists/materialists, a long and complex conversation that for our purposes here can be reduced to a debate between the belief that the mind was the only reality—that humans essentially made their own world and were therefore at the center of it—and the doctrine that there was, in the words of Oxford philosopher Thomas Case, "a reality beyond psychical facts," an *a priori* world of tangible, material objects. Case described realism as the conviction that physical things existed independently of our perception of them and that they were what they seemed to be to our senses, and furthermore that the mediation of scientific analysis and lucubration could help to determine their nature.[75] James, of the SPR, followed a *via media* between these poles and held that the idealist chose the more comforting option: "To say then that the universe essentially is thought, is to say that I myself, potentially at least, am all. There is no radically alien corner, but an all pervading *intimacy*"; the more adventurous materialist (realist), however, possessed "an overpowering desire to escape personality, to revel in the action of forces that have no respect for our ego, to let the tides flow, even though they flow over us."[76]

James was kin to both idealists and realists in seizing on the function of perception, a subject over which psychology gradually gained custody in the period, as the key to proving his theories. For James, the question of identifying the precise source of the sensory experience—and whether it was exterior or mental—was no longer of interest; as he declared, "an hallucination subjectively considered is as good and true a sensation as if there were a real object there. The object happens not to be there, that is all."[77] The belief that the only reality is the one we can perceive permeated philosophy, physiology, and psychology—and, as we will see, art.

Regarding the emergence of psychology between 1850 and 1880, Helmreich cites science historian Bernard Lightman's four-stage model with an eye to tracing the shift away from a "discourse of the soul" to a more medicalized empiricism.[78] She further observes the rise of the newly identified relationship between vision and cognition, beginning with philosopher John Stuart Mill's contention that the physiological function of the eye produced sensations interpreted by the mind. Other philosophers, such as Bain, in *Senses and the Intellect* (1855), and Lewes, in *Physiology of Common Life* (1860), developed the idea that seeing was learned and unique rather than innate and shared; the new emphasis on subjectivity turned vision into a "psychological phenomenon."[79]

Whether the mind generated an idea that the senses then received or the senses caught hold of something in the world and brought it to the mind, hallucinations could, either way, now teach individuals to reorient their definitions of reality and truth, to let go of the need to verify the world's existence, and to replace that compulsion with an acceptance of the fact that one's own perception of the world was the only relevant certainty. For the psychologist, reality was an individual's subjective experience; hauntings made this fact visible. Millais's man, then, both really sees a ghost *and* believes that he sees one, subverting the binary question that the artist must have known would become the most pressing critical concern about the work.

Art and Its Function

Reviewers' anxiety regarding *Speak! Speak!* centered on both the ambiguity of the narrative regarding the ghost's existence and the painter's style, which they found too realist and mired in the material. Commented Claude Phillips, "Where the master fails, it is by reason of a certain curious literalness and insistence on the material side of his conception, which robs it too much of mystery, because it leaves the imagination with little or nothing to complete for itself. . . . [W]e are not carried away into the dim borderland between dreaming and waking, whither the painter would fain transport us."[80] The criticism of *Speak! Speak!* produced a drumbeat of such descriptors: "too tangible and realized"; "curious literalness and insistence on the material side"; "the literalness, the deliberation with which the material side of the subject has been worked out, impairs its power to convince the imagination."[81]

Designating the painting as "realism," both positively and pejoratively, critics focused especially on the specter.[82] Many seemed to feel the need for a more convincing—and by this they meant abstract or painterly—ghost, allowing for the engagement of the viewer's subjective imagination. With a clever reference to *Hamlet*, one carped that "the ghostly visitant . . . seems almost of 'too solid flesh compounded.'"[83] Intriguingly, however, for the experts from the realm of spiritualist study, this very tangibility betokened the woman's status as a visitor from another realm. A review in *Borderland*, a journal devoted to supernatural phenomena, approved of the ambiguity of the woman's identity: "There is a welcome absence of all shimmer and shine about Sir John Millais' apparition, which is true to facts. So far as the evidence goes, ghosts, when they appear, are remarkably like other people."[84] The ghost was therefore the fulcrum upon which the stylistic debate turned: should she be robustly material or ethereal? In her significance as a focus of the gaze and of style, of representation itself, the specter begins to take on another layer of meaning regarding artistic debates.

By portraying the ghost as material and tangible rather than uncertain or visionary, Millais offered a metacritique; the question of *Speak! Speak!* hinged not only on the nature of perception but also on the nature of art itself. Since the Renaissance, art had striven to present the imagined as the real, but at the end of the nineteenth century it was no longer enchained to this directive; it could, rather, function as an emotional expression of the artist's experience. In this way art, too, took up the issues of idealism vs. realism, subjective and objective reality. Helmreich contends that the new psychology emphasizing subjectivity pushed art toward something other than its previous insistence on closely observed detail, constructively interpreting Millais's move away from a detailed early Pre-Raphaelite mode to his later, looser style in relation to shifts in dominant scientific paradigms in light of changing beliefs about perception.[85] Observing how new psychological writings filtered into art

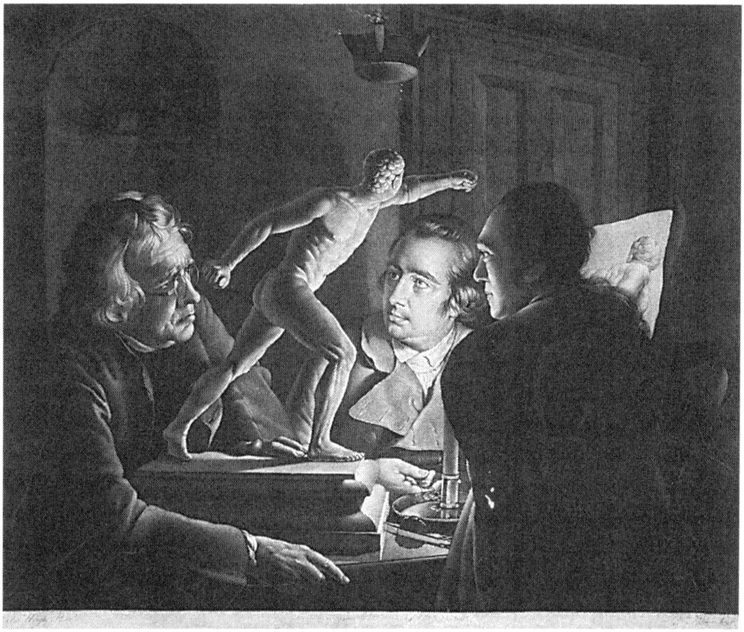

Fig. 4.7. Candlelight pictures became popular in the later nineteenth century, when there was also a renewed interest in the work of Joseph Wright of Derby. William Pether after Joseph Wright of Derby, *Three Gentlemen Viewing the Gladiator by Candlelight*, ca. 1768. Mezzotint. 19 × 29 in (48.2 × 55.5 cm, image/plate). © Victoria and Albert Museum, London.

criticism, she reads art critic Tom Taylor's description of the changing artistic mode over the course of the century as a move from an original goal of "paint[ing] the infinite details of nature as they were" to "paint[ing] the effect of the same details in the eye" by "translat[ing] . . . what they look like by tones and touches in themselves utterly unlike the thing they are to help in representing."[86] The artist, Helmreich observes, was now supposed to "create painted illusions of what the mind saw rather than what the eye saw."[87] The ghost, of course, is both of these things. However, *Speak! Speak!* pointedly does not resemble the brushy impressionism of later landscapes that Helmreich identifies as related to the new dominant model of "what the mind sees." Nor is it typical of Victorian academic painting, or, again, of Millais's early Pre-Raphaelite mode. Rather, the work draws on a wide range of periods and styles, as signaled by the deployment of light and shadow. And the white woman plays a significant role in this summation.

Millais's unusual use of dramatic chiaroscuro, more characteristic of the baroque or the Romantic periods, was part of a minor resurgence of this mode in the late nineteenth century, related to a revival of interest in works such as Joseph Wright of Derby's *Three Gentleman Viewing the Gladiator by Candlelight* of 1765, owned by Millais's friend and patron the Marquis of Lansdowne and engraved by William Pether in 1768 (fig. 4.7).[88] Wright had introduced candlelight scenes to England by reworking baroque Dutch compositions with bright but hidden light sources, such as Gerrit Honthorst's *The Flea Hunt* of 1621 (held at the Dayton Art Institute). A renewed interest in Wright's oeuvre produced publications and an exhibition at the Royal Academy in 1886; he was even compared to Millais by one writer, who proclaimed: "There is no painter, unless it be Sir John Everett Millais, whose merits are more discussed at the present moment than 'Wright of Derby.'"[89]

Millais seems in fact to have been referring directly to Wright of Derby's *Three Gentlemen* in ways that would allow him both to embrace the academy and also to modernize it. The violently physical gesture of the man in *Speak! Speak!* resembles that of a reversed *Borghese Gladiator*, a renowned Hellenic sculpture in the Louvre much copied by art students, as we see in Wright's painting, in which the artist himself compares a drawing with a small replica of the original. Millais, who likewise seems to have quoted the *Gladiator* in an early drawing of wrestlers (ca. 1840, held at Tate Britain), here nods both to the classical tradition upon which the academy was founded and to his own apprenticeship.[90] The RA—and its future president Millais—were in 1895 still invested in the organization's role as the keeper of artistic knowledge and practices handed down through generations, and the pictorial rhetoric of *Speak! Speak!* secured the praise of fellow academician Herbert Herkomer, who commented that "it is the strongest arm that has been put forth for a long time against the fearful (and mad) wave of the modern tendency."[91] In *Speak! Speak!*, however, the "strong arm" of the classical warrior has been transformed into the futile gesture of the desperate young lover, the aggressive stance modified into an endlessly unsatisfied reaching toward.

On the other hand, reviewers recognized in *Speak! Speak!* something more than an academic mode, observing, rather, a renaissance of Millais's Pre-Raphaelite style. One claimed, "In the autumn of his life, [the artist] is reverting to the naiveté and pathos which, not less than its uncompromising realism, serve to distinguish the earlier from the later Pre-Raphaelitism."[92] Following Ruskin, Millais had a keen interest in the close observation of natural world, which accorded with the materialist conception of a "world out there" separate from and indifferent to man; as scholars have found, "Ruskin was a materialist . . . in other words he considered that things produce and confirm ideas, and not ideas, things."[93] Millais's early style confidently claimed that through

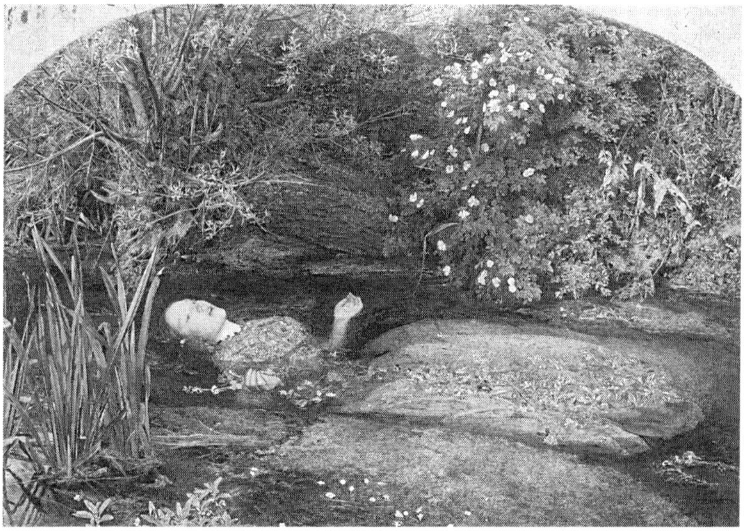

Fig. 4.8. Millais's most famous painting of a woman in white was his image of the drowning Ophelia. John Everett Millais, *Ophelia*, 1852. Oil on canvas. 30 × 44 in (76 × 112 cm). Tate Britain, London. Photo@Tate.

vision the surface character of a thing could be apprehended and that in that surfaceness was true substance. It is relevant that Millais rendered the woman, who offers relief in the possibility of connection and reunion, in the style of Pre-Raphaelite realism; she nostalgically signals his youthful mode while also suggesting that in realism is reassurance. Notably, the ghost's white gown and her gesture that, for one viewer, called attention to itself through its awkwardness, echo that of the figure in what was still, in 1895, one of the artist's most famous paintings, *Ophelia*, of 1852 (fig. 4.8), which portrays another young woman in a liminal condition between death and life; *Hamlet* murmurs multiply as Millais combines various chapters of his career into a grand conclusion.[94]

The painting, then, makes claims for both traditional academic art and for realism. Lit by the drama of chiaroscuro, the man in the bed operates in a traditional idealizing academic aesthetic, while the woman appears in the full flat light of Pre-Raphaelitism; the academic and the Pre-Raphaelite modes are, like the lighting schemes, set in contention—and in complement. In the 1890s, as the Royal Academy was roiled by debates between conventional and more avant-garde camps, a diplomatic painter wishing to be elected president might well take a juste-milieu stance. Millais in fact claimed in 1895: "It is all nonsense . . . to pin your faith to any one school. There is as much room for the old Dutch microscopic painter as for the modern Impressionist."[95] He was, in essence, doing both. The divide between academic art and Pre-Raphaelitism might be very loosely linked to that between idealism and realism in philosophy, which, as we have seen, informed psychology.

In raising these complex artistic issues of his moment, Millais was therefore hitting on key critical debates that, like those in psychology, revolved around prioritizing the ideal or the real: whether art should represent an edited and improved picture of the world or an ostensibly accurate transcription of it. In artistic idealism, the operating principle was that the ordering and refining human mind created masterworks by using its powers to abstract, generalize, and amend nature. The idealist emphasized the spiritual rather than the sensual; as a critic demanded, "Let us show we think there is a soul which is not seen as well as a body which is seen."[96] On the other hand, Pre-Raphaelitism, the realist movement that Millais had helped birth, had long been viewed as messily engaged in the details of the material world, insisting on individual observation and on precise copying of what the eye saw—much like scientists of the day, in fact.[97] In an article on realism in the *Art Journal* of 1880, Millais's work was used as an exemplar of "the notion that a picture must be such as can be easily proved from nature to be correct: there is to be nothing left for the imagination—the fancy must not be free."[98]

Millais seems to be suggesting a merger between artistic styles and philosophical camps, producing a work in which the material and the spiritual are one. Separating the bed hangings to disclose herself, the woman literalizes the metaphorical veil between worlds, comforting the viewer with the idea that lost loved ones may linger just out of sight. The space beyond reassuringly appears as merely a continuum of our own. The otherworldly glow of the woman also suggests a revelation in the manner of Raphael's 1512 Sistine Madonna (held at the Gemäldegalerie Alte Meiste), who also appears behind parted draperies. The ghost becomes a consolatory figure, a divinity for a world in which matter and spirit were newly intermingled. There may be no coincidence in the fact that Millais described the Sistine Madonna as a work that distinctively mingled the material and the spiritual, the real and the ideal: "a picture painted on earth directed by a mind in Heaven." Raphael, for Millais, here "sealed on canvas for all time the nobility which may and does at times breathe through us mortals . . . a combination of the human and the divine."[99] Such a combination may also mark Millais's specter.

The picture operates, then, as a sophisticated career summation; it was also, poignantly, an end-of-life testimony. In light of new scientific information about the fact that things did not disappear but were rather reconfigured, Millais's ghost suggested that souls too can transform

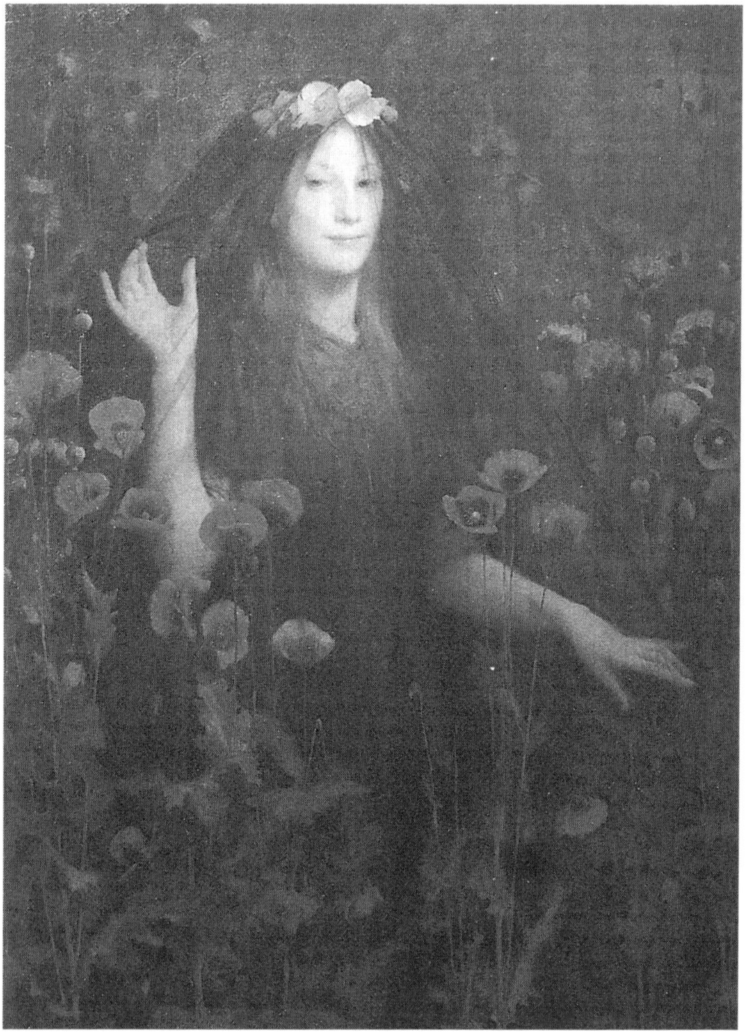

Fig. 4.9. Sex and death intertwine in the popular trope of the ghostly lover come to claim her mortal prize. Thomas Gotch, *Death the Bride*, 1895. Oil on canvas. 60 × 46 in (152.4 × 116.8 cm). Alfred East Art Gallery, Northamptonshire.

yet persist. Dead lovers comfortingly remain present to us just out of view, occasionally even becoming visible. Such reassurance about the persistence of life would have been acutely personal for Millais at this time. Sharply outlined and dramatically pale against the dark shadow cast by the candle, the man's throat—part of the breathing apparatus through which human combustion takes place, generating life as the candle generates light—is prominent. Oddly, Millais drew attention to this passage to more than one interlocutor, one of whom reported, "but for the sight of that throat . . . he might never have painted the picture."[100] The artist would have deeply identified with such a distended neck, as he too was experiencing this condition. His own, however, was due to laryngeal cancer, known to be fatal at the time.[101] By 1895, unbeknownst to the public, the artist was suffering severely and had almost lost his voice; at the RA banquet that year, where he sat in for the dying Leighton, the newspaper reported that he was "pluckily struggling against evidently painful hoarseness," attributed to a cold.[102] Millais, then, was increasingly incapable of responding to the injunction to "speak!"—his own silence, like the ghost's, being that of death.

In my writing of this story I acknowledge my own haunting by the insistent whispers of biography: my own, certainly, but also that of Millais. His forceful statement in the face of his own imminent death is wonderful and moving. In this light, Millais limns himself both as the mortally afflicted man and as the silent female spirit. Simultaneously with *Speak! Speak!* Millais exhibited *Time, the Reaper* at the New Gallery, another work that contrasted moonlight and artificial light. In it, an aged man with Millais's own features enters a house, scythe in hand, astonishingly portraying the artist as the inexorable force he was facing down in a brave cooptation of death itself.[103] In November 1895 he wrote to W. P. Frith, again invoking *Hamlet*, that he would soon be returning to London "to paint, perchance to die."[104] And indeed, despite a dramatic emergency tracheotomy in May, the artist passed in August of 1896.[105]

The spirit bride come for her lover might therefore, like *Time, the Reaper*, be an autobiographical nod. As a revenant lover, Millais's phantom resembled Thomas Gotch's allegorical painting at the same exhibition, *Death the Bride* (fig. 4.9), allegorically conjoining themes of desire and deadliness in the form of a woman. The necro-erotics of the figure approaching the man in bed suggests that death may be a consummation devoutly to be wished.

And the woman comes to stand in for art itself. As the atoms of the burning candle persist, so too does the spirit of the woman persist, so too does art persist, the painting seems to remind us. The overly materialist representation of the ghost asserts not only the fact that spirit and body are one but also perhaps that the material work of art, like the spirit, is persistent. Things—and people—do not disappear but rather became translated into other forms. Things and people do not disappear but rather become translated into art.

Millais makes a grand claim here. Barbara Maria Stafford argues that, from the eighteenth century, "imagery, like the material body, became systematically drawn into a negative analogy with fraudulent apparitions, confounding dreams, and irrational delusions. The Phenomenal, the manual, the somatic were irrevocably divided from the noumenal, the theoretical, the intellectual."[106] Apparitions, delusions, art—touching on psychological discoveries of the day, Millais's work undoes the conventional marginalization of these things by acknowledging that, like hallucination, art negotiates a position between the real and the unreal, being both and neither; in it, the body, the image, and the material are held in equilibrium with the intellect and the ideal.

Millais also insists on Stafford's other disavowed things, the material body and its senses. As we have seen, in this new world ghosts were as real as any other object due to their effect on the sensorium. As the only real access to truth, physiological sense-perceptions were elevated, and as a picture about perception itself, *Speak! Speak!* conveys the process by which the world is unveiled to us. Millais's picture allows us to have it all: it asserts that the world, human beings, and art can be matter and idea both, vindicating both realism and idealism and holding them in equilibrium. Considered in light of several fields of scientific knowledge in the 1890s, then, Millais's painting *Speak! Speak!* erases and makes irrelevant boundaries between reality and fantasy, death, and life. The invisible is made visible: love, art, hallucinations, psychology, chemistry—all realize forms, forces, or materials previously overlooked. Representing a loss salved by a haunting, which then itself becomes a fresh source of trauma, Millais's poignant painting produces a perpetual suspension of opposing elements and forces in a brave and complex statement on the human condition made newly comprehensible through scientific observation.

Yet it would be all too easy to take the intellectually facile path and claim that the picture remains ultimately ambiguous, fluidly able to mean everything and nothing. This may be true, but I would contend that Millais was actually making a case for certain ideas and values over others. That is, he was not merely equivocating about meaning or about art; rather, he was celebrating the way the material nature of painting, its capacity to represent something other than itself with that matter, was in fact persistent: in this picture we see a belief in art as salvific. Art is at core about loss, because an image stands in for something else, something missing. In the same way that a ghostly apparition represents a person who has died, pictures re-present their subject matter. Yet rather than being tragic, this substitution is ultimately triumphant.

And furthermore, that world is metamorphosized into paint: for Millais, what disappears reappears to persist as pigment. Our lost love, our hallucination, our ghost, our perceived image—all are here wrapped up in that impossibly solid, densely material figure of the woman, who is Art itself. Importantly, if the man in the bed cannot hear or touch his love, he *can* see her. Radiant but silent, the woman embodies painting. Vocal, the man stands in for narrative, poetry, or theater. Demanding voice, speech, and words, the title's injunction will remain perpetually frustrated, not only because the person to whom it is directed remains "behind the veil," but also because muteness is an inherent characteristic of oil on canvas.[107]

Art, like ghosts, persists. *Ars longa, vita brevis.* We need to admit more often to a silent but powerful component of our response to painting from previous eras, or even, perhaps, all art: paintings remain, the spirit of the dead silently watching our struggle, reminding us of our own limited existence. Like the man, viewers of Millais's picture engage with an object from a different time and place, drawn to it in a frustrating, never-ending, yearning connection of love, loss, shock, and joy. Millais thus offers representation as compensation. If it is art's job to stand in for what is not there, to figure lack and loss, then representation is a form of death, but also, impossibly, a reparation for that very death. Much as the man is reassured yet disturbed by the appearance of the white woman, so too are we made aware of what is missing behind each solid surface of pigment we encounter. Yet pictures also register visible traces and signs of the lives that gave them life. In what he was aware would be one of his last opportunities to make himself seen, Millais strove to rise to the challenge of producing a work that both acknowledged personal frailty and limitations and offered a powerful vision of how he himself might come to haunt us.

CHAPTER 5

PHOTOGRAPHING ETHER, DOCUMENTING PAIN

REPRESENTING THE CHEMICAL INVISIBLE IN THE DAGUERREOTYPES OF SOUTHWORTH & HAWES

Naomi Slipp

On October 16, 1846, Massachusetts General Hospital (MGH) surgeon John Collins Warren successfully demonstrated the use of ether as a surgical anesthetic. This landmark occasion is a key development of modern medicine, one that fundamentally transformed surgical practice and medical knowledge. This innovation allowed for the rapid expansion of allopathic medicine, which achieved professional hegemony in the postbellum United States, in part thanks to ether. At a local level, the ether discovery launched Boston's reputation as a medical center, a standing that the city—in concert with Harvard Medical School and MGH—cultivated through anniversary celebrations and visual material, including paintings, monuments, and daguerreotypes. These presented the city as a hub of medical innovation.[1] More broadly, the introduction of effective surgical anesthesia dramatically affected contemporary ideas related to physiological pain, its management, and sensorial relation to the human body, in disciplines from medicine to philosophy.

Four extant daguerreotypes and one paper print of a lost daguerreotype by the Boston photographic firm Southworth & Hawes present a reenactment of the first ether demonstration and document two subsequent etherized surgeries in the hospital's surgical theater between 1846 and 1847. Critical reception regards these images—one of which is canonized in the survey text Gardner's *Art through the Ages*—as transparent and uncomplicated records of period surgery, or uses them as illustrations in medical histories.[2] However, by building upon scholarship in the history of photography and medicine and contextualizing these images within dominant period discourses regarding medical

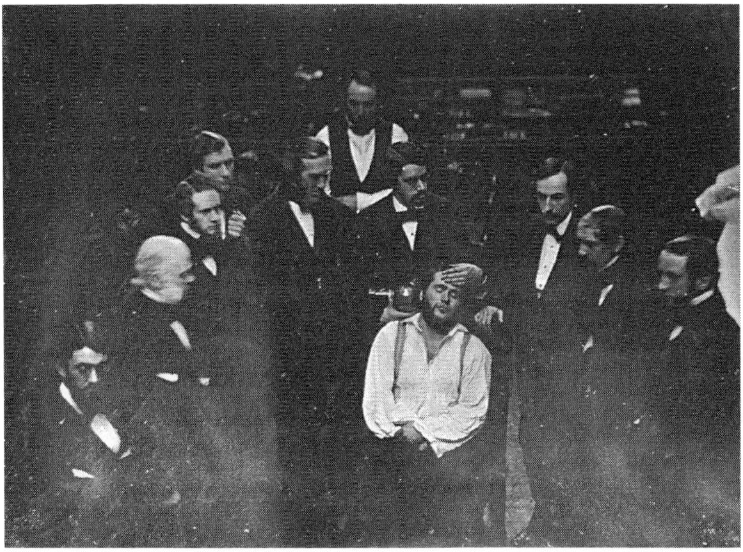

Fig. 5.1. As the first successful surgical demonstration of ether was not documented, the surgeons involved staged a reenactment. Southworth & Hawes, "Untitled (Reenactment of the First Public Demonstration of the Surgical Use of Ether)," 1847. Whole plate daguerreotype, 5 7/8 × 7 7/8 in (15 × 20 cm). Harvard Art Museums/Fogg Museum, Loan from the Massachusetts General Hospital Archives and Special Collections, 3.1979. © President and Fellows of Harvard College.

professionalization; scientific theories on corporeal pain and pain management; and a comparison of the "magical" chemical effects of ether in relation to the relatively new art of daguerreotypy, I instead propose that these five images are highly complex visual objects, which illuminate a number of interrelated photographic and medical concerns.[3] First, by attempting to photograph ether and the absence of pain—two things that could not be visualized, these images make manifest scientific ideas about corporeal interiority and illuminate the challenges faced when attempting to represent the unseen. The images themselves, therefore, indicate a larger uncertainty regarding the purposes and technological limitations of photography. Furthermore, the project of photographing something that could not actually be seen linked medicine with technological and chemical advancements, lashing the presumed documentary effects of the photograph to medical objectivity early in the professionalization of both fields. The daguerreotypes reveal the professional agendas of medical men and photographers who carefully and consciously used the visual arts to advertise medical progress, even when those discoveries were difficult to visually communicate.

I contrast the aesthetic characteristics of the ether daguerreotypes with other Southworth & Hawes images in order to understand the representational flexibility of the new medium of daguerreotypy. This comparison underscores a compelling tension between a subjective photographic practice based on fine art precedents and an objective, mechanical practice increasingly used by the sciences. Finally, by considering the reception of two ether daguerreotypes that hung in John Collins Warren's home near a copy of Rembrandt van Rijn's *The Anatomy Lesson of Dr. Nicolaes Tulp* (1632), I consider how works of art supported the ambitions of medical professionals and question the ways in which certain media better communicated different messages. In other words, what could paintings convey that photographs could not, and vice versa? Ultimately, the ether daguerreotypes visually construct the historical significance of the ether demonstration and telegraph the role of MGH, Boston, and medical professionals, such as Warren, in this scientific discovery.

Documenting an Historic Reenactment

In December of 1846 surgeons at MGH commissioned Southworth & Hawes to document the initial ether demonstration through a staged reenactment in the hospital amphitheater. "Untitled (Reenactment of the First Public Demonstration of the Surgical Use of Ether)" (fig. 5.1) presents a male figure posing as Mr. Gilbert Abbott, the patient in the first ether demonstration. As a direct positive process, the daguerreotype image is laterally reversed; the left of the image is the right of the theater and vice versa. In addition, at 6 1/2 × 8 1/2 inches, the daguerreotype necessitates a concerted physical engagement to decipher; the image slides across the mirrorlike surface, shifting and dissolving at different angles.

The patient is in an operating chair slightly turned to the left of the frame. He appears young, has a full beard, and wears suspenders over a white shirt with the collar opened to reveal his clavicle. His hands are clasped on his lap and his mouth and eyes are closed, implying a sleeping or anesthetized subject. However, he lacks the flushed face and sprawled form of later ether images; his corporeal composure indicates playacting. Mr. Abbott's stand-in does not appear ill, and his clean, white shirt presents him as an unsoiled and orderly, if working-class, patient.

Ten medical professionals in formal attire stand in a semicircle around the seated figure. Although they appear attentive, they do not interact with the patient. Instead, they act as witnesses to an event. The figure behind and to the right of the patient places a hand informally on the top rail of the operating chair and leans upon it, while the man standing directly behind Abbott lays his hand on the patient's forehead and holds a globe for administering ether in his other hand. A bright highlight shines across the convex surface of the glass orb, drawing our attention. The ether globe formally mimics the shape of the patient's head, conflating globe and mind and implying that ether is at work promoting a restful, deep sleep.

MGH's surgical amphitheater was illuminated by a raised clerestory above the central dome. On a clear day, filtered sunlight streaming into the space from above would have facilitated the production of the daguerreotype. While the light was ideal, the steep stadium seating and room's small proportions would have made framing and photographing particularly complex. The image's tilted perspective indicates that the photographers, Albert Sands Southworth and Josiah Johnson Hawes, are in the tiered seating looking down upon the group. The viewer can just make out cases of surgical instruments behind them, along the back wall of the theater. In spite of this, the image is tightly framed around the figures and, while many of the faces are slightly out of focus, the shallow depth of field is focused on the patient, who appears like a still face in a moving crowd. This, along with the white expanse of his shirt and glinting light on the globe, draws the viewer's eye and adds to the sense of calm that the figure projects.

"Untitled (Reenactment of the First Public Demonstration of the Surgical Use of Ether)" presents the successful use of ether as its central subject, witnessed to and attested by the circle of doctors and photographers. Southworth & Hawes depict a contrived, arranged, and theatrical re-presentation of the initial surgery; a representation that was necessary in order to combat public skepticism and claims by other medical practitioners that either emphasized the dangers of ether or supported competing anesthetics. Questions emerge, however, that pertain to the visual legibility of the daguerreotype, its authenticity as a historical document, and its ultimate purpose, and hinge on the fact that this photograph is, in essence, a representation of a reenactment of a historical event. The original surgery two months earlier was not nearly as momentous as it later became. Its history and representation are fraught with controversy.

The Ether Demonstration, Medical Skepticism, and Subsequent Controversy

The first successful demonstration of ether occurred at ten a.m. on October 16, 1846, when dentist William T. G. Morton rendered twenty-one-year-old printer Edward Gilbert Abbott unconscious through Abbott's inhalation of "Letheon," the vapor of pure sulfuric ether. This allowed surgeon John Collins Warren to excise a tumor from his neck while Abbott remained insensitive to pain. To keep the chemical composition a secret, Morton initially masked the scent with oils and added pigments. He intended to patent and charge money for its use. Morton's initial subterfuge, along with his desire to monetize ether in order to profit from the alleviation of physical suffering—something many declared morally repugnant—drew widespread condemnation, spurred declarations of quackery, bolstered competing claims, and fueled doubts over ether's efficacy.[4] Other practitioners, such as Crawford W. Long, Horace Wells, and Charles Jackson, laid claim to Morton's discovery, producing a public contest over who had the rights to ether, stymying Morton's patent application, and jeopardizing the hospital's claim as the site of the breakthrough.[5]

Complicating the historical narrative of "the ether discovery" is the fact that there was no single iconic event. Indeed, Morton's October 16 demonstration was just one of a series of trials performed at MGH following experiments on Morton's dental patients. Later sources claim that Warren recognized the momentousness of the October 16 demonstration *before* it had even occurred, turning to his audience and saying: "I have been 40 years a surgeon in Boston. On every instance when the knife was applied to live tissue, there was pain. I doubt not that every one of my colleagues has had the same distressing experience. And now we have a gentleman here who tells us that he has a liquid preparation by the inhalation of which pain will be entirely done away with in surgery."[6] However, this moving speech is almost certainly a historicizing and mythologizing tribute, written afterward to herald the new anesthetic and to pinpoint October 16, 1846, as a canonical event in medical history. In spite of its questionable authenticity, it is important to consider what textual tributes such as this accomplished and why they may have been necessary. In fact, Warren remained skeptical after the October 16 demonstration. This may have derived from an earlier failed demonstration of nitrous oxide as anesthesia by dentist Horace Wells on January 20, 1845. Not until Warren observed further etherized surgeries, including an amputation on November 7, was he convinced of ether's efficacy.

As professor of surgery at Harvard Medical School and its first dean, cofounder of MGH, surgeon at the hospital, founder of the *New England Journal of Medicine*, sponsor of the 1831 Massachusetts Anatomy Act, and third president of the American Medical Association, John Collins Warren was, without doubt, a giant in American medicine and a highly respected member of his field. Therefore, his approval of the new anesthetic was significant and signaled its potential for professional popularity. At the same time, professionals who could not witness demonstrations firsthand had to be convinced that this anesthetic was legitimate. Because of this, publications and testimonials "streamlined" the narrative of discovery and canonized the October 16 event, marking it as *the moment* when surgical pain ceased to be a surgical concern. They emphatically linked Warren to the ether discovery, highlighting his acceptance of ether and signaling to medical practitioners that it was effective and worth adopting.

Finally, the quotation in question indicates its intended medical audience and speaks to larger professional fissures. In the antebellum United States, pain was the primary hindrance to surgical care; patients did not want to experience it and doctors did not want to cause it. The entanglement between pain and medicine in the public imagination led to the characterization of surgeons as sadists who enjoyed eliciting pain or butchers who would cut first and ask questions later. Medicine was fractured and largely unprofessionalized. Allopathic practitioners or "professionals," homeopaths, quacks, eclectics, and faith healers frequently disagreed over the meanings and causes of diseases and their treatments, especially regarding surgical intervention.[7] The discovery of ether and, with it, the surgeon's ability to dull or eradicate pain created the possibility that more diseased Americans would turn to surgeons, allowing for the subsequent hegemony of allopathy. However, these practitioners needed to actively convince professionals to adopt ether and prove its efficacy to the public. They accomplished this through the creation of documentary evidence, including testimonials and visual materials, that convincingly portrayed ether as a sound anesthetic. Indeed, Morton and hospital staff and trustees used the press and artists to demonstrate their right to patent ether, claim Morton as its "inventor," and memorialize MGH as the location of the "death of pain."[8]

As the newest general hospital in America, MGH had much to prove regionally and nationally. At the time, allopathic medical practitioners looked to Edinburgh, London, and Paris as centers of Western medicine, while New York and Philadelphia were US hubs. Consequently, the role that MGH played in the first successful demonstration of surgical anesthesia, just twenty-five years after its opening, solidified its national and international medical reputation and boosted Boston's reputation as a leading site for medical study and practice. For instance, surgeon Henry Jacob Bigelow announced the event in the *Boston Medical and Surgical Journal* on November 18, 1846; it was republished four times within the week by local papers. In addition, Morton circulated accounts and reports in multiple newspapers and journals, including the *Boston Daily Journal*, the *Daily Chronotype*, and the *Daily Evening Transcript*. Furthermore, all of the hospital trustees signed a petition to the US Congress

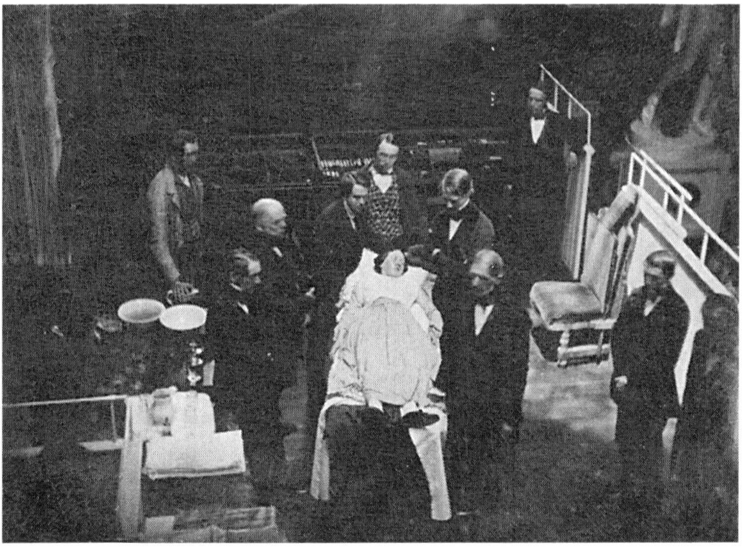

Fig. 5.2. Subsequent surgeries were photographed, demonstrating the efficacy of ether as an anesthetic. Southworth & Hawes, "Untitled (Demonstration of the Surgical Use of Ether)," 1847. Whole plate daguerreotype, 5 7/8 × 7 7/8 in (15 × 20 cm). Harvard Art Museums/Fogg Museuum, Loan from the Massachusetts General Hospital Archives and Special Collections, 2.1979. © President and Fellows of Harvard College.

in support of proclaiming Morton the inventor of surgical anesthesia.[9] Finally, in December 1846, Warren—an avid fan of daguerreotypy and repeat client of Southworth & Hawes—commissioned a daguerreotype to commemorate the initial surgical procedure through a staged reenactment. That image commemorates the events of October 16 as much as it seeks to confirm it. It links MGH, the surgeons, and the new anesthesia, creating a piece of visual evidence. And yet it is a representational fiction, a reenactment. The next step was to photograph a real surgery.

In the spring of 1847, Southworth & Hawes returned to MGH to document an operation on twenty-five-year-old seamstress Athalana Golderman, who had an infected puncture wound in her thigh from scissors. Contrary to the earlier reenactment, "Untitled (Demonstration of the Surgical Use of Ether)" (fig. 5.2) shows a wide view of the operating room, bounded by temporary privacy walls and tables of supplies. Golderman is prone on an operating table surrounded by ten surgeons, including house surgeon Charles Frederick Heywood, who holds an etherized sponge near her face and looks at the camera. The image is devoid of motion; the still surgeons with hands clasped prefigure the rush of the coming procedure. Warren stands to Golderman's right, while a spectator's head invades the left foreground emphasizing MGH's mission as a teaching hospital. The daguerreotype viewer sits in the audience, surrounded by spectators, watching Golderman's anesthetization.

In the daguerreotype, Golderman is clean and neat, her cheeks are flushed, her body orderly, and her skirt chastely tucked at her sides. Her bleached white stockings and smock and carefully pinned hair present her as a good, working-class woman. The way that Golderman is represented signals two entwined missions of MGH. As a general hospital operating under the ideals of Christian benevolence and as an affiliate teaching hospital for Harvard Medical School, MGH aimed to treat regional, upstanding, working-class patients. Patients were used in teaching demonstrations and, at the same time, needed to be understood as morally deserving of public care.[10] This placed contemporary debates about class, ethnicity, healthcare, and morality in tension. In order to determine the moral worth of its patients, MGH required application for admission in writing, proving literacy, and potential patients could be turned away for "bad morals," spitting, drinking, smoking, or swearing.[11]

Finally, Golderman's apparent virtue, as presented in the image, also counteracted period anxieties regarding the hypersexuality of anesthetized female patients, who experienced sexual dreams under the influence of ether. Beginning in 1847, reports in medical journals described sexual dreams in anesthetized women.[12] For example, an 1871 surgical manual cautioned, "Never give ether or chloroform to a woman, unless in the presence of witnesses. A curious but undoubted property of these agents is, that they occasionally produce most vivid erotic dreams and this may happen with a patient whose mode of life and character are above suspicion."[13] While Golderman's flushed face and position lying in the center of the image surrounded by men underscores her vulnerability, the composition gestures toward the relationship between female sexuality, anesthetization, and surgery, and—in the end—affirms her respectability. Such phenomena, however, raised significant questions about the interrelationship between pain, pleasure, and physical sensation. What was actually happening inside the anesthetized patient's body?

Picturing a Body Devoid of Pain?

The ether demonstrations, begun at MGH in October 1846, revolutionized surgery and its relation to pain, both physically and conceptually. It is not hyperbole to declare that the discovery eradicated one of the most significant hindrances for the progression of modern medicine and fundamentally reshaped ideas about the human bodies. To be able to stop pain and alleviate suffering increased questions about where pain originated, how it operated, and what, exactly, ether did.

In a November 1846 letter to William Morton, Boston physician Oliver Wendell Holmes argued that ether and etherization should be given two separate names, one for the thing itself and one for the state produced. He recommended the words *anaesthetic* and *anaesthesia*, drawn from the Greek word *anaisthesia* for a "lack of sensation" or "no feeling." In his choice of terminology, Holmes did not highlight the lack of pain denoted by etherization but instead argued that the name "signifies insensibility, more particularly (as used by Linnaeus and Cullen) to objects of touch. (See Good–Nosology, p. 259). The adjective will be 'Anaesthetic.' Thus we might say the state of Anaesthesia, or the anaesthetic state. . . . The words anti-neuric, aneuric, neuro-leptic, neuro-lepsia, neuro-stasis, etc., seem too anatomical; whereas the change is a physiological one. I would have a name pretty soon, and consult some accomplished scholar . . . before fixing upon the terms, which will be repeated by the tongues of every civilized race of mankind."[14] Holmes makes distinctions between the anatomical versus physiological body. In other words, he is concerned with understanding the taxonomic relationship between corporeal systems and distinguishes anatomy from physiology. Physiology examines the function of and interrelationship between parts of the human body, while anatomy focuses on the structure of parts themselves. In invoking Good, Holmes references an 1817 English publication in nosology, the field of medical science devoted to classifying diseases. In it, Good identifies a class of nervous system diseases he calls "Neurotica," which affect sensation and muscles.[15] Therefore, medical professionals, such as Good, regarded pain as any unpleasant sensation of the nervous system, while Holmes regarded ether as a way of desensitizing the nervous system, thereby inhibiting feeling. While the latter could, perhaps, be conveyed by Golderman's immobile body, how could one reveal an insensitivity to pain—and what was pain really?

A number of historical figures are significant to understanding nineteenth-century conceptions of pain. Holmes's understanding of the neurological basis for pain is connected to the mind/body divide, first theorized by French philosopher René Descartes in 1641. Descartes described the dualism between mind and body, now known as the Cartesian division, wherein the mind controls the body, but the body can also affect the mind.[16] Significantly, Descartes identified the brain as the center of corporeal functioning and argued that pain was a mechanical physiological response removed from the emotional experience of the world. He further theorized that pain was a disturbance transmitted on pathways to the brain. The eighteenth-century Scottish neurophysiologist Robert Whytt also explored the mind/body connection, proving that the nervous system was essential for muscular responses, the spinal cord was integral to reflexes and sensation, and pain was a general "sympathy" of the body's sensations that created a unified experiential whole, thus proving that embodiment was about physical awareness.[17] Finally, in 1811 Scottish physician Charles Bell confirmed the anatomy of sensory receptors and in 1839 Johannes Müller deduced that pain, pleasure, and pressure were physiologically separate neurological sensations.[18] Conflicting contemporary neurophysiological or neuroanatomical theories of pain, however, suggest that we still do not know *exactly* how pain works.

The mind/body divide is implicated in definitions of pain. The *Oxford English Dictionary* defines pain as variously "punishment, penalty, suffering inflicted for a crime," "suffering thought to be endured by souls in hell," and "physical or bodily suffering; a continuous, strongly unpleasant or agonizing sensation in the body, such as arises from illness, injury, harmful physical contact."[19] These meanings identify the two consistent ways pain is identified: first as corporeal *sensation*—feelings localized within the body; and, secondarily as noncorporeal *anguish*—pain as mental suffering. In other words, pain is both a psychological condition and a mental perception related to a physical sensation. While physical pain is related to touch, it is also neurological. Pain is "physical," but it is also mental, conceptual, interpretive, and subjective. Both a biological condition (nociception) and an experiential phenomenon (perception), pain is representationally challenging, as it is highly subjective and difficult to articulate.[20]

While pain is mediated through bodily experience and language, it is also culturally constructed and determined by social and disciplinary discourses. As Steven Bruhn has observed, this provokes "an artificial division," the discarding of which in the eighteenth century allowed for thinkers to "no longer entertain the notion of a physical pain that is not to a significant degree mediated by the patient's culturally determined understanding of pain."[21] In other words, the ways that we are trained to understand pain, hurt, or discomfort as concepts are relative to our social experiences of and cultural relationships to pain. One individual might consider something only slightly painful, while another might find it excruciating. One person may scream loudly and have no compunctions about verbalizing their experience, while another is conditioned not to articulate their pain. At a practical level, the eradication of surgical pain in the 1840s meant that patients were more likely to seek surgery. However, it also complicated morally inflected and shared cultural values.

At a cultural level, a shared understanding about who could (or should) feel pain, and who should be spared suffering, was integral to nineteenth-century debates related to public punishment, sentience, gender, and race, each of which were heightened with the introduction of anesthetic. Individuals were forced to reconcile the experience of pain—whether natural or unnatural—with evolving ideas about cruelty and sympathy and enter into significant ethical debate. If pain could be stopped through anesthesia, who deserved that intervention? For example, nineteenth-century criminal justice demanded that offenders be punished; very often pain came with punishment.[22] As early as 1847, a convict in the United States requested execution under the influence of Letheon.[23] While his request was denied, this prompted a moral inquiry into the shared social goals of corporeal punishment. Scientific racism also attempted to establish biological differences in the perception of pain based upon racial categories in order to support the institution of slavery, which relied upon brutal conditions and physical abuse.[24] In contrast, while white women were generally regarded as delicate and to be protected from physical suffering, many medical professionals considered the pain of childbirth a natural condition. Practitioners who used anesthesia in obstetrics faced scrutiny throughout the century.[25] Finally, the sentience of animals was a key concern in vivisection debates, which polarized the medical community. Such discussions led to the passage of the British 1876 Cruelty to Animals Act, which explicitly required animals used in experiments be anesthetized.[26] The daguerreotype of Golderman directly participates in such period debates by underscoring her moral virtue and highlighting the hospital's role in alleviating her physical suffering. The implication is that she deserves to be spared pain.

In addition to these ethical dilemmas, the introduction of anesthesia reinforced the fact that pain, as a subjective, corporeally bound experience, was difficult to communicate. For patients like Golderman and Abbott, the only way that MGH surgeons knew whether the ether had worked or not was by asking the patient after the surgery if they had felt pain. Bigelow noted that Abbott "muttered, as in a semi-conscious state, and afterwards stated the pain was considerable, though mitigated . . . as though the skin had been scratched with a hoe," and describing another surgery, the patient "professed not only to have felt no pain, but to have been insensible to surrounding objects, to have known nothing of the operation."[27] Yet insensibility is not necessarily the same thing as insensitivity to pain. Troublingly, each individual's experience of pain is variable, subjective, and mediated by personal experience and cultural conditioning. How could surgeons be sure that a patient felt no pain and that the next patient would have the same experience? And, how could photographers communicate consistent painlessness in an image? The

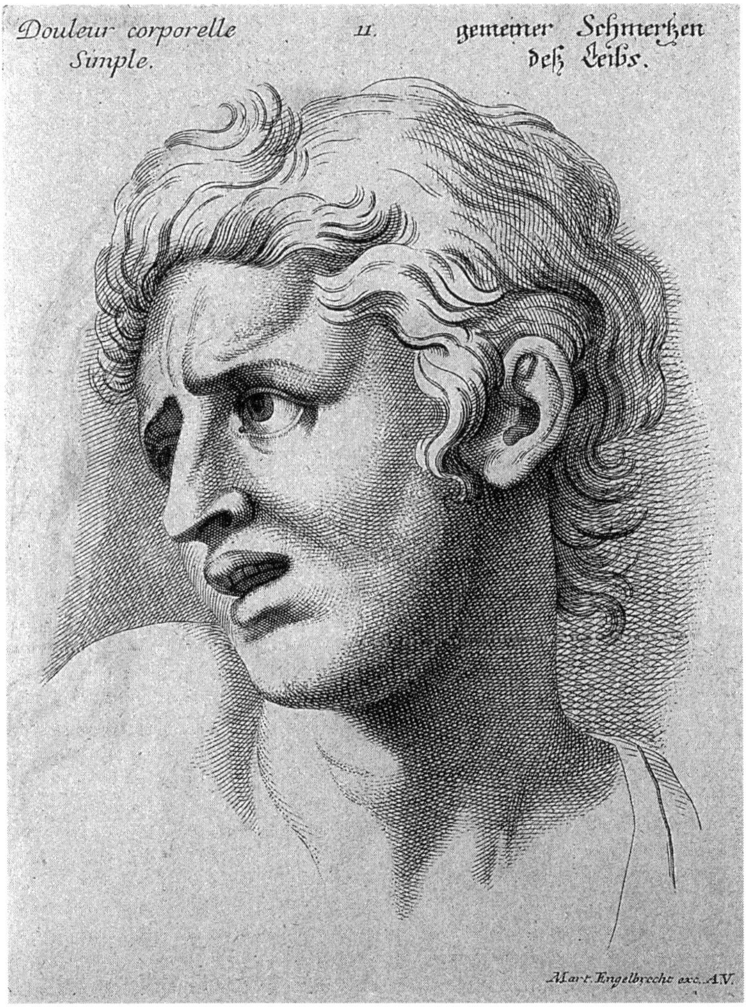

Fig. 5.3. Drawing manuals and art treatises concretized how facial expressions communicated emotions, suffering, and pain. Engraving by M. Engelbrecht, after Charles Le Brun, "The face of a man expressing simple bodily pain (*douleur corporelle simple*)," from *Expressions des passions de l'âme* (1732). Courtesy of the Wellcome Collection.

first step toward communicating pain, or a lack of pain, was the development of a shared cultural language surrounding this condition.

In the late eighteenth century, an epistemological transformation occurred, wherein pain came to be understood as the physical signification of unseen internal disease or pathology. The articulation of pain in a diagnostic setting offered a significant, although imperfect, vehicle of communication between patient and practitioner. In verbalizing pain, the patient and doctor agreed that it was a mutually affective sensation, wherein the body is a unified organism that translates internal symptoms of disease or injury into sensations then described by the subject in a way that makes sense relationally to the doctor. In *The Birth of the Clinic*, French theorist Michel Foucault identifies a disciplinary turn in Western medical practice toward clinical methodologies and the description of pain as a process of linguistic signification between body and mind. For Foucault, this epistemic shift is signaled linguistically by the transition from the question "What is the matter?" to "Where does it hurt?" While the first indicates an emphasis on visible, external symptoms, the latter signals a fundamentally different notion of diagnostics focused on the internal and invisible. Foucault argues that physicians used the first question in the eighteenth century and then adopted the second question in the nineteenth. He observes, "From then on, the whole relationship of signifier to signified, at every level of medical experience, is redistributed: between the symptoms that signify and the disease that is signified, between the description and what is described, between the event and what it prognosticates, between the lesion and the pain that it indicates." The clinical emphasis on pain—of isolating and articulating a response to *where does it hurt*—marks an empirical diagnostic process mediated by the professional, who can understand what the pain signifies. It also highlights the difficulties faced by the professional, who had to read the surface of another's body as a marker of interiority. In this new phase of medical diagnostics, the subjective articulation of pain takes the place of or works in concert with visible symptoms and physical examinations. The disciplinary structures of diagnostic medicine rely on touch, sight, and sensation, on objective observation and subjective articulation; this is what Foucault beautifully describes as "the figures of pain . . . redistributed in the space in which bodies and eyes meet."[28]

If pain was difficult to articulate, it was equally difficult to visualize. In art, images of pain and suffering are most commonly religious in nature, and depict execution and martyrdom or hell and eternal damnation.[29] By the seventeenth century, suffering was pictorially codified within academic spheres via drawing manuals.[30] Pain and anguish were conveyed through facial expression and bodily contortion. For example, French artist Charles Le Brun included images of "douleur corporelle simple" (fig. 5.3) and "douleur aigue" in his *Expressions des passions de l'ame* (1732). In both, the mouth is agape, eyes are wide, and eyebrows knit together.[31] Le Brun believed that passions acted upon the muscles of the face; emotion (and pain) could therefore be read via expression. Between 1770 and 1781, Austrian sculptor Franz Xaver Messerschmidt

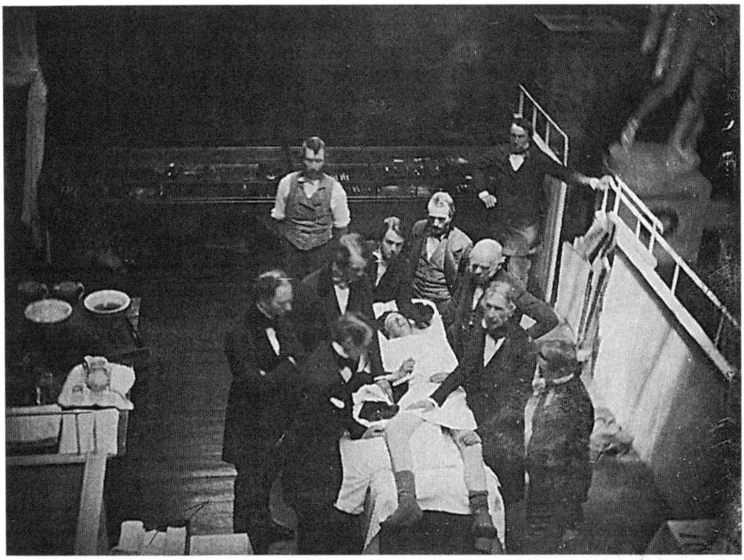

Fig. 5.4. Photographers struggled to demonstrate the durational efficacy of ether in a single image. Southworth & Hawes, "Untitled (Early Operation Using Ether)," 1847. Whole plate daguerreotype, 5 7/8 × 7 7/8 in (15 × 20 cm). Harvard Art Museums/Fogg Museum, Loan from the Massachusetts General Hospital Archives and Special Collections, 1.1979. © President and Fellows of Harvard College.

created sixty-nine character heads showing extreme facial expressions; to prompt these looks, Messerschmidt would pinch himself in the ribs.[32] Later, French neurologist Guillaume-Benjamin-Amand Duchenne and photographer Adrien Tournachon used electrical probes to stimulate the facial muscles and create similar extreme expressions, which were then photographed and published in *Mécanisme de la physionomie humaine* (1862).[33] These examples demonstrate the ways that artists and scientists sought to visually convey the physical expression of bodily pain. In the ether daguerreotypes, however, the problem Southworth & Hawes faced was the opposite of their forbears. How did ether work, how could one be sure etherized patients were without pain, and how could that be communicated to an audience? In other words, how could one photograph and visually telegraph a *lack* of sensation or physical pain?

A Surgical Progression over Time

Sometimes described as "The Anatomy Lesson Daguerreotypes," the final three Southworth & Hawes images considered here answer this question. They present a teenage boy anesthetized on an operating table. In the first, "Untitled (Early Operation Using Ether)" (fig. 5.4),

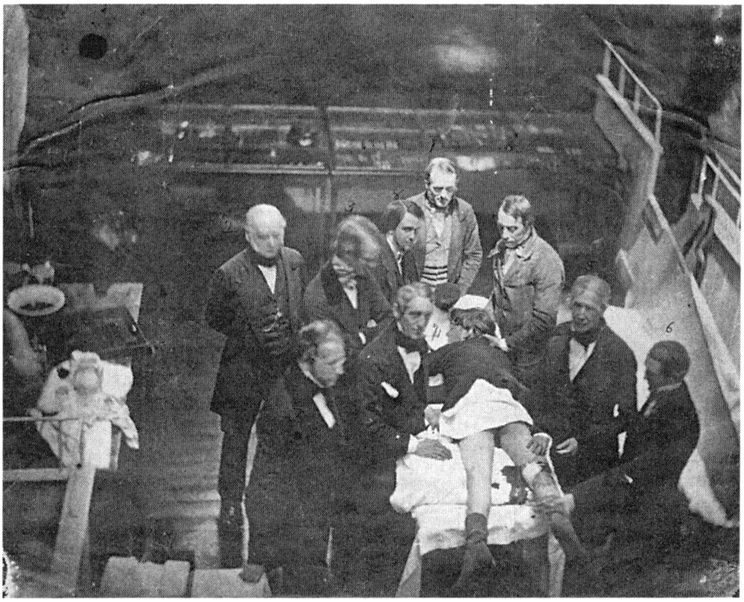

Fig. 5.5. In this case, photographers took three images of the surgery showing before and after. Southworth & Hawes, "Young Man Receiving Ether from a Sponge," 1847. Paper print after a whole plate daguerreotype, 5 3/4 × 7 7/8 in (14 1/2 × 20 cm). Center for the History of Medicine, Francis A. Countway Library of Medicine.

Heywood holds an ether-soaked sponge next to the patient's face. The boy's hands rest on his midsection and his left leg is elevated on a pillow, while his feet—clad in long, dark socks—dangle from the table's edge. His mouth is open and his face is flushed. With the tipped vantage point of the camera, the viewer looks up the patient's hospital gown, enhancing the voyeuristic experience. The patient's unconscious body thus becomes an object of scrutiny.

Similar to the Golderman image, sunlight floods the middle ground and the room is presented at an angle from above. Surgical tools are laid out at the left and back and a temporary wall encloses the scene on the right. Basins, pitchers, and folded towels stand ready. A waiting attendant leans casually by the case of spare instruments. The preparation foreshadows the coming operation. A copy of the *Apollo Belvedere* and the Egyptian mummy Padihershef are in the upper right, while a skeleton appears in the upper left. All three were permanent fixtures of the MGH surgical amphitheater.[34] Unlike the Golderman image, however, here the doctors crowd around the patient and physically engage with his body. Multiple hands are present and all of the doctors except Warren look at the patient. Warren touches the boy's elevated leg to indicate the location of the coming surgery to treat necrosis of the tibia and looks at the viewer. This demonstrative gesture underscores the educational action of the surgical theater—like the last image "we" are presumably among a student audience—and emphasizes the boy's presumed *insensibility* to the surgeon's touch.

In the next image, "Early Operation Using Ether for Anesthesia," many figures remain in the same positions, while the attendant has left the frame. This implies that only a few minutes have elapsed. It is possible that the surgeons had, by this time, become comfortable with the prolonged effects of ether and were willing to allow Southworth & Hawes to take a second image. The presence of multiple hands is minimized, while Warren rests his hands on the boy's right thigh. Warren looks up at the viewer, the only one to directly connect with the audience. The figure to his right has stepped back, either to allow him to speak or to clear the composition.

In the final image, "Young Man Receiving Ether from a Sponge" (fig. 5.5), the viewer is presented with the conclusion of the successful surgery via a paper print made after a lost daguerreotype. Heywood again holds the etherized sponge near the boy's face; however, his position has shifted, and an orderly holds the back of the boy's neck, perhaps monitoring his pulse. The patient, still unconscious, is on his stomach with a bandaged left thigh. A stretcher is laid against the right wall to transport him from the operating theater. Blood soaks the pillow—a literal sign of the internal reality of the operation. The patient's stockinged feet hang off the edge of the operating table and his flushed face contrasts to the whiteness of the sheets. The table at left has been disturbed during the procedure, the water pitcher moved and towels crumpled. In this image, the participants stare at the viewer and touch the patient, proud of a successful surgery. The patient, however, remains insensible to the doctors' touch, even throughout the prolonged, invasive, and presumably painful surgery. His relaxed body visually expresses his obliviousness to the operative procedure just performed.

In portraying a surgery across three images, Southworth & Hawes create a temporal experience that counters the difficulties faced in representing a prolonged occurrence as a single image. Taken together, they present surgery as an ongoing action, portraying its beginning and end. While the viewer is not shown the surgery itself, the middle, the place of action, is left to the imagination. In this way, we realize that the subject is not the operation; instead, the images reveal the durational efficacy of this new anesthesia. Therefore, the daguerreotypes seek to make clear that the patient has been desensitized to pain.

Visual Legibility and Medical Objectivity

The period understanding of pain created a difficulty for the artist who sought to represent the pain-free body. Photographing the invisible effects of ether also raised questions about the efficacy of photography, the goals of representation in the medical context, and the signification and legibility of a medicalized body for a professional audience. If the daguerreotypes cannot visualize the invisible process of ether anesthetization occurring inside the patient's body, then what work are these images doing and to whom do they communicate?

A problematic assumption of photographic transparency taints scholarship on the ether daguerreotypes. For example, Bates Lowry and Isabel Barrett Lowry write: "Not a single aspect of this scene is the result of the artist's imagination . . . since nothing could be omitted by the camera in favor of a more artistic presentation, this image has preserved for us the precise details of what a surgical operation looked like in the late spring of 1847."[35] As this quotation demonstrates, these images have been interpreted as representations of a medical procedure, devoid of artistic intervention or aesthetic intention. Their use as referents to a historical event implies both a problematic understanding of the medium's transparency and an implicit misunderstanding of the images' context. These daguerreotypes are complicated cultural objects that participate in and support institutional dynamics of power and cultural ideas about corporeal representation. As explored in the formal analysis of these images, Southworth & Hawes made decisions about framing, lighting, and composition, and collaborated with medical professionals at MGH to create a mediated, composed, and highly self-conscious representation.

In some ways, the photographic medium itself becomes complicit in the misreading of these daguerreotypes through the mid-nineteenth-century epistemic success of mechanical objectivity. Historians of science Lorraine Daston and Peter Galison describe the development and eventual dominance of mechanical objectivity—which they define as "the insistent drive to repress the willful intervention of the author-artist, and to put in its stead a set of procedures that would, as it were, move nature to the page through a strict protocol"—as a disciplinary episteme, one that empowers visual representations through the purported erasure of human mediation and validates, especially, scientific endeavors.[36] The abiding dominance of mechanical objectivity in photographic discourse allows these images to obfuscate their own content. Daston and Galison expand upon this when they write: "The machine seemed to offer images uncontaminated by interpretation . . . [it] stood for authenticity."[37] In other words, the fact that viewers look at these images and claim that they show something "as it is" underscores the hegemony of "mechanical objectivity" as an ideological construct. Therefore, the very success of the ether daguerreotypes resides in their effective deception. They convince audiences they see something they do not: first, that the images visualize ether working inside the body, and second, that the bodies are without pain.

As Tanya Sheehan argues in *Doctored: The Medicine of Photography in Nineteenth-Century America,* American photographers borrowed the language of American medicine to undergird professional agendas and lend credibility to their practice. Sheehan perceptively connects the discomfort of studio portraiture with surgical anesthesia: "Like the surgeons to whom they were so often compared, photographers needed to equip themselves with the rhetorical and material means . . . for rendering sitters' bodies insensible to pain. Out of this logic emerged fantasies of sitters' bodies as anesthetized, docile objects, and of operators as agents of their control."[38] Sheehan identifies a shared vocabulary invested in "insensibility" and "immobility" between the two disciplines, citing Bigelow, who used this language when he described etherization.

In "Insensibility of Surgical Operations Produced by Inhalation," published in the *Boston Medical and Surgical Journal* November 18, 1846, Bigelow described how: "The character of the lethargic state, which follows this inhalation, is peculiar. The patient loses his individuality and awakes after a certain period . . . entirely unconscious of what has taken place."[39] For Bigelow, it is not pain that is done away with but an inability to remember such sensation upon waking. The etherized patient becomes unconscious, unaware of whether the body feels pain. Therefore, the efficacy of anesthesia is not to eradicate pain per se, but to immobilize and desensitize, thereby throwing into relief the ambiguities of the etherized state, which itself remained troublingly mysterious. Reversing the directional emphasis of Sheehan's study allows us to consider how medicine used the flexibility of photography to visually communicate physiologically complex concepts. In other words, did the medium of photography itself operate to validate the painlessness of etherized surgery?

The subject of the ether daguerreotypes is the invisible process of anesthesia; the photographs therefore seek to make clear that the patient is desensitized to pain. The photographer relies on visual cues that together communicate effective anesthesia by revealing "insensibility" in a medical setting. These signals include the sleeping figure, the etherized sponge or globe, the surgical theater, and the surgeon's touch on the patient's body. The final three daguerreotypes rely on their seriality to convey a progressive narrative of successful, continuous anesthetization. Due to these signifying details, these images were legible to a period audience familiar with ether anesthesia. Trained in reading the body as a series of diagnostic signifiers, professional viewers would have understood these clues and recognized that what was depicted, but not actually seen, was ether working inside the patient's body and the corresponding eradication of pain.

The challenge of depicting the unseen—the ether, or lack of pain—situates interpretive pressure within the surgical subjects' bodies, which advertise anesthetization. The use of photography to "document" etherized surgery was inherently a fraught endeavor and required the medium to reveal something the camera could not "see"—the effects of an invisible chemical inside an unconscious patient's body. Furthermore, the objectivity of photography semantically offered a metaphorical bond between the relatively new process of daguerreotypy and the equally mystifying process of ether anesthesia, conferring scientific legitimacy upon ether as a surgical anesthetic. Significantly, both anesthesia and daguerreotypy are conceptually bound, as they prompt chemically induced transformations through the use of similar compounds to induce invisible effects.

Magic or Science?

Introduced in 1839 by Louis-Jacques-Mandé Daguerre, the daguerreotype process offered a revolutionary way to capture a permanent image by directing the sun's rays through a camera and fixing the resultant image onto a plate through chemistry. The daguerreotype revolutionized the world, offering a new way of conceiving of vision, technology, and the arts. Daguerreotypy was quickly popularized in the United States by Samuel F. B. Morse and others, who adopted the new medium as thoroughly democratic and advertised it as perfect for the burgeoning nation, enterprising entrepreneurs, and middle-class buyers.[40]

A simple process, daguerreotypy used a camera, constructed of a wooden box and lens, to capture the image seen through the lens on a chemically treated plate. Daguerreotypes are a direct positive, unique, laterally reversed image, fixed on a polished silver-plated copper plate treated with bromide and/or iodine vapor before exposure. The image is developed by mercury vapor and fixed with ordinary salt (sodium chloride).[41] The resulting objects are highly reflective, forcing the viewer to shift the plate back and forth to make sense of the image. Alan Trachtenberg notes that this difficulty in visual comprehension operates as a metaphorical sign relating to cultural anxieties about the unfixedness of the photographic image itself, indicated by the numerous terms coined for the new medium.[42] In 1855, Philadelphia photographer James E. McClees despaired of this, writing: "Daguerreotype, Crystalotype,

Talbotype, Calotype, Crystalograph, Panotype, Hyalograph, Ambrotype, Hyalotype . . . the sooner the art becomes divested of the quackery of terms the better."[43] By referencing the "quackery of terms," McClees adopts the language of medical chicanery to highlight public confusion over the proliferation of photographic practices and underscore widespread skepticism over their function and efficacy. The chemically transformative effects of daguerreotypy were technologically progressive and defamiliarizing—combining an alchemical mystery with the science of optics and modern chemistry.

By midcentury a relentless stream of inventions led many to praise the magic of the age. Significantly, both daguerreotypy and ether rely on chemical processes to enact what were perceived as magical effects. Regarding daguerreotypy, Antoine Claudet explained in the *British Journal of Photography*: "By pouring a solution . . . upon his photographic paper, still perfectly white on its removal from the camera-obscura, but containing the latent image which in an invisible state had been impressed upon it . . . as by the power of a magic wand, the vanished picture is evoked to light—'Apparet imago!'"[44] Likewise, in 1845, Claudet wrote in "The Progress and Present State of the Daguerreotype Art": "Two centuries ago the daguerreotype art would have been looked upon as the work of witchcraft, but in our age of improvement we are accustomed to extraordinary discoveries; we are capable of admiring and appreciating these huge efforts of genius, and nothing surprises us."[45] This conflation between witchcraft, technological progress, and chemical experimentation typify responses to the medium, and are evident in examples ranging from the magic lantern, to Henry Fox Talbot's description of his "magic pictures," to the 1861 text *The Magic of Science: a Manual of Easy and Instructive Scientific Experiments*.[46] Douglas Nickel demands that we take this slippage seriously, arguing that the linguistic conflation between photography and magic was predicated upon and reflective of the ways in which period science was deeply rooted in pre-Darwinian, post-Enlightenment beliefs in the divine and marvelous in nature.[47] That we may be surprised by the ways in which magic and photography were linked signals that our scientific episteme is radically different from that of Bigelow and Southworth & Hawes.

The alchemical transformations and enigmatic effects of daguerreotypy were confusing and mysterious; at the extreme, daguerreotypy was regarded as confounding, and downright dangerous. In *The House of the Seven Gables* (1851), Nathaniel Hawthorne conflates the magic of .daguerreotypy with chicanery.[48] The protagonist daguerreotypist, Holgrave, misrepresents himself as an innocent lodger, while secretly seeking revenge upon his hosts. By extension, the daguerreotype is presented as dangerous and untrustworthy, a product of imperfect, mechanical mimesis. Similar to Holgrave's elusive motives, daguerreotypes were not always easy to read.

Like daguerreotypy, the intangible effects of ether were met with wonder and doubt.[49] For example, the *Edinburgh Review* declared: "The scientific discoveries of recent years, and their marvellous applications in the arts, have been of such a nature and magnitude, as to astonish the most sober philosophers. . . . Surgeons have not employed ether and chloroform as anaesthetics for three years."[50] Others described how "operations were performed by the surgeons of Boston, in hospitals and in private practice, upon patients rendered unconscious of pain by inhaling the letheon. . . . What wonder, in view of effects so magical, that those great teachers should break out in strains of rejoicing."[51] Finally, journalist Margaret Fuller wrote of her own etherization, "It was a sensation like wandering in long garden-walks, and through many alleys of trees . . . I quitted the body instantly, and cannot remember any detail of what I saw. . . . Suddenly I seemed to see the old dentist . . . in the twilight the figure had somewhat of a Faust-like, magical air."[52] In these examples, the surgeon, like the daguerreotypist, is seen as a magician and ether as magical.

Yet if ether was mysterious, it was also sinister and disturbing. Ether rendered one unconscious and disembodied—caught in a dream state between life and death. Fuller's conflation of the ether-administering dentist with Faust, the German scientist who trades his soul to the devil for earthly knowledge, suggests the ways in which ether stimulated widespread concerns regarding subjectivity, consciousness, and embodiment. Indeed, ether proved that medicine was probing the very nature of life, death, and reanimation, like the titular character in Mary Shelley's 1818 novel *Frankenstein*, in spite of theological or cultural hesitations regarding the human soul or sanctity of the body.

The term *ether* comes from the Greek for "upper air."[53] While it scientifically denotes an organic compound in which two hydrocarbon groups are linked to an oxygen molecule, it also describes an invisible medium wherein light and particles move about; refers to something intangible with no precise location; and characterizes the essence of the universe or the sublime. Ether refers to air above the atmosphere, a place beyond what is known: the heavens, and is ethereal, celestial, or otherworldly. As a scientific compound, ether acts upon the subconscious to induce a dream state, a place beyond comprehension and knowing. Morton took his name for the new anesthetic (Letheon) from the Greek *lethe*, referring to the river in Hades—the "waters of oblivion"—from which the dead drank to forget their mortal life.[54] The dead slept through their afterlife, insensible to what came before it. Morton's choice was apt; after its introduction ether was increasingly regarded as dangerous, a step from "the sleep of death."[55]

Indeed, concerns grew almost immediately over patient safety and ether's unpredictability, alongside discussions regarding the benefits of painlessness. While ether was used in France and England beginning in December of 1846, responses were marked by skepticism related both to affective experiences and representational veracity.[56] As late as 1852 Pennsylvania Hospital refused to use surgical anesthesia, because of a lack of clarity over how it worked and concerns related to ether-induced deaths.[57]

"The Finest in Every Respect"

Much of the consternation over daguerreotypy and ether related to the ways in which epistemology was bound with vision in the mid-nineteenth century. If sight was associated with knowledge, daguerreotypy and ether both confounded that paradigm.[58] In the 1840s the camera captured things beyond human vision: the moon's craters, fleas magnified hundreds of times, and ghosts and spirits. Revelations of the strangeness of the world were unceasing. In light of this, what else could the camera "see"?

The material complexities of ether anesthesia relate to the representational difficulties of the ether daguerreotypes, whose medium was likewise elusive, and which purported to picture something unseen and poorly understood. In analyzing the ether daguerreotypes, it becomes apparent that the representational agenda of the images is contradicted by the challenges of the subject matter. They oscillate between objective, scientifically oriented, "transparent" documents and obscure, hard-to-read images that are thick with discipline-specific details. A consideration of their authorship indicates that, early on, daguerreotypy was highly flexible. Practitioners embraced both objectivity and idealization, drawing from fine arts conventions; indeed, these were not mutually exclusive aesthetic programs. Southworth & Hawes's photographic output reveals their aesthetic range and the medium's adaptability for different subjects.

The authorship of the five ether daguerreotypes is well established. In 1852 the firm stated, "We have made several pictures of the Surgeons of the Massachusetts [General] Hospital with the patient under the influence of ether, all accurate likenesses."[59] Operating in Boston from 1843 to 1863, Southworth & Hawes were known for whole-plate daguerreotypes with spectacular tonal quality and detail. With his first partner Joseph Pennell, Albert Sands Southworth set up shop as a daguerreotypist in Cabotville, Massachusetts, before moving the firm to Boston by the summer of 1841. Josiah Johnson Hawes joined the partnership in 1843,

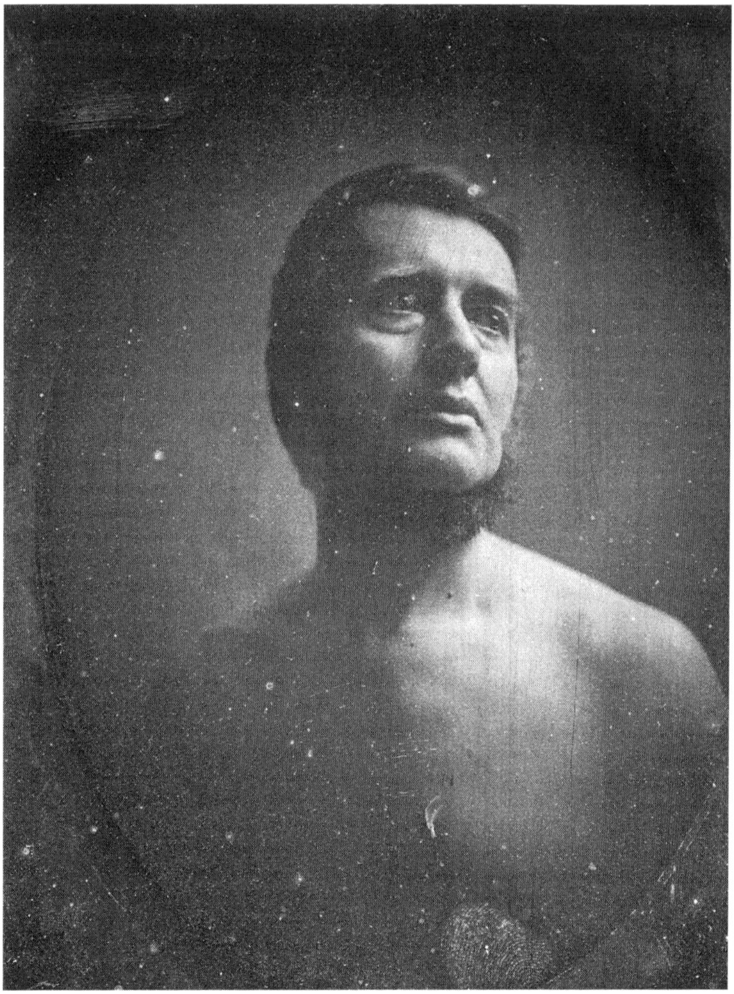

Fig. 5.6. The photographers maintained an active studio practice that included idealized portraiture. Southworth & Hawes, "Albert Sands Southworth," c. 1845–50. Daguerreotype, overall: 11 1/8 × 9 3/16 in (28 × 23 cm). Courtesy of Metropolitan Museum of Art, Gilman Collection, Gift of The Howard Gilman Foundation, 2005.

following Pennell's departure. As one of only two daguerreotype studios initially operating in Boston, Southworth & Hawes quickly made a name for themselves.[60] Daguerreotypes and artworks were regularly on display in their studio and clients could look at examples before having their own likeness taken.

Southworth & Hawes proclaimed quality over economy, drawing elite clientele. In 1853, an advertisement read: "As no cheap work is done,

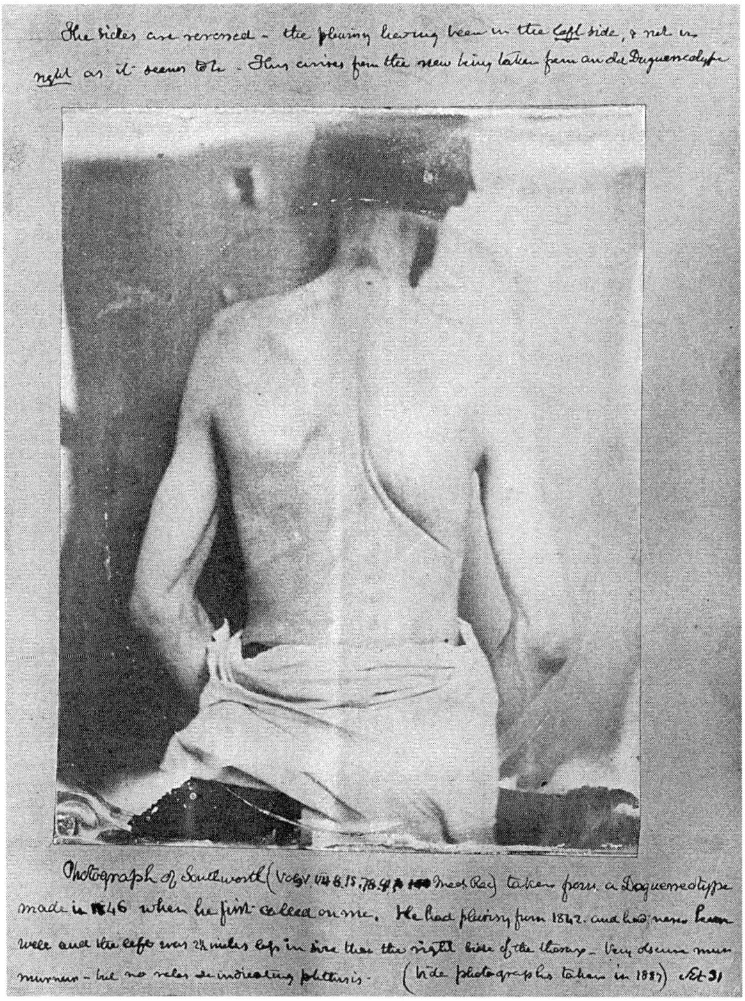

Fig. 5.7. Southworth was the subject of medical photographs, as in this example taken for his physician. "Southworth as he looked in 1846," in "Collection of Letters and Portraits, Dr. H. I. Bowditch," portrait vol. 4. Boston Medical Library in the Francis A. Countway Library of Medicine.

we shall spend no time in bantering about prices; and wish to have all understand that ours is a one-price concern."[61] While Matthew Brady charged two dollars per plate, a Southworth & Hawes daguerreotype cost customers five dollars.[62] In comparison, a skilled Massachusetts carpenter made one dollar per day.[63] In 1847, the year of the ether daguerreotypes, Southworth & Hawes began ordering French plates that were smoother and more reliable than American plates.[64] This superior material increased the aesthetic and commercial appeal of Southworth & Hawes's products.

Southworth & Hawes were known for their style, quality, elite patronage, and use of light, careful compositions, and—in portraiture—references to classical models. Southworth self-consciously elevated daguerreotypy to an art form, describing the role of the daguerreotypist as more than mere operator: "The artist, even in photography, must go beyond discovery and the knowledge of facts. He must create and invent truths, and produce new developments in facts."[65] Here, he emphasizes that a photographer is not just endowed with technical prowess or chemical mastery, but must also understand artistic conventions to create works of refinement.

The flexibility of the medium and manner in which Southworth & Hawes approached picturing the human subject is exemplified by a comparison between two images of Southworth. The first shows Southworth shirtless and rendered as if he were a marble classical bust (fig. 5.6). The camera has been manipulated to maximize the fall-off and create an oval vignette. The shallow depth of field draws attention to the sitter's face, while his chest and shoulders melt into shadow. Southworth tilts his chin forward to catch the light from the right of the frame. This illuminates his face, catching the bridge of the nose, pursed lips, and left cheekbone, and creates a subtle highlight in each eye. His expression is melancholic. Borrowing from academic traditions—including formal portraiture, classical sculpture, and nude figure study, this daguerreotype demonstrates Southworth & Hawes's familiarity with fine arts discourses. In contrast, the latter image (fig. 5.7) reveals a medical representation of the nude human body. Taken for Dr. Henry Ingersoll Bowditch, Jackson Professor of Clinical Medicine at Harvard Medical School, the photograph shows the effects of pleurisy of the left lung on Southworth's back and shoulder. The image is a visual document of disease, made for his physician—Southworth bares his back for the camera and reveals himself as physically marred by prolonged illness.[66] Lacking the lush lighting and classical pose of the previous image, this showcases an ailment and is explicitly diagnostic in function.

These two images—one using the language of fine art and the other diagnostic practice—demonstrate the versatility of the firm and medium, and contribute to contemporary debates over daguerreotypy as either a fine art or a science.[67] Both represent the same subject; one adopts a familiar aesthetic language that was based upon classical precedents and used in academic portraiture, while the other relies upon the novel reformulation of conventions of portraiture to fit the developing needs of medical science via the new medium of photography. Southworth &

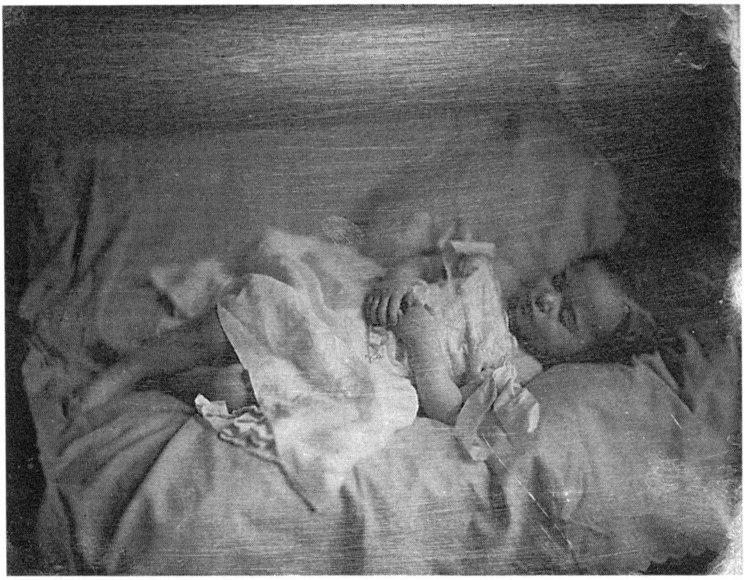

Fig. 5.8. The photographers excelled at post-mortem photography, giving their subjects a life-like appearance. Southworth & Hawes, "Post-mortem of Child," ca. 1850. Whole plate daguerreotype, 6 1/2 × 8 7/16 in (15.25 × 21.5 cm). George Eastman Museum, gift of Alden Scott Boyer. Courtesy of the George Eastman Museum.

Hawes seem to embrace the expansive capabilities of their medium, instead of regarding their practice as exclusively for science or art. In other words, this comparison reveals that the epistemic regime of "camera objectivity" was still being negotiated in the 1840s.

Southworth & Hawes also excelled at postmortem photographs. "Post-mortem, unidentified young girl" (fig. 5.8) depicts a child with arms folded softly over one another. Her ankles are crossed and her eyes closed, as if in permanent sleep. An 1848 Southworth & Hawes advertisement states: "We take great pains to have miniatures of deceased persons agreeable and satisfactory; and they are often so natural as to seem, even to artists, in a quiet sleep."[68] Here, they liken death to sleep, a reference that resonates with the sleep of etherized surgery. Aside from the difference in setting, little distinguishes the subject in the postmortem photograph from the stillness of patients in ether images. Representationally then, how can viewers distinguish between the dead figure and the anesthetized one?

Anxieties about the imprecision of the identification of death were widespread, as indicated by the fear of premature burial.[69] The visual ambiguity of postmortem photographs, then, when placed alongside images of anesthetized patients, evoked nineteenth-century anxieties over the slippages between anesthetization, sleep, and death. This is evident in the linguistic confusion over sleep. When the anesthetist says, "I will put you to sleep," they mean that you, the patient, will be rendered insensible to external stimuli, while "We put our cat, Fluffy, to sleep," means Fluffy died. Contemporary anesthetists have begun to question the use of *sleep* to describe anesthesia, because of the conflation between sleep and death.[70] The porousness of language, the distinctions between sleep and death, and the inscrutability of images picturing anesthetized and dead subjects, reveal the ways in which imagery and words prove incapable of translating some physical states of being. In spite of this difficulty, Southworth & Hawes's photographic practices—their lifelike postmortem images and familiarity with fine arts and medical precedents—undoubtedly affected how they approached photographing etherized surgeries.

John Collins Warren, Rembrandt, and the Legacies of Medical Art

Thus far, I have focused on the representational challenges of photographing etherized patients, arguing that details in the images denote the subject and communicate a lack of pain, but are also contradicted by the images' aesthetics and rhetorical slippages. Yet, one question remains. Who was the intended audience for these daguerreotypes? How were they disseminated and where might they have been seen? Unique versions of five ether photographs are extant; however, Southworth & Hawes made multiple daguerreotypes at single sittings and created daguerreotype copies of originals. They would display these in their showroom, making it plausible that other etherized surgery images were taken and exhibited at their studio on Tremont Row. At least one daguerreotype of an ether surgery was in Hawes's possession in 1897.[71]

However, the production of these daguerreotypes was also one component of a citywide commemorative campaign to establish Boston as a medical center. Displayed in black, carved frames with gold-foiled mats, at least two of the daguerreotypes, "Use of Ether for Anesthesia" and "Demonstration of the Surgical Use of Ether," were exhibited in Warren's residence, along with family antiques, medical specimens, a mastodon skeleton, and a Gilbert Stuart portrait of the young Warren.[72] In this house, Warren conducted private surgeries and patient consultations, prepared anatomical specimens, and entertained students and colleagues. In this environment, the two daguerreotypes confirmed Warren's role in the ether demonstrations, cementing his reputation at the vanguard of American medicine.

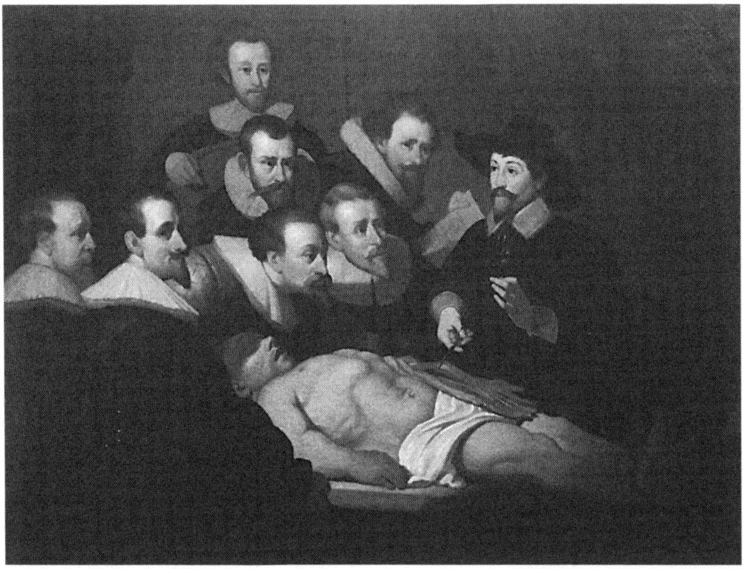

Fig. 5.9. This copy of a famous medical painting connected American practitioners and institutions to their European forebears, establishing a professional genealogy. Artist unknown, copy after Rembrandt van Rijn's *The Anatomy Lesson of Doctor Nicolaes Tulp of 1632*, undated. Oil on canvas, 63 × 84 in (160 × 213.25 cm). Warren Anatomical Museum in the Francis A. Countway Library of Medicine.

In addition, Warren commissioned two copies of Rembrandt van Rijn's *The Anatomy Lesson of Doctor Nicolaes Tulp of 1632* from an unidentified artist. One version hung in the front hall of his home; the other went to Harvard Medical School (fig. 5.9).[73] This brightly colored, stylized copy of Rembrandt's depiction of Nicolaes Tulp's annual anatomical demonstration in Amsterdam is faithful to the original in terms of composition and perspective. Tulp was widely regarded in Western medicine as a great orator, anatomist, and teacher. In the painting, Tulp stands over a cadaver demonstrating the action of tendons of the forearm and hand, while consulting Andreas Vesalius's *De humani corporis fabrica* (1543), a canonical illustrated anatomical text. Warren regarded himself as an inheritor of Tulp and Vesalius's anatomical and pedagogical legacies, writing: "I was necessarily compelled, before lecturing, to go over every part which had been dissected, to remove its imperfections and supply its deficiencies. The arrangements preliminary to the presentation of each part to the student's eye required a good deal of thought, time, and practical tact."[74] In this quotation, Warren identifies the artistry of anatomical dissection, highlighting how the instructor was responsible for manipulating the body in order to clarify relevant information for student viewers and present an idealized and therefore "typical" subject. In effect, the anatomist was similar to the artist, who "sculpted" the cadaver in order to represent a uniform visual lexicon of information that would align with textbook illustrations, or the photographer, who framed the ether daguerreotypes in order to make their subjects clearer.

By commissioning and giving a copy of Rembrandt's portrait of Tulp to Harvard Medical School, Warren drew a genealogical significance between anatomical study at Harvard and the teachings of Tulp. The copy establishes a visual precedent for nineteenth-century American medicine. While Harvard Medical School was young—it was founded in 1782 by Warren's father—the painting indicates that the institution was built upon European pedagogical precedents. By also installing a copy of Rembrandt's portrait of Tulp in the entryway of his home alongside two of the ether daguerreotypes, Warren connected himself to earlier medical and pedagogical history. In his home, where he entertained colleagues, taught students, and received and treated patients, his commission—of both daguerreotypes and paintings—underscored Warren's role as an anatomical demonstrator and medical orator, advertised his preeminent surgical position, and indicates how he used the fine arts in order to further American medicine at large.

The ether daguerreotypes created by Southworth & Hawes use the objective characteristics of the new photographic medium to lend factual weight to the efficacy of ether as a surgical anesthesia. Medical professionals relied upon the presumed transparency of the photographic medium to present proof of the surgeries performed at the hospital; this also masked the ideological work these images perform. As ideas about vision, pain, and photography actively evolved, proving to be slippery and elusive, the images, upon closer examination, similarly prove representationally challenging. The photographs seek to capture that moment when, as Holmes described, the patient is "lulled by the strange magic of the enchanted goblet, held for a moment to their lips, into a repose which has something of ecstasy in its dreamy slumbers."[75] Long regarded as transparent images, the daguerreotypes in fact use a highly specialized vocabulary to communicate two key facts that were difficult to visualize: anesthetization and painlessness. That viewers look at the ether daguerreotypes and see things that are not actually pictured, indicates that the work that these images enact is rhetorically complex and stunning in its efficacy, erasing any question about what is pictured.

CHAPTER 6

DRAWING RACIAL COMPARISONS IN NINETEENTH-CENTURY BRITISH AND AMERICAN ANATOMICAL ATLASES

Keren Rosa Hammerschlag

In this chapter I take as my point of departure a largely unknown illustration in an almost entirely overlooked atlas by a mostly forgotten artist. The illustration, which appears as plate 5 of Joseph Maclise's *Surgical Anatomy* (1851), depicts the "The Surgical Dissection of the Sterno-Clavicular or Tracheal Region, and the relative position of its main blood vessels, nerves [, &c] etc." (see plate 1). What makes it significant is that it juxtaposes the tracheal regions of a Black man and a white man, inviting readers to compare them. Historian Nancy Stepan highlights the role of comparison in establishing racial hierarchies. "Race, in short, was a contrastive concept," she states. "Any representations of one race could only be understood in explicit or implicit relation to the representation of another, through a process that highlighted the signs of difference."[1] In the preface to *Surgical Anatomy*, Maclise describes his desire to apply the principle of comparison to the study of the human figure: "Comparison may be fairly termed the pioneer to all knowledge. . . . Every anatomist is a comparer in a greater or lesser degree; and he is the greatest anatomist who compares the most generally."[2] "The comparison of facts already known," he writes, "must henceforward be the scalpel which we are to take in hand."[3] To reinforce

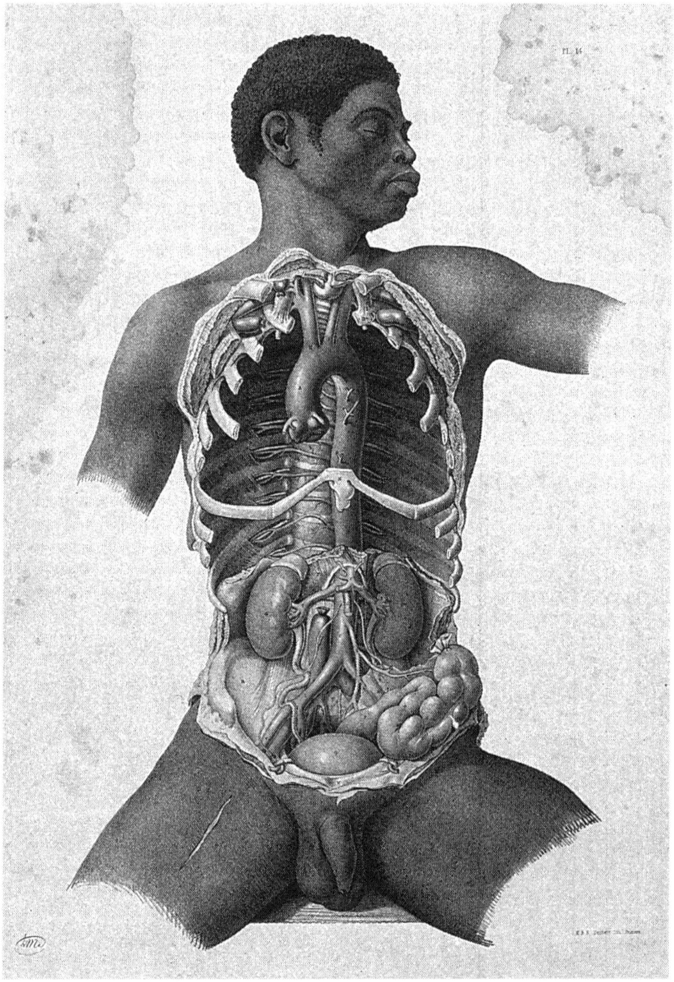

Fig. 6.1. It is rare to encounter an illustration of the anatomy of an idealized Black figure for general anatomical instruction. Joseph Maclise, "Dissection of the trunk of a seated black man, showing major blood-vessels," colored lithograph, sheet 21 1/2 × 14 3/4 in (54.5 × 37.7 cm), 1851. Courtesy of Wellcome Collection.

this position, Maclise included in *Surgical Anatomy* a series of striking visual comparisons as a way of training the eye of surgeons and future surgeons, thereby making clear the importance of comparison to medical (as well as art historical) looking. What makes these illustrations unique is the way that comparison is utilized to highlight the similarities rather than the differences between the anatomical structures of men of different races.

The appearance of a Black man in an anatomical atlas like Maclise's is highly unusual. The same Black figure reappears in numerous other illustrations in the atlas, including in plate 14 (see fig. 6.1), an illustration of the dissection of the major blood vessels. The normative race for anatomical models in illustrated atlases dating back to Andreas Vesalius's *De humani corporis fabrica* (1543) was the European (i.e., white) man.[4] Illustrations of the anatomy of nonwhite people were generally the reserve of comparative anatomy and the emerging nineteenth-century race sciences of ethnology and anthropology. By situating plate 5 in the context of the atlas as a whole and Maclise's oeuvre more generally, and by putting it into dialogue with the work of Joseph Maclise's brother, the Royal Academician Daniel Maclise, I seek to reveal the ways in which this image challenged dominant narratives of white racial supremacy in the developing discourses of race, evolution, and culture in mid-nineteenth-century Britain. A comparison between the work of Joseph and Daniel Maclise reveals the intersections of "high" art and scientific illustration during the mid-Victorian period, with Joseph's anatomized corpses striking poses familiar from classical art and Daniel's painted figures appearing on occasion like the dissected corpses of anatomical illustration.[5]

Even more intriguing than the appearance of the Black figure in *Surgical Anatomy* is his disappearance from the American editions of the same atlas (the accompanying text remained the same; race is mentioned in neither).[6] The inclusion of the Black figure in the British edition of *Surgical Anatomy*, and his omission, or "whitewashing," for the American edition illustrates how, in the words of Ludmilla Jordanova, "it is perfectly possible for an image to transgress only at some times or in the eyes of particular groups."[7] I use the term *whitewashing* purposefully. It is a term that refers to the erasure, masking, and bleaching out of color.[8] It speaks to a particular kind of "whiteness" that is, according to David Batchelor, "*generalized* . . . abstract, detached and open to contamination by terms like 'pure.'"[9] On the one hand, whitewashing exposes the superficiality of color by showing that color can be erased. But the act of erasure is rarely superficial. In Maclise's *Surgical Anatomy* the puzzle of the disappearing Black man brings to the fore that which has been excluded, elided, or denied in medicine's visual culture—namely the use and abuse of nonwhite bodies in the production of anatomical knowledge.[10] It reminds us that, as historians of art and medicine, we need to be attentive to both the presences and absences in the archives, the voices and the silences in the record.

The Arteries

I have been able to find but one precedent for Joseph Maclise's illustration of the anatomy of a Black man in a surgical anatomical atlas: plate

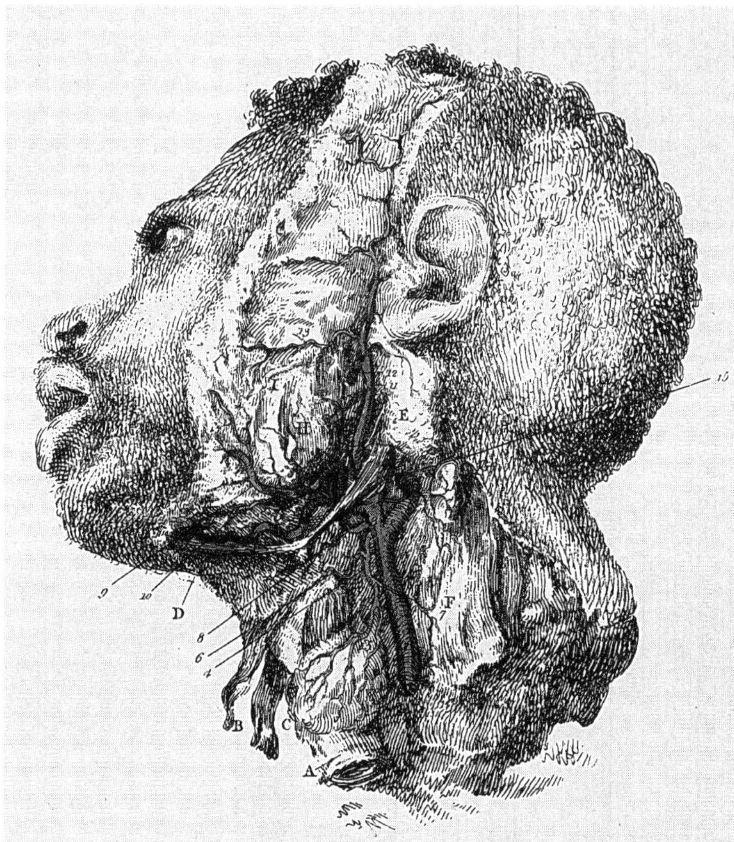

Fig. 6.2. Bell claimed to have found in the head of this Black man "the most common and regular distribution of the branches of the Carotid Artery." Plate 4 in Charles Bell, *Engravings of the Arteries*, 3rd ed. London, 1811. Courtesy of Wellcome Collection.

4 of Charles Bell's 1811 *Engravings of the Arteries* (fig. 6.2). Like Maclise, Bell was a surgeon who illustrated his own medical publications, and he here portrayed a Black figure with his head and neck cut open to reveal the carotid artery.[11] Bell's original ca. 1797 ink and watercolor sketch (held in the National Library of Medicine, in Bethesda, Maryland) includes details of the figure's dark face and hair set against the bright red of his internal anatomy.[12] The figure made his first public appearance in the third edition of Bell's *Engravings of the Arteries*—he is absent from the first (1801) and second (1806) editions—and was included in every British edition thereafter (his absence from American editions will be addressed in due course). In the preface to the third edition, Bell states: "I have added some other Plates, and I have etched most of the Plates

with my own hand, preferring accuracy to elegance. I have often had to regret in former editions of my work, that my intentions were entirely mistaken by the publishers and the artists employed."[13] As evidence of Bell's particular interest in the subject, plate 4 is signed "Etched by Charles Bell."

The pictures that furnish the third edition of Bell's *Engravings of the Arteries* are noticeably more detailed—and, in some cases, more dramatic—than those that appear in the two earlier editions. And, of all the illustrations in the third edition of Bell's text, plate 4 is the most detailed. The other figures depicted in the book have their skin cut away, but the Black man alone retains his flesh. The positioning of his head in profile effectively reveals the branches of his carotid artery, while at the same time presenting to the viewer a carefully rendered ear, open mouth, nose with flared nostril, and eye with delineated eyelashes. The accompanying text offers a justification for the inclusion of this figure: "Finding in the head of this black the most common and regular distribution of the branches of the Carotid Artery, I took this sketch from it."[14] In what constitutes a striking contradiction, the image of the Black man as rendered by Bell is the most individualized figure in *Engravings of the Arteries*, and yet, in the text, the Black man's anatomy is described as the "most common and regular distribution."

Maclise paid homage to Bell in one of his illustrations for Richard Quain's influential 1844 anatomical atlas, *Anatomy of the Arteries*.[15] Quain, surgeon and professor of descriptive anatomy at University of London (now University College, London) from 1832 to 1850 (Charles Bell held the post of professor of general anatomy and physiology at the University of London until 1832), enlisted his former student Maclise to produce the illustrations for his treatise, stating in the preface: "In reference to that gentleman's [Maclise's] labours, it may be allowed me to say, that while I have had the co-operation of an anatomist and surgeon, obviously a great advantage, the drawings will, I believe, be found not to have lost in spirit and effect."[16] In plate 66 of Quain's *Anatomy of the Arteries*, Maclise included a depiction of a book with "C. Bell" inscribed on its spine, thereby connecting Bell's text on the arteries to Quain's. Furthermore, there are stylistic similarities between Maclise's and Bell's illustrations, especially in their shared repertoire of poses, as, for instance, the similar position of the legs and feet in plate 9 of Bell and plate 82 of Quain.

But Maclise's illustrations claim more of a "high art" classicizing style through the incorporation of idealized bodies assuming poses that resemble classical statuary. This generalizing and idealizing style of anatomical illustration enjoyed a final colorful flourish during the first half of the nineteenth century, before it was superseded by the clinically pared down style of *Gray's Anatomy* (1858). In both his illustrations for Quain and in his own *Surgical Anatomy*, Maclise lavished attention on the depiction of lustrous hair, fabric turned drapery, and the contours of undulating musculature encased within blemish-free skin. Some figures, including the Black cadaver in plate 14 of *Surgical Anatomy*, are seated upright without the assistance of ropes and wooden blocks, their bodies gently cropped at the top of their arms and thighs—much like that embodiment of the academic bodily ideal, the *Belvedere Torso* (held at the Vatican Museum).[17] The example of plate 14 makes clear the degree to which Black figures in Maclise's atlas—despite their race—represent an idealized, even classicized anatomical specimen.

The aestheticized aspects of Maclise's illustrations indicate the artist's desire to produce beautiful figures that elevate the atlas (and anatomical illustration more generally) to the status of high art. Academic art required artists to study nature, but then improve on it through a process of abstraction and generalization. In Joshua Reynolds's influential *Discourses on Art*, a series of lectures delivered between 1769 and 1790, and republished during the nineteenth century, the Royal Academy of Art's first president considered the relationship between variety and unity—terms, as we shall soon see, that were of seminal importance to Maclise's theory of evolution. Reynolds stated that "variety reanimates the attention, which is apt to languish under a continual sameness."[18] However, he cautioned against variety being "carried to excess" at the expense of "the pleasure proceeding from uniformity and repetition."[19] Similarly, atlases were the product of selection, standardization, and generalization. Theorizing the rise of "objectivity" as an epistemic virtue during the mid-nineteenth century, Lorraine Daston and Peter Galison in their foundational study *Objectivity* classify scientific atlases as "systematic compilations of working objects." "No science can do without such standardized working objects, for unrefined natural objects are too quirkily particular to cooperate in generalizations and comparisons," they write.[20] In both high art and scientific illustration a balance needed to be struck between the typical and the particular, unity and variety.

In the case of Maclise's anatomical illustrations, variety or a degree of quirky particularity persists, adding a sense of veracity to the figures depicted. Individualizing details—chapped lips, whiskers, foreskins—testify to the particulars of particular dissections of specific individuals, such as Maclise undertook at University College, London, and the École Pratique and School of Anatomy adjoining the Hôpital de la Pitié in Paris.[21] In the case of the figures in plate 5, close attention is paid to the rendering of the Black man's tightly curled hair, arched eyebrow, flared nostril, full lips, shadowing along his cheek and under his chin, and his

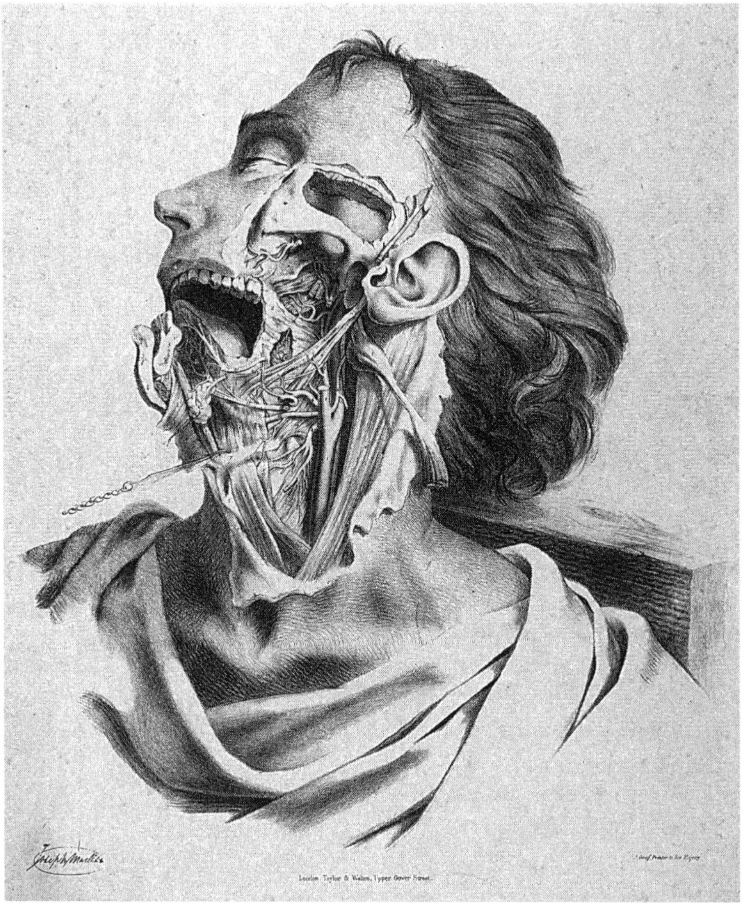

Fig 6.3. Despite the inclusion of classical elements, the horrors of dissections prevail in anatomical illustrations such as this one. Joseph Maclise, "The circulatory system: partial dissection of the neck, jaw and face of a man, with arteries and blood vessels indicated in red," lithograph with watercolor, sheet 22 1/2 × 17 3/4 (57.5 × 45.4 cm) 1844. Courtesy of Wellcome Collection.

bulging jugular. Even eyelashes, whiskers, and an ear piercing are visible. The white man boasts prominent muttonchop sideburns and a lick of hair at his temple. Shading and hatching indicate the contours of his face and neck, with highlights catching the white of his eye, his forehead and cheekbone, the line and crease of his nose, and the inner folds of his ear.

There are also moments in Maclise's illustrations where classical idealization gives way to the monstrosity of dissection, revealing the inability of the artist-anatomist to escape fully the horrors of dissection, and hinting at the underlying sadistic pleasure of dissecting.[22] In plate 10 of

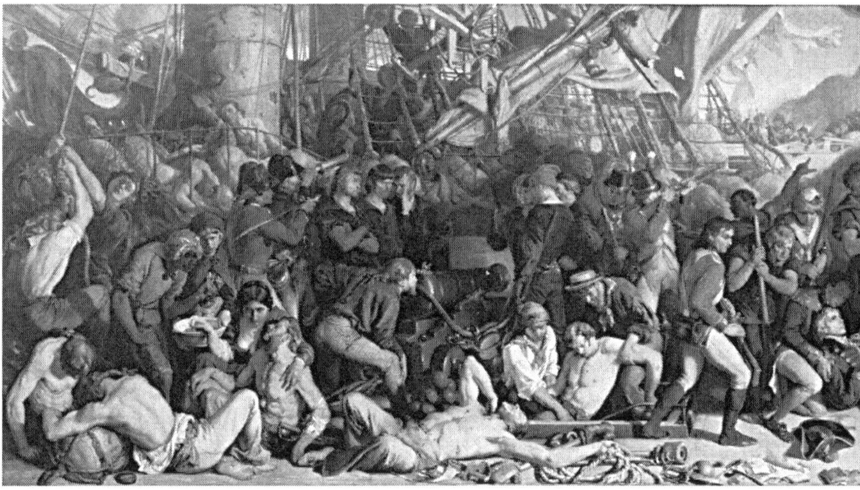

Anatomy of the Arteries (see fig. 6.3), for example, an illustration of "the lingual, the pharyngeal, and the internal carotid arteries," the cadaver reveals the relevant anatomy, as well as a delineated profile, ear, and lustrous hair, with fabric draped across his shoulders. But, in turning his head with a hook attached to the cadaver's throat, the figure is revealed to have exposed bone that resembles an eye socket, with an open cavity in the place of a mouth, complete with a row of teeth. The impression is of a ghoulish second face—a macabre foil to the figure's refined profile. This example visualizes the degree to which Maclise's illustrations resist easy categorization through the incorporation of contrasting elements: living and dead; preserved and decomposing; classical and gothic; ideal and real.

The Muscles

The high art aspirations of Maclise's illustrations connect them to the work of his brother, Daniel Maclise. Biographies of Daniel only ever mention the work of Joseph in passing—if at all—revealing the continued art historical prioritization of certain kinds of images (history painting) over others (medical illustration). It is worth quoting, however, one of only two references to Joseph in Nancy Weston's *Daniel Maclise: Irish Artist in Victorian London*: "In addition to the learned medical commentary, he [Joseph Maclise] drew meticulous finely finished medical illustrations which are not only informative and beautiful but bear a striking similarity to Maclise's *Waterloo.* The similarities are so strong that it is impossible not to imagine a lifetime of these brothers drawing side by side with the only line blurred being that between the poet and the

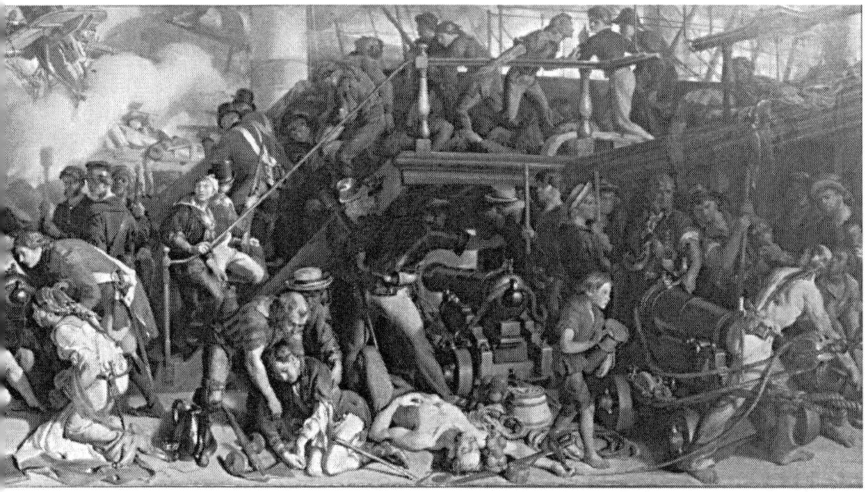

Fig. 6.4. Daniel Maclise's frescos for the Royal Gallery of the Palace of Westminster provide evidence of the artist's engagement with medical subjects. Daniel Maclise, *The Death of Nelson* (finished study), oil on canvas, 388 × 1390 in. (985 × 3530 cm), 1859–1864. Courtesy of National Museums Liverpool: Walker Art Gallery, Liverpool.

practitioner."[23] Weston's imagining of the Maclise brothers drawing side by side is evocative of the kinds of exchanges that took place between artists and surgeons during the Victorian period.

It is also easy to imagine the brothers—artist and surgeon—collaborating in the study of anatomy, drawing together around the dissecting table.[24] We know that Daniel attended anatomy lectures and participated in dissections as part of his art training at the Cork Academy of Arts between 1822 and 1826.[25] In 1844, the year that Quain published *Anatomy of the Arteries*, Daniel and Joseph traveled together to Paris. Between 1861 and 1863, the brothers shared a residence: 4 Cheyne Walk in Chelsea, putting them at the heart of the literati and intellectual circles of the moment.[26] And in 1866 Daniel exhibited at the Royal Academy a painted portrait of Quain, his brother's teacher and collaborator. But it is Daniel Maclise's frescos for the Royal Gallery of the Palace of Westminster—namely, *The Meeting of Wellington and Blücher after the Battle of Waterloo* (completed 1861) and *The Death of Nelson Supported by Captain Hardy on the Victory at Battle of Trafalgar* (completed 1865; see fig. 6.4)—that provide the best evidence of the impact of Joseph's medical training on Daniel's art.[27]

Waterloo is a somber picture of the meeting of the Duke of Wellington and his Prussian counterpart Field Marshal Gebhard Leberecht von

Blücher after their joint victory over Napoleon. A soldier left of center with an impressively muscular torso has attached to his shoulder what has been identified as a "Petit-type" tourniquet, an instrument designed by the French surgeon John-Louis Petit to apply pressure in order to stop heavy bleeding, especially during amputations.[28] Toward the bottom left of the scheme a medical man takes a pulse. Additionally, a surgeon with a leather medical pouch containing three medical instruments—a pair of bone nippers, a lancet, and a pair of spring forceps—bandages a leg.[29] These details illuminate the attention the artist paid not only to likeness, costume, and weaponry but also to the rendering of medical procedures and instruments (it is also worth noting that a third Maclise brother, William, was a surgeon in the army).

The foreground of both *Waterloo and Trafalgar* (the most visible portions of the designs) feature groupings of dead and dying bodies, stretched out and bent over, some fully dressed but many with their bodies exposed, in poses not unlike those in Joseph's illustrations for *Anatomy of the Arteries* and *Surgical Anatomy*.[30] Like *Waterloo*, *Trafalgar* shows a crowded but solemn scene, this time with the dying General Nelson at its center. Daniel was not the first artist to depict the death of Nelson on a monumental scale; he did, however, produce a more somber version than, say, Benjamin West's iconic rendering of the event in *The Death of Nelson* (1806; held at the Walker Art Gallery, Liverpool). West devoted substantially more space to Nelson than did Maclise. In fact, Maclise's Nelson risks being overlooked among the many dying men depicted.

Daniel's *Trafalgar* features an array of exposed muscular dead or dying male bodies, arranged in positions familiar from Joseph's anatomical atlases. Heads turn and tumble, torsos extend and contract, arms reach and hands grasp in gestures that convey the suffering of noble British soldiers and highlight their exemplary anatomy. A figure in the foreground just left of center, with a bullet wound in his chest and an expression of excruciating pain on his face, lies on his back with one arm bent and raised, and the other arm extended above his head to grasp a rope, revealing to the viewer the undulations produced by strained bones, muscles, veins, arteries, and ligaments. A figure diagonally behind him is propped up by an older man, his head hanging and blood dripping down his face and chest. His torso is slightly concave, while his front arm is held horizontally with its lower half dangling to produce a right angle, the reverse position of the arm of the male cadaver in plate 7 of Joseph's *Surgical Anatomy* (fig. 6.5). The cadavers in Daniel's paintings and Joseph's atlases appear to have been maneuvered into the sorts of positions assumed by life models at the academy, revealing a shared

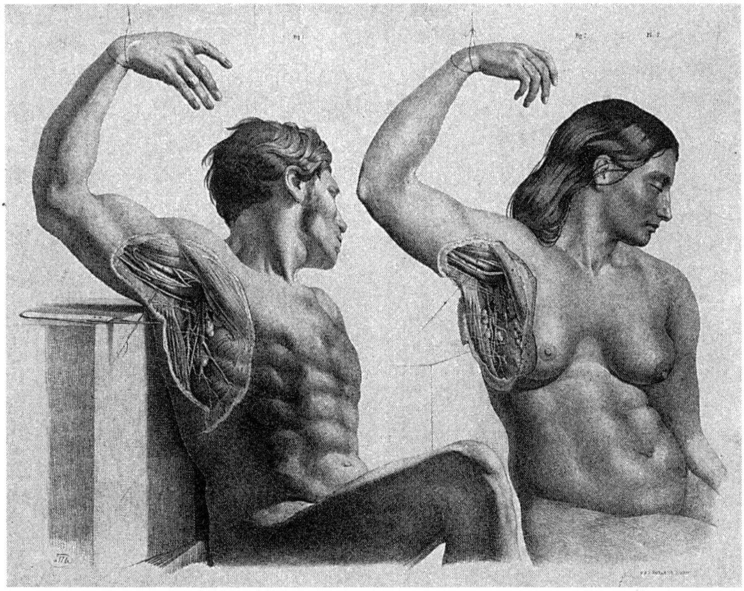

Fig. 6.5. Maclise utilized comparison to emphasize the similarities between differently gendered bodies. Joseph Maclise, "Dissection of the muscles and blood-vessels of the axilla of a seated man and a seated woman," lithograph with watercolor, sheet 21 1/2 × 15 in (54.5 × 37.7 cm), 1851. Courtesy of Wellcome Collection.

vocabulary of poses across high art history painting and anatomical illustration during the period.

Another way in which Maclise's *Trafalgar* is distinguished from other renditions of the death of Nelson is through the inclusion of two Black figures. Both Maclise brothers, it seems, were invested in the depiction of nonwhite bodies. As in John Singleton Copley's *The Death of Major Peirson, 6 January 1781* (1783; held at the Tate, in London), a Black man in Maclise's *Trafalgar* plays a central role.[31] In Copley's painting the Black man avenges Major Peirson's death by shooting his master's killer; in Maclise's fresco the Black man performs the important task of pointing out Nelson's killer for the soldier beside him to shoot.[32] Maclise's rendering of the figure with head turned and arm outstretched directs the viewer's eye into the action of the painting, in a gesture reminiscent of the Black man at the apex of Théodore Géricault's *The Raft of the Medusa* (1819; held at the Louvre), a painting that Maclise saw on his first visit to France in 1830.[33] One might also recognize in the mass of corpses at the base of *Waterloo* a visual reference to the pile of corpses at the base of *The Raft of the Medusa* and other French Romantic paintings.[34]

The gesture of the Black man in Maclise's *Trafalgar* reveals to the viewer the figure's noticeably muscular arm and thick neck. He is presented as noble in terms of both his important role in the drama and his exemplary musculature. Weston recounts how, "while interviewing survivors at Greenwich, Maclise met a man named Pollard who claimed to have deliberately shot the sniper whose bullet killed the Admiral." Maclise then made a point of championing the forgotten hero.[35] Not so with the second Black figure, described in the *Athenaeum* as "an old Negro," who appears to the left of the scene, hunched over and offering a glass of medicinal rum to a wounded soldier.[36] In contrast to the evident muscularity and masculinity of the individualized Black man at the center of the painting, this second Black man might be recognized as a denigrated racial "type," positioned amid a group of women. The latter figure's rounded shoulders, protruding neck, and furrowed brow make him appear, by the visual language of the day, less intelligent and more "primitive" than the white people in the scene, thereby reiterating stereotypes of African racial inferiority, about which more will be said shortly.

The Bones

The complexity that characterizes Daniel Maclise's depiction of racial difference in *Trafalgar* is evident in his brother Joseph's *Comparative Osteology: Being Morphological Studies to Demonstrate the Archetype Skeleton of Vertebrated Animals* (1847), the text that preceded *Surgical Anatomy*. Maclise wrote and illustrated *Comparative Osteology*, a comparative study of vertebral structures, especially those of humans, the principal contribution of which was Maclise's identification of what he termed the "archetype" skeleton, or the complete form from which different skeletal structures derived.[37] Historian of science Nicolaas Rupke has shown that Maclise claimed to have come up with the concept—or at least this particular use of the term *archetype*—prior to the influential comparative anatomist and opponent of Darwin's theory of evolution by natural selection Richard Owen.[38] Owen never disputed Maclise's claim; he did, however, provide a divergent account of the structure of the "vertebrate archetype" in his own *On the Archetype and Homologies of the Vertebrate Skeleton* (1848).[39]

While Maclise's contribution to mid-nineteenth-century comparative anatomy has been largely eclipsed by Owen's, certain illustrations in *Comparative Osteology* offer surprising presentations of, to use Maclise's phrase, "unity in variety." According to Maclise in the book's introduction, the "archetype form" was the "prime model," and the vertebrae of different species—human and animal—were the product of a process of constant modification. He claimed that the vertebra was not a perfect whole but was rather part of some other complete or whole form that was once, but is no longer, in existence. In other words, all extant vertebrate animals exhibited some part—"plus or minus qualities"—of the original archetype form.[40] This reveals the impact of French "philosophic anatomy," developed by Étienne Geoffroy Saint-Hilaire, professor of zoology at the Muséum national d'histoire naturelle in Paris, on Maclise's account of the relationship between the anatomical structures of humans and other animals. As Adrian Desmond has shown, by 1830 Geoffroy and his disciples came to accept "a unitary composition for all animals," in opposition to George Cuvier's division of animal life into discrete "divisions."[41] Desmond writes of Geoffroy's theory of "unity of composition" that "not only were all vertebrates built to the same blueprint; in its most extreme formulation, the theory allowed insects, molluscs, and man to be reduced to common organ components."[42] Geoffroy's leading disciple was Étienne Serres, senior physician at the Hôpital de la Pitié, where Maclise studied. Desmond describes Maclise as "a partisan morphologist, even in the late 1840s, who saw Geoffroy 'shake the throne of Cuvier' and sweep aside his 'tottering fabric.'"[43] When, in plate 29 of *Comparative Osteology*, Maclise shifts focus from the human skeleton to the study of the vertebrae of "lower animals," he makes direct reference to Geoffroy and Cuvier: "Animal bodies are one and all products of the same law of development [as humans]. A Newton, a Leibnitz [*sic*], a Geoffroy, or a Cuvier, could not deny that their own corporeal frameworks bore an unmistakable similitude to that of an orangutan, a chimpanzee, a mandril, and a baboon."[44]

Applying his theory of "unity in variety" to the vertebrae of men of different races, Maclise included as plate 47 a comparative study of the skeletons of "a Chinese" and "a Carib," and as plate 48 a comparison of the skeletons of "a New Zealander" and "a Negro." In the text accompanying his first racial comparison, Maclise effectively counters Pieter Camper's eighteenth-century system of racialized facial measurements, stating that "in whatever other respect these skeleton forms may be said to manifest distinctive characters, it is at once rendered evident that no mode of measurement, by the facial angle or minute examination of special character, can help us to an interpretation of the law which planned them, as they are, the proportional of archetype unity."[45] Camper had compared the skulls of different racial types by drawing "a line along the forehead and the upper lip" and measuring the angle. He used this measurement to rank races from those exhibiting the most "beautiful" lines and angles, through to the opposite end of the scale.[46] Camper's study of different human and animal facial angles marked the beginning of physical anthropology, and, as historians of the visual cultures of race have shown, was all too easily used in nineteenth-century anthropology and ethnology to associate the "lower" races with primates, and to establish the intellectual and moral supremacy of privileged white races.[47] The skeletons of humans of different races, argues Maclise, cannot be placed on an evolutionary scale organized according to Camper's measurement. Rather, comparison of the skeletons of "the Caucasian, Mongolian, American, and Ethiopian" illustrates the plus and minus development of archetypal unity.

Maclise writes that "while we compare fig. A, the skeleton of a Chinese, with fig. B, the skeleton of a Carib, we see every reason to name them homologues."[48] In his description of the New Zealander and the Negro, Maclise observes: "The same number and relative position of elementary parts are visible in the two figures." He continues: "So, also, are the like parts found in the skeleton qualities of a baboon, a mandril, an orang, or a chimpanzee."[49] By comparing the skeletons of an Indigenous New Zealander and a Black person with those of the lower primates, Maclise reiterates racial hierarchies that put the "lower" races, in particular Black people, next to apes on the evolutionary ladder. Yet he also likens the skeletons of "a Newton, a Leibnitz, a Geoffroy, or a Cuvier" to those same primates, and describes humans and other vertebral animals as the product of the same evolutionary processes. Hence, while histories of Victorian science tend to emphasize the conflict between Owen and Thomas Huxley over Darwin's theory of evolution, Maclise's *Comparative Osteology*, with its emphasis on the commonalities between humans and other primates, offers a significant but largely overlooked precedent for Huxley's *Evidence as to Man's Place in Nature* of 1863, which made clear the shared anatomical structures—especially skeletons—of humans and other primates.[50]

Flesh and Blood

Surgical Anatomy was initially released as fascicles, starting in London in 1848. In 1851, Blanchard and Lea of Philadelphia and John Churchill of London published the atlas as complete first editions. In 1856 Churchill published a second expanded edition, and in 1859 Blanchard and Lea did the same. Henry C. Lea published a smaller, cheaper version in 1870. In the American context, Maclise's illustrations were reproduced in watercolor for pedagogical use and now form part of Jonathan Hutchinson's pathological print collection in the William H. Welch Medical Library at Johns Hopkins Medical School. The atlas contains thirty-five color plates (fifty-two in the 1856 British edition) showing surgical dissections, standard dissections, and the relative position of anatomical structures that would have been of particular use to surgeons, such as those regions relevant to surgeons performing a lithotomy or the surgical removal of a stone from the bladder, kidney, or urinary tract, referred to at the time as "cutting for stone." While the large folio edition would have been a beautiful collector's item, the illustrations may also have been hung on the wall during dissections to direct medical students.[51]

Overall, the illustrations offer striking depictions of supreme physical specimens, opened up for the instruction of medical students and medical practitioners. They focus on the dissected areas—the veins, arteries, and organs conveniently color-coded and numbered for easy identification—while at the same time showing the jagged edges produced by the surgeon's scalpel, and the surrounding enclosed areas of the body. The exposed arteries are organized into systems devoid of messy viscera and blood. The only suggestion of blood is offered by the red color of the arteries. But it is not crimson; it is a uniform red appropriate to a medical diagram. In the second British edition the illustrations become increasingly dramatic—bordering on theatrical—with striking contrasts of light and dark and a more varied palette, including pastel pinks and purples. Cadavers not seen in the first edition have their bodies turned inside out and exhibit expressions suggesting that they have fainted from excruciating pain, complete with backward-rolling eyeballs. More attention is paid to the rendering of ropes, which hold cadavers' arms in place, reminding us of the use of ropes to keep the tired limbs of life models steady, as seen in Johann Zoffany's 1771–1772 painting *The Academicians of the Royal Academy* (held in the Royal Collection Trust).

As historians of anatomy and its attendant visual culture have shown, despite the "Enlightenment" emphasis on direct observation of human dissections as the path to scientific truth, anatomical illustration continued to include irrelevant aesthetic elements associated with Renaissance anatomical illustration, such as putti, drapery, landscapes, and classical ruins.[52] Maclise's illustrations, with their attractive animated cadavers, can be seen as part of this hybrid visual tradition. Distinguishing Maclise's illustrations for *Surgical Anatomy* from earlier anatomical illustrations, including those by Bell and Maclise's own illustrations for Quain, however, is their emphasis on comparison and the degree to which he organizes the anatomy of the body according to geometry. In image and text Maclise draws comparisons between healthy and diseased bodies; the body's regions in relation to the whole; the various organs contained within these regions; and the dissected and nondissected areas of the body. This is, to use Maclise's phrase, "relative anatomy."[53] Utilizing a topographical analogy, Maclise informs us in his preface that "the unbroken surface of the human figure is as a map to the surgeon, explanatory of the anatomy arranged beneath," and that he has therefore "left appended to the dissected regions as much of the undissected as was necessary," with the professed objective of indicating "the interior through the superficialities."[54]

It is the comparative aspect of plate 5 in *Surgical Anatomy* that makes it such a compelling and potentially controversial image. The illustration features a white man on the left and a Black man on the right, both with eyes open, their shoulders facing forward and heads turned in opposite directions to reveal the dissection of their tracheal regions. Unlike the more intimate encounters that Maclise sets up in some of his other illustrations (other illustrations have the all-white figures facing in the same direction), in plate 5 he depicted the men facing away from each other, as if unrelated or even adversarial. And yet, when one penetrates the surface of the skin—through what appears to be a window or portal—to the anatomy beneath, one finds the same universal anatomical structures, rendered in the same schematized colors. This calls to mind the American physician-educator William Andrus Alcott's argument in his popular guide to anatomy and physiology, *The House I Live In, or, The Human Body, for the Use of Families and Schools*, first published in full in 1836, that "the skin itself, properly so called, is alike in the whole human race . . . and that the color might be removed . . . leaving one individual as white and as dark as another."[55] Michael Sappol explains that, according to Alcott, "skin color . . . is only a superficial difference and its removal reveals universal anatomy."[56] The radical nature of the concept of universal anatomy is described by Darcy Grimaldo Grigsby in her stomach-churning essay about the relationship between colonialism, slavery, and cannibalism: "But Europeans' queasiness derived in the end not from the alienation inspired by racial difference but from the perceived sameness—the ways violence and consumption and coerced passivity, the dissembling of bodies into meat and buttocks, levelled difference. . . . And black bodies once butchered resembled white ones. This fact was discomforting. It was also radical."[57] In the version of plate 5 that appears in the second British edition of *Surgical Anatomy*, Maclise added a focused study of the men's shared anatomy, shown front-on, and strategically positioned between them on the page.[58]

Nineteenth-century race scientists can roughly be divided into two camps; those who believed that the different races derived from a common origin and were therefore all of one species (monogenists), and those who believed that the different races of men constituted entirely different species (polygenists). Both monogenists and polygenists relied on drawing comparisons between the anatomical structures of different races—putting them into close proximity—in order to establish their relative positions on the racial scale. A monogenistic account of the

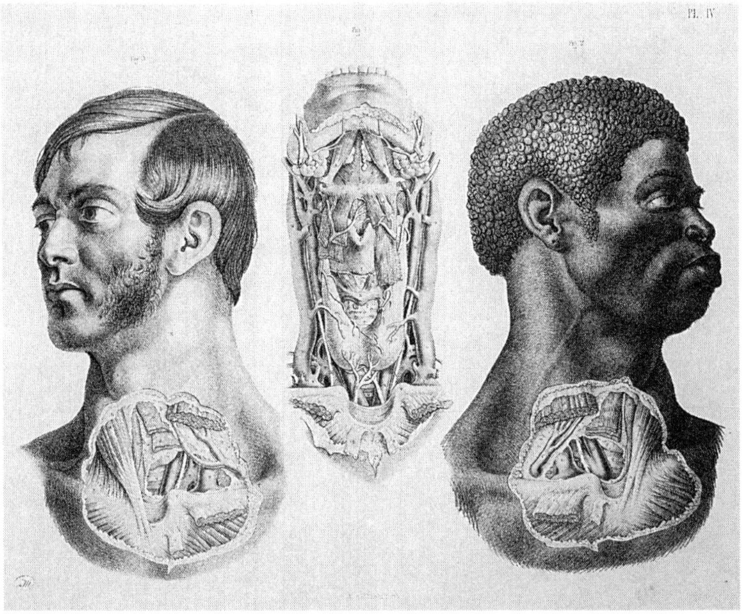

Fig. 6.6. At the center of this illustration is a detailed study of the shared anatomy of a Black man and a white man. Joseph Maclise, *Surgical Anatomy*, 2nd ed. (London, 1856), plate 4. Courtesy of National Library of Medicine, Bethesda, MD.

nature of racial difference might emphasize the biological similarities between people of different races as signs of their common origin and shared humanity. Many comparative anatomical studies used skulls or crania to perform this ideological work (others, of course, emphasized hips, buttocks, and genitals). The justification for the study of skulls was that skulls housed brains, and brains were the seat of intelligence. Consider Josiah C. Nott's statement in his chapter "Comparative Anatomy of Races," in his highly influential American polygenist work on race, *Types of Mankind*: "A minute and extended anatomical comparison of races, in their whole structure, would afford many curious results; but such detail does not comport with the plan of this work, and would be fatiguing to any but the professed anatomist. It is indispensable, however, that we should enter somewhat fully into a comparison of crania; and it may be safely assumed, as a general law, that where important peculiarities exist in crania, others equally tangible belong to the same organism."[59] Nott furnished this chapter with almost one hundred illustrations of heads and skulls, overwhelmingly depicted in profile and mostly obtained from Samuel George Morton's famous skull collection, in order to illustrate his polygenist position. The skulls are shown either facing forward or in

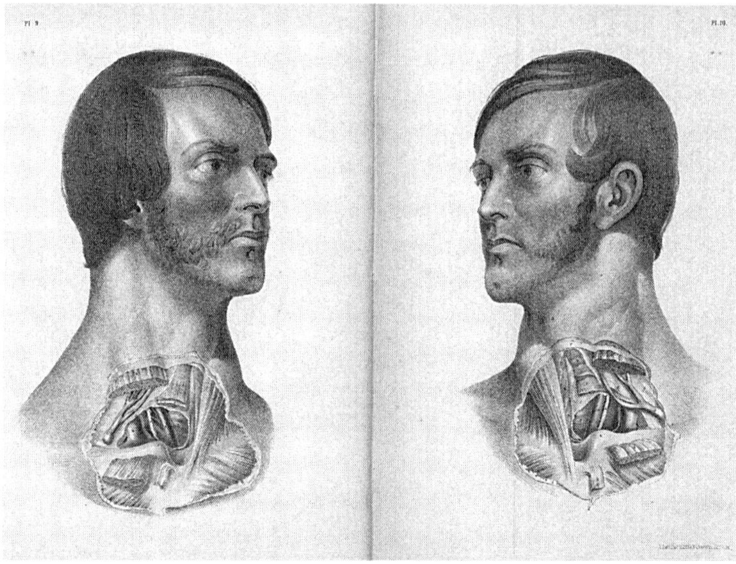

Fig. 6.7. Maclise's *Surgical Anatomy* was whitewashed for American audiences. Joseph Maclise, *Surgical Anatomy* (1851), plates 9 and 10. Courtesy of Rare Books Collection, Booth Family Center for Special Collections, Georgetown University Library, Washington, DC.

profile, allowing for the measurement of the face according to Camper's angle, culminating in a comparative illustration of an American Indian and European brain.

Continually working against the trend of using comparison to highlight signs of racial difference, Maclise used visual comparisons to emphasize the similarities between the anatomy—bones in *Comparative Osteology*, organs in *Surgical Anatomy*—of the different "races of men." Maclise also used comparison to emphasize the similarities between differently gendered bodies. The aforementioned plate 7 is an illustration of the anatomy of the male and female axillae or armpit, which, Maclise concludes, "are anatomically homologous or similar at basis." The subtle distinctions between male and female anatomy are, writes Maclise, a product of "the plus or minus condition"—a description that recalls his theory of the diversity produced from the vertebral "archetype." In the first version of the illustration from 1851 (fig. 6.6), the admirable and noticeably "ripped" musculature of the male figure's abdominal region and raised arm is revealed. The figure's bent knee conveniently obscures his genitals from view, only narrowly avoiding touching the thigh of his female companion. The female figure similarly exhibits a defined, muscular abdomen, but with fleshed-out contours that represent her

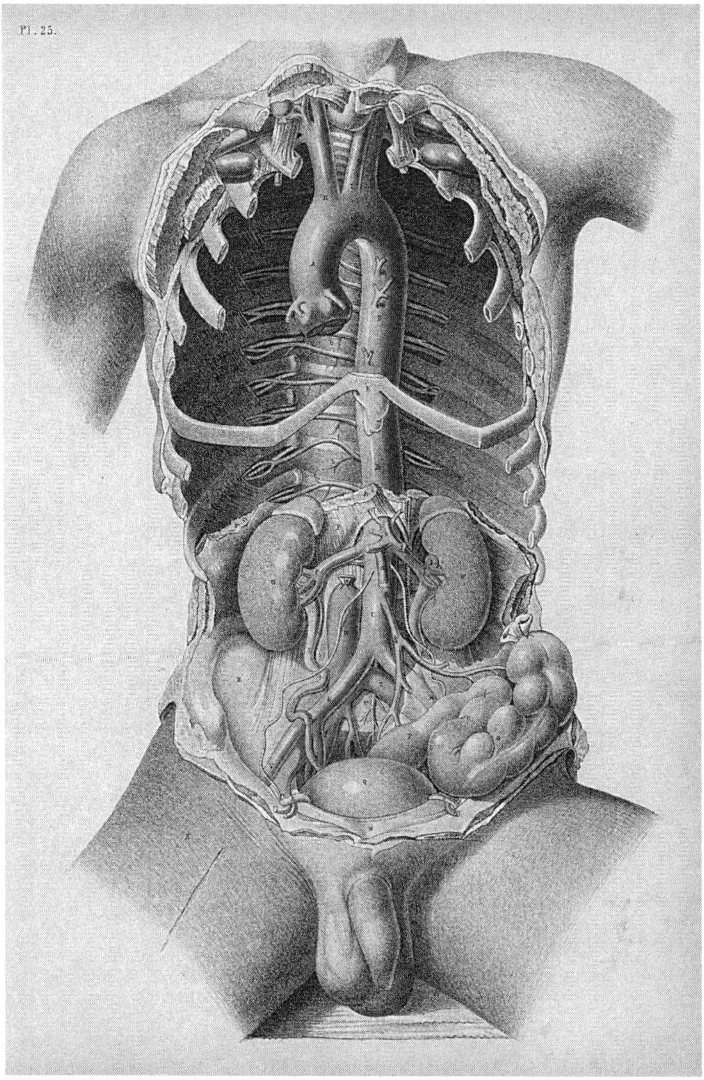

Fig. 6.8. The original version of this image was altered to give the impression that we are looking at the body of a white man. Joseph Maclise, *Surgical Anatomy* (1851), plate 25. Courtesy of Georgetown University Library, Washington, DC.

perceived natural fecundity. She is also in possession of an unnaturally hard-looking breast and nipples that, in their perfect regularity, resemble bolts. The symmetry in the positioning of the two figures draws attention to their homogeneity and puts them into a sort of dance formation. Their shared anatomy is further emphasized in the altered version of the illustration that appears in the second British edition of *Surgical Anatomy*. Here the placement of the figures is switched, and they are brought into even closer contact. On the one hand, this allows us to compare their anatomy more easily; on the other hand, it serves to heighten the intimacy between them. The face of the woman is now obscured, while her right-hand forefinger gently lingers on her male companion's elbow, her left breast highlighted by the shadow cast by his raised arm.[60]

In the case of plate 5, what alerts us to its problematic nature is the conspicuous absence of the Black man from all the American editions of the atlas. Similarly, Bell's illustration of the dissection of the Black man was never reproduced in American editions of *Engravings of the Arteries*.[61] In the 1851 American version of Maclise's *Surgical Anatomy* (fig. 6.7), the Black man's anatomy is shown as a white man's anatomy. The men now face each other, their similar appearance indicating that we are being given different views of the same man, although there are subtle differences in the styling of the hair around the ears, the amount of facial hair shown, and the shape of their noses. In case of plate 14, the Black figure is depicted with closed eyes, protruding lips, and a distinctive cut on the inside of his thigh, which appears to have been sutured in plate 48 of the 1856 British edition. In the American version of plate 14, which appears as plate 25 (fig. 6.8), the figure's skin tone is lightened and his head is cropped out of the picture. We can, nevertheless, tell that this is the same figure from his pose, the form of his genitalia, and, more tellingly, from the cut on his inner thigh.

One possible explanation for this omission lies in the particularly high-stakes nature of debates over racial difference in America at the time. In Britain the theological orthodoxy of the day was monogenism. By contrast, as Stephen Jay Gould has argued, polygenism dominated American race science during the nineteenth century, so much so that it was referred to in Europe as the "American School." "It is obviously not accidental," writes Gould, "that a nation still practicing slavery and expelling its aboriginal inhabitants from their homelands should have provided a base for theories that blacks and Indians are separate species, inferior to whites."[62] Indeed, there were more polygenists in the United States, for the reasons laid out by Gould, and they often occupied influential positions: Louis Agassiz, for instance, was professor of zoology and geology at Harvard, and founder and director of the Museum of Comparative Zoology. But polygenism also had its opponents, foremost among them the influential American Lutheran minister and naturalist Reverend John Bachman, who argued that the claims of polygenism were contrary to the account of creation in the Bible and were therefore blasphemous.[63] Hence, it was not the depiction of the dissection of a Black man that would have caused offense in the American context. After all, as medical historian John Harley Warner has shown, cadavers dissected in nineteenth-century American medical schools were disproportionately African American.[64] Rather, it was the depiction of a Black man in possession of ideal or, worse, universal anatomy that was so disruptive to the perceived racial order. Maclise, or more likely his American publishers, must have calculated that plate 5 in its original form would have been too inflammatory for the American scientific community (why it was admissible in the British context is a topic for another day).

The work of Joseph Maclise makes clear that nineteenth-century anatomical illustration engaged with the most contentious aesthetic and scientific debates of the day. When it is put into dialogue with the work of his brother Daniel Maclise, we see the degree to which Joseph Maclise drew on the visual language of high art to sanitize, generalize, even classicize his dissected cadavers. Joseph Maclise sought to present, as was characteristic of nineteenth-century academic art, ideal physical specimens, and, significantly, he used Black cadavers to do some of this aesthetic and ideological work. The racial implications of this cannot be overstated. The version of plate 5 that appears in the British edition of *Surgical Anatomy* uses the body of a Black man to teach anatomy. With knowledge of the racial origins of the original source, it is the American version—in which a Black man's anatomy appears in a white man's body—that, ironically, is the most radical because it more effectively illustrates the superficiality of racial difference when faced with the deeper truth of our shared anatomy.

There is a lot that we do not know about plate 5. Despite an exhaustive search, I have not been able to identify the Black man. Nor have I been able to ascertain how his corpse came to be dissected. Since the publication of *Surgical Anatomy* postdates the Anatomy Act of 1832, we can assume that he died, like the other men and women in the atlas, in a state of destitution and was unclaimed after his death. Perhaps by virtue of his race he was a particularly desirable specimen and dissection was a more likely fate. One need only think of the well-documented case of Sarah Baartman, the so-called "Hottentot Venus."[65] Or, perhaps, as for Bell, the Black man simply offered the best example of the relevant anatomy. We also do not know who was responsible for the erasure of his skin color for the American edition. Extrapolating from this example, one might be led to wonder how many of the white male anatomical specimens furnishing iconic anatomical atlases past started out as Black corpses on the dissecting table. For now these questions remain unanswered. But, by following plate 5 as it moved across different locations and was reworked for different audiences, one can start to recover a sense of the complex and contested relationship between anatomy and race in the developing discourses of evolution in midcentury Britain and America. Conversely, by working backward and exposing the racial origins of the whitewashed American version we can start the messy process of inserting the flesh and blood back into the history of nineteenth-century anatomical illustration.

CHAPTER 7

THE POST-DARWINIAN EYE, PHYSIOLOGICAL AESTHETICS, AND THE EARLY YEARS OF AESTHETICISM, 1860–1876

Barbara Larson

Aestheticism came to public attention in 1877 with exhibitions at the prominent Grosvenor Gallery and the infamous Whistler trial the following year, and this formerly marginalized movement became popular among the cultural elite.[1] While distinctly modern and even subversive in approach, the artists of aestheticism neither felt bound to subjects of the current period nor followed a unified style. They ranged from artists like James McNeill Whistler and his friend Albert Moore, who explored color harmonies, to Edward Burne-Jones, Dante Gabriel Rossetti, and Lord Frederic Leighton, who often used a warm, sensual palette to depict voluptuous, languorous female figures. They became known for their promotion of well-appointed decorative interiors that confirmed elite sensibilities, as well as for their love of spectacular aspects of the natural world such as peacock feathers and sunflowers, which were repeated to the point of becoming virtual symbols of the movement. Some of the artists of aestheticism were once part of the Pre-Raphaelite school, but now ventured away from the kinds of generally descriptive scenes and narrative art with which they had once been associated. Instead, they became engaged in considerations of the specifics of beauty (color and form) and in producing harmonious and indistinct scenes. The unifying credo of aestheticism was "art for art's sake," which seems to reject by its very declaration any adherence to the world, let alone science, but, as I argue here, aestheticism is actually close in principle to developments in psychology and physiology in the 1860s and early 1870s that were tied to evolutionism. This includes Darwin's arguments about the function of the eye—its attentiveness to attractants ("beauty") on the one hand and inability to grasp overall detail on the other, and it includes the current and influential idea promoted by Darwin supporter Herbert Spencer that art, though an eventual outcome of the evolutionary and physiological history of the body, indeed had no purpose other than itself.

The Scientific Background of Aestheticism

The field of psychology as it emerged in Great Britain in the 1850s was tied to the work of empirical philosophers engaged in mental studies like Alexander Bain, who emphasized the mind's response to physiological processes.[2] Grounded in the body, psychological response was directly tied to corporeal operations; even "spontaneous thought" had its basis in physiology. After the publication of Darwin's *On the Origin of Species* (1859), physiological psychologists became attuned to Darwin's ideas regarding holistic corporeal response to the environment and a fallible, limited, yet highly responsive eye, a perspective on sight and experience of the world that diverged from that upheld by contemporaneous natural theologians. When the new movement of aestheticism emerged in the 1860s, its characteristics included the foregrounding of sensory experience from the artist's and viewer's perspective that strongly suggests the influence of the physiological theories of Bain, Herbert Spencer, and Darwin.

Spencer was one of the most influential of the psychologists on developing physiological aesthetics of the period. Like Bain, Spencer tied physiology to psychology in his *Principles of Psychology*, published in 1855, but followed Jean-Baptiste Lamarck on historical and linear evolutionism, believing the brain was layered with ancestral memories. After the publication of the *On the Origin of Species*, Spencer emerged as a Darwin supporter, but maintained certain Lamarckian applications of evolutionism as a teleological process, as opposed to Darwin's undirected theory of natural selection (survival of the fittest). He began to issue his second expanded edition of *Principles of Psychology* by 1868, and then in complete text form in two volumes by 1872. In it he acknowledged that his original publication, steeped in "the Doctrine of Evolution," was ridiculed by most in the mid-1850s, but in the last ten years, presumably since the publication of *Origin*, applications of evolutionary physiology to the mind were now taken seriously.[3] In *Principles of Psychology* Spencer demonstrated an interest in the evolutionary implications of aesthetics, adding an entire section in the second edition that directly addressed art.[4] According to Spencer, art as practiced by the civilized was a pursuit uncoupled from issues of survival. Energy that had once been necessary for survival from an evolutionary and physiological perspective was now expended by neurological "effervescence" through play or art. Art went hand in hand with the advancement of culture and race—it was pursued for its own sake; and the greatest art produced the most pleasurable sensations.[5] Though the phrase "art for art's sake" has a history predating aestheticism in Great Britain, both Spencer and the British physiological aesthetic theorists who followed him gave greater scientific credibility to this position.[6]

In terms of what pleases the eye, Spencer followed Bain, who believed we prefer winding movements that are curvilinear—waxing and waning forms by which the eye is stimulated, then sinks briefly into repose.[7] Aestheticism, with its curvaceous compliant female forms, as in the paintings of Frederic Leighton (such as the famous *Flaming June*, 1895, held at the Museo de Arte de Ponce), is replete with these shapes. Spencer, like many of the physiological psychologists, added the older concept of associationism to aesthetics—that is, positive and negative associations related to the experience of pleasure and pain invoked by memory in relation to the current moment. In *The Senses and the Intellect* of 1855, Bain too had drawn upon associationism and explicitly addressed art. The role of the artist was to emotively suggest rather than replicate, a principle of aestheticism.[8] According to Bain, the viewer then draws upon associations and "the mind supplies from the past what the eye does not distinctly see at the time, so that the picture actually realized is not the bare optical impression of the moment, but a much fuller picture which that impression suffices to suggest."[9]

New scientific ideas on subjectivity were readily available to the artists of aestheticism through popular journals that combined science and the arts, such as the culturally progressive and influential *Fortnightly Review*, the *Westminster Review*, and the *Contemporary Review*. Aesthetic artist Rossetti, for example, was also a poet who published in the *Fortnightly Review*, perceived as a leading organ of modern physiology as well as evolutionism by the 1860s. Its first editor, philosopher George Henry Lewes, was an amateur physiologist, and he published no fewer than four articles on "Mr. Darwin's hypotheses," beginning in April of 1868.[10]

The evolutionary history of the organism was central in explicating physiological theories of response to the environment, and the imperfection of the eye (of interest to Lewes) underscored the importance of a coordinated sensorium in successfully experiencing the world. Before the publication of *On the Origin of Species*, the eye, though understood to be subject to imperfections of vision, was nevertheless thought of in early Victorian Britain as an ideally constructed organ. For more than fifty years prior to the appearance of *Origin*, natural theologians had identified the eye as the exemplary organ of divine design. For example, William Paley, in his central text *Natural Theology* (originally published in 1802), used the eye as the ultimate example of God as Creator of the human body, comparing the eye to instruments such as the camera obscura or telescope. It is the reaction of this British scientific audience that Darwin anticipated when he addressed the structure of the eye in *Origin*:

> To suppose that the eye, with all its inimitable contrivances for adjusting the focus to different distances, for admitting different amounts of light, and for correction of spherical and chromatic aberration, could have been formed by natural selection, seems, I freely confess, absurd in the highest possible degree. . . . It is scarcely possible to avoid comparing the eye to a telescope. We know that this instrument has been perfected by the long-continued efforts of the highest human intellects; and we naturally infer that the eye has been formed by a somewhat analogous process. But may not this inference be presumptuous? Have we any right to assume that the Creator works by intellectual powers like those of man? If we must compare the eye to an optical instrument, we ought in imagination to take a thick layer of transparent tissue, with a nerve sensitive to light beneath, and then suppose every part of this layer to be continually changing slowly in density, so as to separate into layers of different densities and thicknesses, placed at different distances from each other, and with the surfaces of each layer slowly changing in form. Further we must suppose that there is a power always intently watching each slight accidental alteration in the transparent layers; and carefully selecting each alteration which, under varied circumstances, may in any way, or in any degree, tend to produce a distincter image. We must suppose each new state of the instrument to be multiplied by the million; and each to be preserved till a better be produced, and then the old ones to be destroyed. In living bodies, variation will cause the slight alterations, generation will multiply them almost infinitely, and natural selection will pick out with unerring skill each improvement. Let this process go on for millions on millions of years; and during each year on millions of individuals of many kinds; and may we not believe that a living optical instrument might thus be formed as superior to one of glass, as the works of the Creator are to those of man?[11]

Having laid out the history of the eye through natural selection, Darwin further discussed its evolutionary aspects, citing cases in which there was rudimentary sight or where there apparently had once been sight but which was significantly reduced or existed no longer, as in moles or certain crabs.[12] And it was not the human eye that Darwin found to be the most able, but rather that of the eagle.[13] Moreover, on the imperfection of the human eye he added, "Natural Selection will not produce absolute perfection, nor do we always meet . . . with this high standard under nature. The correction for the aberration of light is said, on high authority, not to be perfect even in that most perfect organ, the eye."[14] The "high authority" for his thoughts on the imperfection and, ultimately, the limits of the human eye was identified in later editions of *Origin* as "Muller," by whom he no doubt meant Johannes Müller, who wrote two books on the eye in the 1820s and whose 1833 *Handbook on Human Physiology* became the central text on the senses, including sight, in the developing science of human physiology through the middle of the nineteenth century.[15]

Darwin may have known of Hermann von Helmholtz's important work on the eye by the 1860s even though, despite the fact that Helmholtz invented the ophthalmoscope (an instrument used to examine the retina) in 1851, his physiological work was less celebrated in Great Britain at midcentury than his research in physics.[16] Lewes, whom Darwin knew personally, was well versed on Helmholtz, having read Helmholtz's three-volume *Handbook on Physiological Optics* (1856–1867) in the original German in the 1860s.[17] However, Helmholtz on the eye was popularized around 1870 in England to the scientific elite through lectures and translations by fellow physicist John Tyndall. In the sixth edition of *Origin*, Darwin gratefully acknowledged Helmholtz's willingness to find the eye imperfectly formed.[18] In the second edition of *The Descent of Man* of 1874, Darwin demonstrated greater familiarity with Helmholtz, his *Handbook on Physiological Optics* and other writings having been translated into English in 1873. Darwin was blunt and perhaps somewhat defensive: "And we know what Helmholtz, the highest authority in Europe on the subject, has said about the human eye; that if an optician had sold him an instrument so carelessly made, he would have thought himself fully justified in returning it."[19] In fact, as detailed in the *Handbook on Physiological Optics*, Helmholtz felt the eye was so compromised in its ability to accurately gather visual information that he posited that we must be making inferences from incomplete data on an ongoing basis.

The years that followed the initial proposal in *On the Origin of Species* in 1859 of the evolutionary eye developed by natural selection were rife with criticism, culminating in 1871 with St. George Jackson Mivart's *The Genesis of Species*, in which the eye was the clinching case against Darwinian aspects of evolutionism. Mivart pointed out that rudimentary structures within the eye did not in and of themselves produce sight. Thus, evolution according to natural selection was nonsensical since there would have been no useful purpose in such structures to begin with.

The eye as defined by Darwin was not only imperfect, subject to illusions in the natural world (thus, the value of camouflage), but varied considerably among individuals. In that same year, 1871, Darwin elaborated on the human eye in the first edition of *Descent of Man*. He claimed that "savages" had better eyesight than Europeans and that this was so through use and inheritance. Indeed, he claimed for this group a far superior development of all the senses, even citing dissections that seemed to prove his case. According to Darwin, "The inferiority of Europeans in comparison with savages, in eyesight and in the other senses, is no doubt the accumulated and transmitted effect of lessened use during many generations."[20] Moreover, nearsightedness and farsightedness among Europeans also tended to be hereditary as well as attached to certain professions. Engravers and watchmakers, according to Darwin, tended to be nearsighted. The Darwinian eye was historical, situated, and not privileged among Europeans.[21] And it was as much a product as an object of perception.

From Darwin's perspective, however good or bad one's own sight may be, vision was part of a coordinated sensorial response in which sensibility was tied to self-preservation—averting harm on the one hand and positive excitation of tissue on the other. Based in the vital qualities of cell structures, in which both plants and animals shared protoplasm (as was now recognized), all organic beings responded to sensory stimuli in the environment in a holistic manner. Pleasure and pain, principles that had been explored in the seventeenth and eighteenth centuries by empiricists, were central to Darwin's evolutionism.[22] The eye had developed as it had to the extent that survival and reproduction were ensured. Thus, in the animal world the eye responded strongly to stimulants such as markings, coloration, secondary sexual characteristics like antler size, or the overall impression of an animate form as larger or potentially threatening. These stimulants varied among species and through time. And physiological processes evolved along with mental faculties.

Given the circulating discourse on the eye, then, Darwin's observations on embodied vision and its evolutionary history should be considered in light of the rise of aesthetic painting. His mechanism of sexual selection (competition for mates), wherein visual attractants are prioritized, is also highly relevant to aesthetic painting. In a brief section on "secondary sexual characteristics" in *On the Origin of Species*, Darwin gave but a couple of examples of what was meant by this, one of which was the difference in coloration, display, and tail feathers of male gallinaceous birds, which include peacocks.[23] His explanation for the sometimes extravagant plumage of male birds was that it was used to attract the female of the species during courtship display. It was an issue to which he would return in subsequent editions of *Origin* in the 1860s and in *The Descent of Man* of 1871.

Male peacock feathers had long been considered a symbol of beauty with erotic overtones. Caitlin Silberman's essay in this volume on Whistler and the Peacock Room explores the multiple meanings the peacock had acquired as a decorative device by the 1870s. Given Darwin's new scientific explanation behind the reason for the peacock's train, this bird became the creature around whom questions of aesthetics (beauty) and sexual selection consolidated even before the publication of *The Descent of Man*, the subtitle of which is *And Selection in Relation to Sex*.[24] The peacock became the creature, next to the ape, that most invoked Darwinian thinking in the public eye.

In the 1860s, Darwin's interest in that which stimulates the eye had become increasingly engaged in the concept of beauty, and he expanded upon his ideas in the fourth (1866) and fifth (1869) editions of *Origin*. The aesthetic preferences of those who select mates and, therefore, provide another engine of evolutionism (notably females in the nonhuman world) could be arbitrary and capricious.[25] In 1869 Darwin wrote, "I willingly admit that a great number of male animals have been rendered beautiful for beauty's sake."[26] A sense of beauty could be found in all living creatures and was tied to pleasure given by form, color, and sound. Here he may well have been influenced by Bain, whose contention that beauty gives pleasure in his *The Senses and the Intellect* of 1855, though deriving from sensationist philosophy of the eighteenth century, provided a scientific updating of the idea.[27] By 1871 the role of beauty for humans was now made explicit in Darwin's *Descent of Man*. For example, he believed that the reduction of body hair was a sexually selected trait and made frequent comparisons between birds and humans in terms of aesthetics.[28] Attention given by Darwin to attractants (such as color and form), explanations of the workings of the imperfect eye by scientists, and specifics of the aesthetics of pleasure and ideas concerning a "suggestive" art developed by psychophysiologists should be considered in light of the emergence of aestheticism.

Aestheticism

Sensory effects of color and form and a foregrounding of beauty over content among the artists of early aestheticism were discussed by a supporter of the movement, Sidney Colvin, in the *Fortnightly Review* in

1867. Rossetti, Burne-Jones, Leighton, Whistler and Moore, among others, were found to be above all interested in form and color. They were among "those artists whose aim, to judge by their works, seems to be pre-eminently beauty."[29] Colvin was sensitive to the subjective response of individual artists. He was part of the same milieu in which art and science mixed and was a friend of Lewes, frequently visiting the philosopher's home.[30] Lewes had written about sensory response to the world as early as his *Physiology of Common Life* (1859–1860), and in the prolegomena of his *History of Philosophy* in 1867 he detailed his thoughts on physiology and individualized mental perception. Visual impressions stimulated local or generalized bodily sensations, which then via sensory nerves transmitted the message of the impression to the brain.[31] Colvin shared the conviction that specific impressions stimulated sensations, an idea famously discussed by Walter Pater as well.

In 1873, Pater, the best known of the literary critics of aestheticism, wrote passionately about the importance of the impression as a sudden sensory moment profoundly meaningful to the artist (poet) in *Studies in the History of the Renaissance*. Having already maintained that the route for inspiration was "the desire for beauty, the love of art for art's sake" in an 1868 review of William Morris's poetry, he now called upon artists to "burn" with a "hard, gem-like flame."[32] Familiar with the new science on the experience of the impression, he wrote of the "whirlpool" of inner forces of the natural world (perceptions, feelings, thoughts, and memories) brought about by a swarm of impressions—colors, odors, and textures "unstable, flickering, inconstant" as opposed to solid, defined objects.[33] Pater also underscored the importance of the individual personality of the artist in selecting a single, sharp sensory impression. Here Pater seems to be working under the influence of Bain, who had elaborated on perception and phenomenology in *The Emotions and the Will* (1859). Consciousness was stimulated by a series of impressions, which were best distinguished from one another through novelty or surprise. Bain wrote, "The brain is more sharply stimulated . . . by reason of novelty of the impression . . . distinct discrimination . . . of different things that strike us . . . is the very foundation of our intellectual development."[34] In his article, Pater noted the difference between the earlier division of body and mind, now replaced by the "law of life of the human spirit" and the passage from passions to mind. From Pater's perspective, impressions, whether visual, musical, or olfactory, should not be automatic but ardent, and one should then reflect upon them.

Pater also discussed the idea of flux in life, and here Billie Inman has argued that Pater was influenced by Lewes's third article on Darwin in the *Fortnightly Review*, in which he engaged in a discussion on cells and protoplasm as the basis of life.[35] The idea of flux, along with transformations in mental life, can also be directly tied to other Darwinian ideas. Pater might have had evolutionism itself in mind in *Plato and Platonism,* possibly alluding to Darwin by discussing aesthetics according to the universality of "man's kinship to the animal, the material."[36] He also stated: "Darwin and Darwinism for which 'type' itself is not but is only always *becoming.* . . . And the Darwinian theory that 'species,' the identifying forms of animal and vegetable life, immutable though they seem now are fashioned by slow development, while perhaps millions of years go by: well! every month is adding to its evidence. Nay, the idea of development . . . is at last invading one by one as the secret of their explanation, all the products of mind, the very mind itself."[37] In a later essay on Dante Gabriel Rossetti's poetry Pater upheld the importance of corporeal response to beauty and sensuality. Anything mental or spiritual was "sensuous also, or material."[38] These writings, given Pater's central importance to aestheticism, are examples of the familiarity with new scientific ideas among the critics and artists associated with the movement certain of which are discussed below.

Rossetti, Leighton, and Moore

As early as 1859 Rossetti began to diverge from the detailed, drybrush literary watercolors with narrative associations of his earlier Pre-Raphaelite years and started to paint voluptuous women at half length, often in glossy oils. Whether or not titles associate them with the historical past, these women are rendered lifelike, with flushed cheeks, and are so tightly compressed against the picture plane that they appear (in some cases) to nearly burst into the viewer's space. Aestheticist scholar Elizabeth Prettejohn has argued that Rossetti was now "implicitly arguing for a full acknowledgement of the claims of the senses, of the body, and sexuality," which she relates to Rossetti's new interest in the sensuousness of Venetian art, particularly that of Titian, Paolo Veronese, and Giorgione.[39] Tellingly, however, Rossetti's former Pre-Raphaelite colleague William Holman Hunt found a different explanation—he referred to Rossetti's sensual female figures as "advocating as a principle the mere gratification of the eye and if any passion at all—the animal passion to be the aim of art."[40]

The notion of animal passions and gratification of the eye brings Darwin to mind in terms of his arguments on attractants. Jane Munro has discussed a perceived correspondence between Rossetti's new erotic figures and Darwin's interest in the motivating factor of beauty and sexual signaling within evolutionary theory by contemporary critics.[41] Rossetti's former mentor, the eminent art critic John Ruskin, was one of Darwin's earliest detractors in the British art community, calling attention to the amoral and even ruthless implications of the scientist's ideas.[42] Paradoxically, Ruskin's protests explicated evolutionary ideas to Pre-Raphaelite and other artists who moved in a direction that he did not approve of; namely, aestheticism, and this may well have been a motivating factor for Rossetti as he increasingly turned to sensual female figures.

By the mid-1860s Rossetti's new paintings often juxtaposed women and flowers in ways that seemed to move beyond the merely symbolic, although his brother, the writer and critic William Michael Rossetti, referred to them as "beautiful women with floral adjuncts."[43] Beginning in 1862 Darwin's botanical books had outlined sexual reproduction in plants as well as the adaptive strategies of alluring color and form in flowers as attractants for birds and insects in terms of pollination and seed dispersal. Though Darwin's ideas drew many enemies, his botanical publications helped make the case for evolution and sexual signaling in the natural world, ideas alluded to in Rossetti's paintings. Rossetti too emphasized the sexual parts of plants. In *Blue Bower* of 1865 (plate 6), free-floating white passionflowers to the right and left of the sensual head reveal the stigma, or female part, of the flower at its greatest extension—in fact, distended to an almost impossible degree, ready for pollination. The very notion of a "bower" has romantic and botanical implications; humans and flowers are found together in an arena of erotic associations. The blue of the backdrop in this painting is as intense as the blue cornflowers in the foreground, a flower which may be a play on the name of the model, Rossetti's mistress Fanny Cornforth.[44]

The originary connection between flowers and females is suggested by the background of *Blue Bower*. It is imprinted with the prunus blossom of blue-and-white china, popular during the period of aestheticism, but the specific treatment of the pattern suggests fine membranes dividing floating blossoms as if they are contained in cells infused with protoplasm.[45] Unifying protoplasm (in plants and humans) was a recent discovery popularized by Darwin follower Thomas Huxley and others, and its reference in this painting functions as a reminder that visual stimulants—whether of flowers or women—had an underlying organic basis throughout all of nature.[46]

Venus Verticordia (1863–1868, Russell Coates Museum), a rare nude by Rossetti, includes a profusion of flowers. In the honeysuckle at bottom, bright stamens and anthers—sexual parts—project clearly. Ruskin, who considered himself an amateur botanist, characterized the honeysuckle as "licentious" in his botanical *Proserpina*.[47] He also wrote negatively to Rossetti about the appearance of the flowers. According to Ruskin, the flowers were "awful—I can use no other word—in their coarseness:

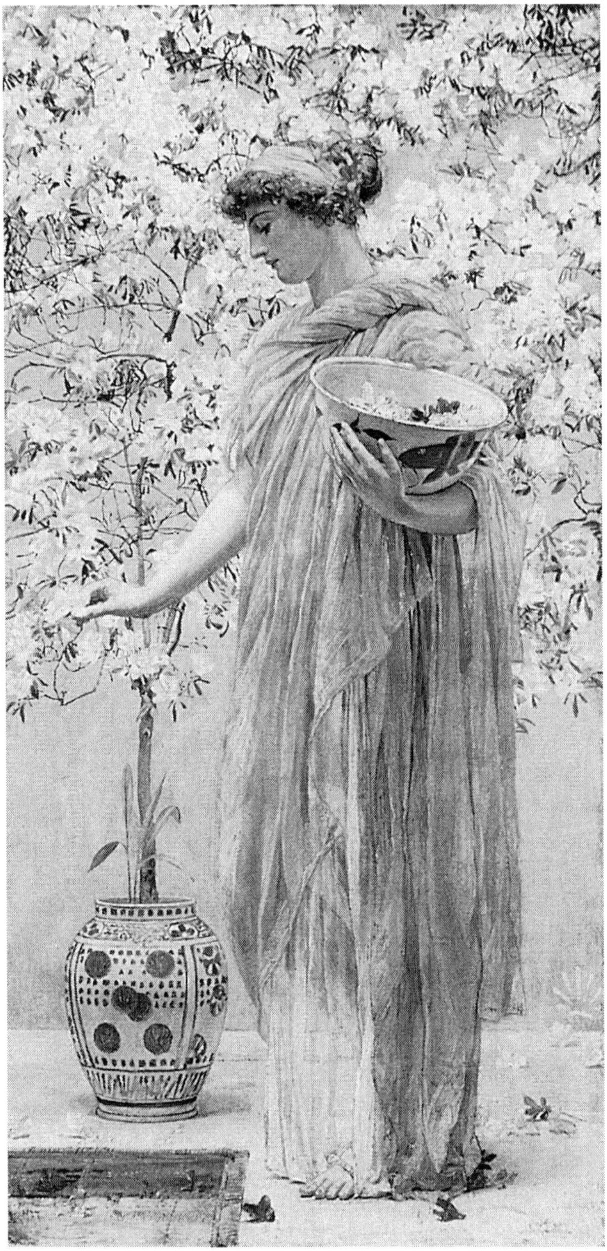

Fig. 7.1. In *Azaleas* harmonies of color and form were interpreted in the contemporary period as exercises in the beautiful. Painterliness and variability of brushstrokes contribute to sensory effects. Albert Moore, *Azaleas*, 1868. Oil on canvas, 78 × 39 1/2 in. (198.1 × 100.3 cm). Collection and image © Hugh Lane Gallery, Dublin (Reg. No. 5).

showing enormous power; showing certain conditions of non-sentiment which underlie all you are doing now."[48] Ruskin had written extensively about plants in the first (1846) and last (1860) volumes of his *Modern Painters*.[49] Flowers were to be connected to a morally healthy art, to spiritual and physical truths, and not to sensory pleasure.

Frederic Leighton included references to that most aesthetic and much discussed beast, the peacock, in paintings that foregrounded the sensory potential of physical allure. In *Nanna* (*Pavonia*) (*pavonia* being Italian for peahen) of 1859 (held in the Royal Collection), a fan of peacock feathers surrounds the head of the lovely bust-length sitter. Four years later he painted *A Girl Feeding Peacocks* (1863, private collection), in which the physical surface, including various textures of satin and feathers, is carefully explored. This is one of the paintings William Michael Rossetti had in mind when he praised Leighton's work as representing "beauty for beauty's sake, colour, light, form, and choice details for their own sakes, or for beauty's."[50]

The aesthetics of allure in animal and plant life was one aspect of the evolutionary framework of the eye that arguably influenced aesthetic artists, but another was a general sensory environment in which the ability of the eye to grasp specific details is demoted. Moore, who preferred complete forms of full-bodied, statuesque, or languorous females in diaphanous classicizing gowns, juxtaposed his figures with flowers or fruit, but here the tones of the entire environment and the figures themselves are often muted and similar. In this case, color harmonies in garments and the organic world echo one another and blend. The eye is not drawn to particular attractants; instead, the entire painting is equally sensory.[51] One such example is *Azaleas*, of 1868 (fig. 7.1), which features a single large-scale figure next to a Chinese vase containing the titular azaleas. Algernon Swinburne, an aesthetic poet and art critic, wrote of *Azaleas* that the "painting is . . . an exclusive worship of things formally beautiful. . . . The melody of colour, the symphony of form is complete . . . its meaning is beauty; and its reason for being is to be."[52] Despite the harmony of colors and lines, the variation in painterliness across the painting calls attention to itself. The azaleas are made up of multiple, broken brushstrokes that gives them a sense of liveliness, whereas the woman's drapery is created with elongated strokes.

The lusciousness and materiality of Rossetti's brushstrokes in his new paintings and his glazing and layering techniques; the attentiveness to fabric, feather, and flesh by Leighton; and a general sensory environment in paintings by Moore can be identified as part of the transformation from Pre-Raphaelite painting to that of aestheticism. The materiality of paint calls attention to sensory experience as opposed to the highly

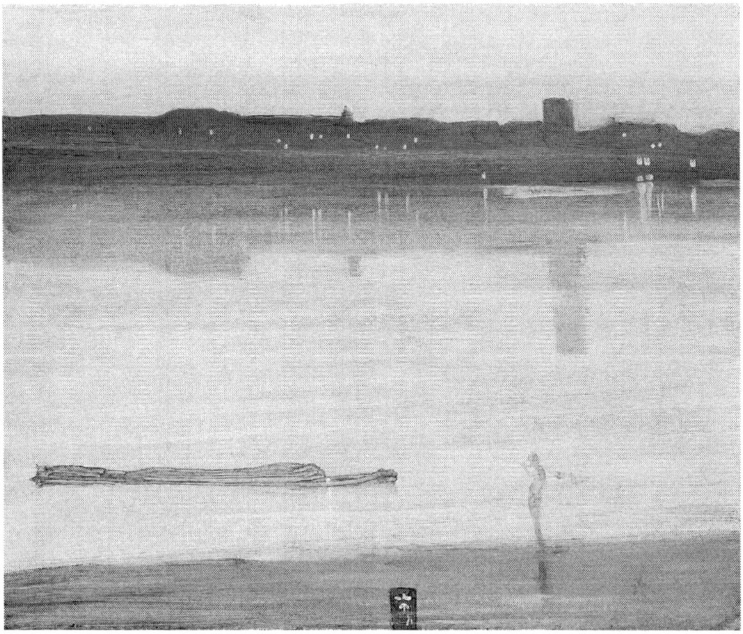

Fig 7.2. *Harmony in Blue-Green—Moonlight* captures atmospheric effects along the Thames. A growing interest in musical analogies led the artist to change the title to *Nocturne—Blue and Silver – Chelsea*. James McNeill Whistler, *Harmony in Blue-Green—Moonlight (Nocturne: Blue and Silver, Chelsea)*, 1871. Oil on Canvas. 19 3/4 × 24 in (50.2 × 60.8 cm). Tate Britain. @Tate, London / Art Resource, NY.

detailed and narrative content of earlier Pre-Raphaelite or nonmodernist Victorian painting.

Whistler

Whistler, like his close friend Moore, was interested in color harmonies and obscurity, though unlike Moore he often turned to the landscape. Art historian Christopher Newall cites the 1860s as the period in which landscape painting once attached to the Pre-Raphaelite school moved from detail to the elusive, as in John Inchbold's *Venice from the Public Garden* (1862–1863, private collection).[53] Whistler, who moved to Great Britain at the beginning of the 1860s, began to create works of art wherein forms were simplified and general atmospheric effect foregrounded, such as *Brown and Silver: Old Battersea Bridge* (1865, Addison Gallery of American Art, Andover, Massachusetts). Such paintings are closer to the idea of the process of seeing and the production of effect in a specific situation than to a factual record of that which was momentarily present. In 1866 Whistler began to paint veiled, atmospheric harbor scenes at Valparaiso, Chile, termed "moonlights." By 1871 he was applying atmospheric effects to scenes along the Thames with even more reductive means. These effects were often based on the perceived beauty of a sooty London fog at night from various locales along the river.[54] Thinned pigment and broad brushstrokes seem to envelop the river, barge, and lone spectator in vapor in *Harmony in Blue-Green—Moonlight* (1871, fig. 7.2). Beyond demonstrating a deep appreciation for the harmonious, Whistler's vaporous new style, in which one can almost imagine the dampness of the fog on the skin, is consistent with the increased visibility of theories of physiology regarding corporeal effect and resulting states of mind.[55]

Whistler's foregrounding of the atmospheric and his obscuring of detail in his outdoor scenes, along with the suggestion of musical modes of experience in his titles, may demonstrate the kind of physiological awareness in art promoted by James Sully. In April of 1871 Sully, an emerging physiological aesthetic theorist who followed Bain and Darwin, published an article on the aesthetics of pleasure in the *Fortnightly Review.* Sully considered, among other things, the value of obscurity in art. He wrote of the artist's love of the indefinite, including haziness, for it allowed for a play or freedom of the imagination for the individual personality and at the same time a connection of minds.[56] He also referred to Bain and the variety of impressions experienced in a given day, along with discernment, ideas that could be readily connected to Whistler. With regard to Whistler himself, however, Sully claimed later in the decade that his "misty nocturnes" were *too* obscure for the "urgings of the eye" in perception.[57]

Having found out about Helmholtz through the lectures of Tyndall, Sully went to Berlin at the end of 1871 to work with the German scientist, studying the cognitive and sensory psychology of hearing and vision and the aesthetics of musical theory and color. By this time Helmholtz had demonstrated his interest in art in "On the Relation of Optics to Painting," published in English in 1871.[58] This essay largely focuses on color, light, and optical experience (such as binocular vison), but Helmholtz approaches the subject from "the physiological manner in which our senses originate, how impressions from without pass into our nerves, and how the condition of the latter is thereby altered."[59] He too foregrounded the importance of the artist's preliminary visual impression. Like his mentor Helmholtz, Sully was interested in musical harmonies, which he likened to color harmonies in an article on music in the *Fortnightly Review* in April of 1872.[60] Musical comparisons to aesthetic art were frequently mentioned by artists and critics of aestheticism, since in both forms, narration, and detail were subordinated to abstraction. Indeed, Sully's theories are not unlike Whistler's own aesthetics, in which he emphasized the effects of color harmonies with musical implications. Although Whistler's synesthetic titles date back to 1867, he first began to give his moonlit paintings the musical title of "Nocturnes" in 1872 and reconsidered the titles of other paintings as well. The Thames painting *Harmony in Blue-Green—Moonlight* (1871), for instance, was renamed *Nocturne: Blue and Silver—Chelsea.*[61]

Katharine Lochnan has focused on Whistler's interest in Helmholtz after his infamous trial of 1878. She has discussed his reworking of earlier etchings of the Thames (1859) in the late 1870s according to Helmholtz's theory of direct and indirect vision as postulated in the *Handbook on Physiological Optics.*[62] Accordingly, Whistler portrays objects as if in "peripheral vision," fading away from sight. Later in life Whistler told his follower Mortimer Menpes that he left areas around the periphery of his main point of interest less finished, putting them in with "one broad sweep."[63] In "The Eye as an Optical Instrument," Helmholtz stated:

> The image which we receive by the eye is like a picture, minutely and elaborately finished in the center, but only roughly sketched in at the borders. But although at each instant we only see a very small part of the field of vision accurately, we see this in combination with what surrounds it, and enough of this outer and larger part of the field, to notice any striking object, and particularly any change that takes place in it . . . this we may call direct vision, applying the term indirect to that exercised with the lateral parts of the retina.[64]

While certain of Whistler's experiments based in Helmholtz's discussions might date to the later 1870s, the importance of the scientist's ideas for artists was made known to artists through Sully throughout the 1870s.

Sully published the book *Sensation and Intuition* in 1874, which elaborated upon the ideas of Helmholtz to his British audience along with those of Herbert Spencer; it was well received by Darwin, Bain, and Lewes.[65] In his writings, Sully explored a neurological loop from eye to mind via the sensations thereby produced, resulting in perception and involving memory (associations). He followed Spencer on the evolution of aesthetic taste. Refinement of feeling needed to be upheld over "primitive and universal gratifications," which are in essence mimicry of the natural world, a lesser pursuit.[66] Complex, imaginative artworks corresponded to refined sensibilities, which had developed over time as the result of intellectual and emotional activities. The person with the greatest mental development experienced the most pleasure—as opposed to "savages," who responded to the momentary pleasure brought by color.

Sully followed Spencer and Bain on the equation of the quality of the aesthetic with the amount of pleasure given.[67] Moreover, harmony, which involved the subordination of detail to a central emotion or idea, was an important endeavor, a concept that could be applied to Whistler or Moore.[68] Sully believed that qualities of "simplicity and natural force" were important to artists, writing in *Sensation and Intuition*, "Such emotions as the joyous elation which is begotten of morning light [or] dewy air . . . have a high aesthetic value as old, widely-diffused, and permanent attributes of human nature," other ideas that could be readily applied to Whistler.[69]

Sully followed Darwin's evolutionism, but he disagreed on many points with the scientist on psychology, feeling that his approach to the animal world was too anthropomorphic, and that heredity and evolutionism were perhaps overshadowing the importance of individual psychology. In the final chapter of *Sensation and Intuition*, Sully addressed art and aesthetics, arguing that while art was the result of evolution and a way of dispersing surplus energy, as Spencer had maintained, individual experience was also important. The individual brought a range of experiences and even a hereditary or national bias to aesthetics, and thus "aesthetic impression presents itself as eminently inconstant and relative."[70] For Sully, aesthetics were based on all of the following: sensory response, such as to color or form; Bain's multiple impressions and the individual's response to novelty; aspects of Darwin's evolutionism; and Helmholtzian ideas on vision. To this he added the stimulating intellectual activity of intuition; Spencer's advance of civilization and mind in terms of an increasing appreciation for visual complexity and refinement of feeling; and associations through memory.

Aestheticism emerged at a time when the Pre-Raphaelite idea of "truth to nature" no longer seemed like a viable pursuit, given the psychological and physiological complexity of impressions of the world on the individual—now understood as entirely subjective. By the 1860s discussions surrounding Darwinian ideas on vision and its implications regarding aesthetics were circulating. The eye had developed as it had in the context of the overall sensory system and according to older ideas concerning the poles of pleasure and pain. Beauty and sexual selection increasingly became preoccupations of Darwin in his discussions of evolutionism and its application to humans. Physiological psychologists thought about not only beauty in the abstract but art as well, theorizing about sensory response to the world and how this could be understood in terms of the art object. And Spencer's evolutionary ideas on the pursuit of art for its own sake, which resonates with and indeed underscores the new direction in art, is in keeping with the dandified atmosphere surrounding the aesthetes. While subjects of aestheticism do not moralize or have a narrative function, they can be either modern (Whistler's views of the industrialized banks of the Thames) or historical (Moore's classical women), the unifying principle having to do not with content but with sensory aspects of form, line, color, or the environment, especially concerning the experience of "the beautiful." In following new and even radical directions in science on physiology, psychology, and the visual, the artists of aestheticism responded to a reconceptualization of the relationship of body to mind and of the sensory eye to its environment that set a course for modernist experiments to follow.

CHAPTER 8

DARWINIAN AESTHETICS AND AESTHETICISM IN JAMES MCNEILL WHISTLER'S PEACOCK ROOM

Caitlin Silberman

Visiting James McNeill Whistler's nearly completed Peacock Room (1876–1877) in February 1877, a reviewer for the journal *London* described the six large birds painted on the walls and window shutters as "not the common fowls of the farmyard, but creatures compounded of pheasant and phoenix—such things as peacocks may come to be in the course of natural and sexual selection" (fig. 8.1, plate 7).[1] That is, the artist had transformed the mundane reality "of the farmyard" into something beyond or above nature: in a literal sense, supernatural. The contrast between "natural" birds and Whistler's unnatural or supernatural ones is reinforced by the reviewer's invocation of "natural and sexual selection" as transformative forces that might someday generate something like Whistler's peacocks, but as yet have failed to do so. In linking peacocks and Darwinian evolution, moreover, the review attests to the pervasive connection between the two subjects in late Victorian culture.

Peacocks, and especially the ornate feathers in the trains of male peacocks, figured prominently in a range of cultural, artistic, and scientific discourses beginning in the second half of the nineteenth century. An enormous number of British artists employed peacocks and their feathers in painting and the decorative arts from the late 1850s onward.[2] These birds were activated as symbols of beauty, vanity, aristocracy, and the resurrection, as Orientalist invocations of India, China, and Japan, and as emblems of the aesthetic movement and, later, art nouveau.[3] Alongside

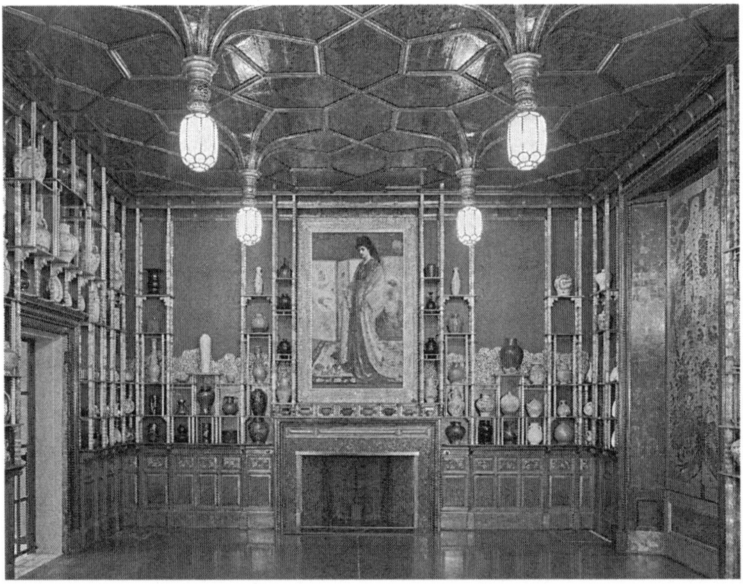

8.1. View of James McNeill Whistler's Peacock Room, facing north. *Harmony in Blue and Gold: The Peacock Room*, 1876–1877. Oil paint and gold leaf on canvas, leather, mosaic tile, and wood. 166 × 241 1/2 × 404 in (421.6 × 613.4 × 1026.2 cm). Freer Gallery of Art, Smithsonian Institution, Washington, DC: Gift of Charles Lang Freer, F1904.61.

its rise as an artistic motif, the peacock played a prominent role in debates over the plausibility of Charles Darwin's mechanistic theories of evolution: natural selection, first proposed in *On the Origin of Species* in 1859, and sexual selection, mentioned briefly in *Origin* but fully elaborated in *The Descent of Man, and Selection in Relation to Sex* (1871). It was the peacock's prominence as an example of a species shaped by sexual selection that cemented the association between peacocks and Darwinism from the 1860s onward. This association not only pertained among men of science but spread quickly to popular culture, such that magazines like *Punch* might use peacocks and peacock feathers as visual shorthand for Darwinian theory in cartoons.[4]

While the twenty-first century has seen a scholarly turn toward examining Darwin's impact on American and European art and visual culture—an impact described by Linda Nochlin as the "Darwin effect"—attempts to assess the place of Whistler's Peacock Room within a culture shaped by Darwin have been cursory.[5] The omission is surprising, as the Peacock Room may be the single most famous and influential work of late Victorian interior design, a touchstone for the aesthetic movement and later for art nouveau. If we are to take seriously Barbara Larson's assertion in *The Art of Evolution* (2007) that shifting "philosophies about human behavior, society and its institutions, and politics in light of Darwinism are embedded in much of the art and graphic production of the modern period," the Peacock Room merits analysis as an example, even a central example, of the ways in which these shifts played out in visual culture.[6] In chapter 7 of this volume, Larson implicates Whistler's misty landscapes in the context of connections between aestheticism, shifting theories about the evolution of the eye, and Victorian explorations of the limits of objective vision. In this chapter, I suggest that James McNeill Whistler's Peacock Room, along with related artworks created for Whistler's great patron Frederick Richards Leyland, can be productively read as the artist's intervention in contemporary debates over the origin and use-value of an aesthetic sense.

Whistler's decorative program in the Peacock Room asserted the unique creative power of the artist in the face of a cultural moment in which the origin of beauty was increasingly sought elsewhere. In the 1870s, two competing ways of thinking about the origin and use-value of the aesthetic sense were supremely visible: theologies of nature, epitomized by the writings of the eminent critic (and Whistler's eventual opponent) John Ruskin, in which a divine Creator apportioned beauty for the purpose of delighting and instructing humankind, and Charles Darwin's new theory of sexual selection, in which human and nonhuman animals' aesthetic sense arose as a tool for judging the reproductive potential of prospective mates. For Darwin and his adherents, animals' impulse to reproduce with suitable mates drove the development of an aesthetic sense, which thus served a purely instrumental purpose. For Ruskin, as for many of Darwin's critics, beauty could not be explained by mindless and undirected reproductive pressures, but only as a gift from God. Like other artists associated with the aesthetic movement, Whistler sought to decouple the fine arts from any moral or instrumental purpose, a notion encapsulated in the aestheticist rallying cry of *l'art pour l'art*, or "art for art's sake." In pursuit of this goal, Whistler reworked Ruskinian and Darwinian uses of the peacock motif to argue for beauty as an end to itself, and to stake a claim for artists—not God, not Nature—as the supreme creators of beauty.

Historically, sources for the Peacock Room have been sought primarily in Whistler's engagement with Asian art and the art of his British contemporaries. Here, I attend rather to Whistler's departures from these sources to suggest the ways that the artist engaged contemporary debates over Darwinian sexual selection through the figure of the peacock. In the Japanese artworks with which Whistler was familiar, multiple peafowl appeared together almost invariably in male/female pairs, while Darwin cited peahens' choice of mates as the origin of the peacock's elaborate feathers. Whistler, however, depicted a band of male peacocks displaying for each other. I suggest that the Peacock Room, taken in conjunction with another main element of the room, a painting in progress called *The Three Girls* that Whistler envisioned as the manifesto of his aesthetic theory, models a supernatural world in which beauty is generated by acts of attraction, fertilization, and reproduction divested of Ruskinian moral lessons and Darwin's shackling of beauty to a heterosexual reproductive drive. Moreover, the room—at least in its early stages—also implicated Leyland as the artist's partner in a queer variation on sexual reproduction, in which beauty resulted from the union of artist and patron.

This interpretation draws in part upon the work of Alison Syme, who has argued that fin-de-siècle artists and critics responded to growing scientific understanding of "the intricacies of plant reproduction through pollination" by adopting a "metaphorical equation of painters as pollinators," whether bees, birds, or butterflies.[7] Artists pollinated by hand and brush, like butterflies and bees with their proboscises and birds with their beaks. The artist's act of pollination/fertilization produced not offspring but art. As Syme's research has demonstrated, the art world of late nineteenth-century Europe and America was saturated with metaphorical comparisons between artistic creation and animal fertilization and reproduction, with the artist-as-pollinator metaphor appearing not only in artists' own writings and artworks but also in the popular and specialist press. Syme also notes the wide variety of pollinators—bees, bats, birds, and butterflies—that nineteenth-century artists took as their emblems.[8] Whistler had begun signing his work with a stylized butterfly in the 1850s, but in the Peacock Room the eponymous bird becomes another animal avatar for the artist's acts of fantastic fertilization.

The Evolution of the Peacock Room

To suggest new readings of Whistler's iconic decoration is to enter an already crowded field, as Whistler is among the best-known and most-studied artists of the nineteenth century. The American-born expatriate, trained in France in the 1850s, had established himself as an avant-garde character in the London art scene by the end of the 1850s. He would reside primarily in London through his death in 1903. Today Whistler occupies a place of honor in the art historical canon of three countries that he called home: France, Britain, and America. The Peacock Room has commanded scarcely less scholarly attention than the artist who created it. As Linda Merrill has observed, writers began apologizing for the surfeit of writing on the Peacock Room as early as 1878.[9]

The origins of the space gave little indication of its outsized influence. The Peacock Room began life as the dining room in 49 Prince's Gate, South Kensington, a newly built townhouse in a rising neighborhood that had attracted the attention of Leyland, a shipping magnate seeking a more spacious home in London.[10] Leyland was born in Liverpool in 1831 in obscure circumstances, but had risen to great wealth as a ship owner and stakeholder in the transatlantic shipping trade.[11] In common with many nouveau riche Victorian industrialists, Leyland became a patron of music and the visual arts. In a self-conscious imitation of Renaissance merchant princes, Leyland drew a veil over his humble origins and accusations of mercantile avarice through lavish patronage and displays of taste. Leyland's initial forays as an art collector focused on Old Masters and Romantic watercolor landscapes, but in the 1860s he turned to collecting contemporary Pre-Raphaelite and aesthetic art on a large scale, buying extant works and commissioning new ones. He was especially partial to Edward Coley Burne-Jones and Dante Gabriel Rossetti, purchasing large numbers of their oil paintings, drawings, and watercolors.

It was Rossetti who introduced Leyland to Whistler, around 1866–1867. Leyland, Rossetti, and Whistler moved in many of the same networks of taste and consumption. Beyond their preference for Pre-Raphaelite and aestheticist painting and design, the men shared a passion for Chinese blue-and-white porcelain, a collecting craze that came to be known as "chinamania." Around 1870, Leyland purchased one of Whistler's earliest works of Japonisme, *The Princess from the Land of Porcelain*, also known as *La Princesse du Pays de la Porcelaine* or simply *La Princesse* (1863–1865). Leyland also commissioned many new works from Whistler. Among the earliest and most important of these commissions was a painting to be titled *Symphony in White No. 4: The Three Girls*. This painting was to be the fourth, culminating canvas in Whistler's series of Symphonies in White, formally innovative paintings featuring languid young beauties in white dresses.[12] Whistler built up lofty expectations for *The Three Girls* in himself and his patron that the painting would be the summation of Whistler's artistic theory. It was intended to be the final word in Whistler's career-long argument that Nature, unaided, creates beauty only accidentally and imperfectly, and that true, complete beauty could only be achieved by the intervention of an artist. This view was most fully articulated in Whistler's *Ten O'Clock Lecture* (1885), in which the artist declared: "Nature contains the elements of color and form of all pictures—as the keyboard contains the notes of all music—but the artist is born to pick, and choose, and group with science, these elements, that the result may be beautiful—as the musician

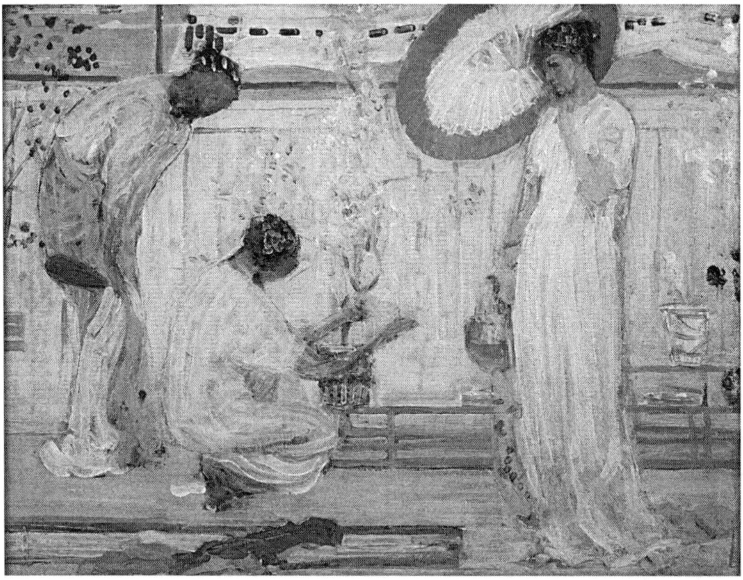

8.2 In 1868, Whistler gave Leyland an oil sketch for *The Three Girls*. James McNeill Whistler. *The White Symphony: Three Girls*, 1868. Oil on millboard mounted on wood panel. 18 1/4 × 24 1/4 in (46.4 × 61.6 cm). Freer Gallery of Art, Smithsonian Institution, Washington, D.C.: Gift of Charles Lang Freer, F1902.138.

gathers his notes, and forms his chords, until he brings forth from chaos, glorious harmony."[13]

Accordingly, *The Three Girls* pits inadequate Nature against the superior creative power of Art. The small oil sketch of *The Three Girls* (fig. 8.2) that Whistler presented to Leyland in 1868 indicates the intended composition. The scene takes place on a sheltered balcony, set up as a greenhouse. White fabric screens protect a collection of potted plants from the bright blue skies beyond; these blocks of white also flatten the picture plane such that the scene is almost entirely composed of a frieze-like foreground. The white-robed three girls are lined up in this space. The girl on the left leans forward to observe the middle figure, who is crouching to prune a potted cherry tree. The righthand figure stands and observes the action, a Japanese parasol in one hand and a water jug in the other. As Merrill and Robyn Asleson have observed, "the hot-house imagery" of *The Three Girls* conveys the message that "if art is to flourish, it must be sheltered from the real world, an environment inhospitable to ideal beauty."[14] The blossoming cherry tree, a well-known symbol of youth and beauty across cultures, is attractive enough as a raw material, but *The Three Girls* suggests that only human intervention can perfect what nature began. The girls, picking and choosing among the choicest branches and sprays of flowers, are thus stand-ins for the artist—and perhaps for the patron, whose largesse allows the artist to pursue his goals.

Whistler found himself overwhelmed by the task of completing *The Three Girls*. Whistler's mother, Anna McNeill Whistler, wrote to Leyland in the spring of 1869 to apologize for the delay in delivering the painting, explaining that her son "tried too hard to make it the perfection of Art, preying upon his mind unceasingly it has become more & and more impossible to satisfy himself."[15] In anticipation of the final canvas, or as a distraction from it, in 1872–1873 Whistler created a custom frame for *The Three Girls*. Measuring an imposing 73 by 55 inches, it is adorned with hawthorn blossoms, Whistler's butterfly signature, and a musical passage. The inclusion of lines from Franz Schubert's *Six moments musicaux* (1828) was a nod to Leyland's love of music and an homage to the collaboration between patron and artist.[16] Yet the frame remained empty, even as Whistler continued to collect Leyland's advance payments for *The Three Girls* for years on end.[17]

By 1873 Leyland's collection had outgrown his homes in London and Liverpool. Expanding to South Kensington not only eased this strain but also offered Leyland the opportunity to craft a consummately artistic environment in the aesthetic mold, fulfilling his "dream of living the life of an old Venetian merchant in modern London."[18] The exterior of 49 Prince's Gate was Italianate, but the interiors would be decorated in the eclectic style that was the highest mark of contemporary artistic fashion. The design of the dining room that would later become the Peacock Room was initially entrusted to Thomas Jeckyll, an expert in the fashionable Anglo-Japanese idiom.[19] Jeckyll was charged with designing a functional dining room that would also showcase some of Leyland's most prized possessions: his collection of porcelain and two of his Whistler paintings. *La Princesse* would hang on the north wall, while the south wall was reserved for the ever-forthcoming *Three Girls*.

Jeckyll modeled the dining room after early modern European porcelain display rooms (*Porzellenkammer*) and chinoiserie garden pavilions.[20] In accordance with aesthetic interior design, Jeckyll installed a walnut-wood dado, or wainscot, over the lower third of the walls. The north, south, and west walls were covered with antique leather wall hangings that were gilded, embossed, and painted with twining red roses, while the east wall (facing the garden) was dominated by three tall windows, each with a pair of floor-to-ceiling shutters in the same plain walnut as the dado.[21] Along all available wall space, Jeckyll installed an

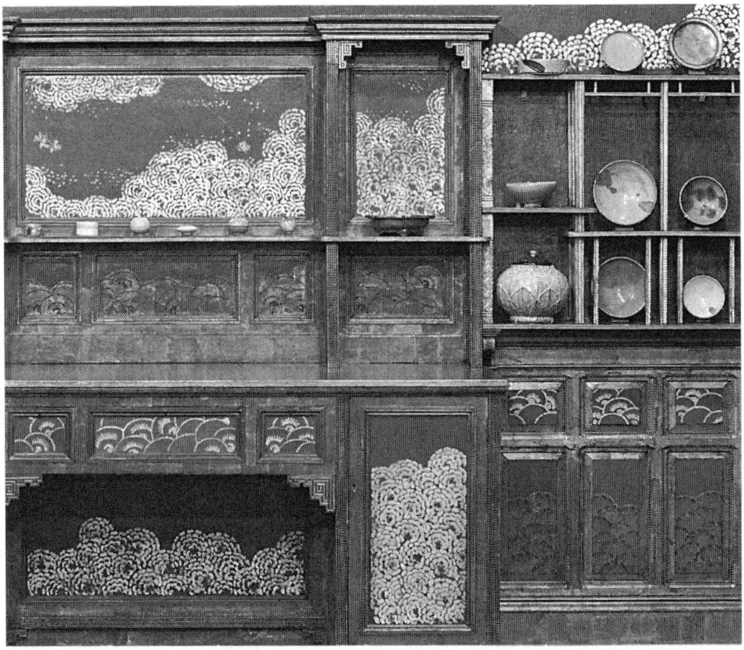

8.3 Designs on the Peacock Room's wall panels imitate peacocks' eyespots and breast feathers. James McNeill Whistler. *Harmony in Blue and Gold: The Peacock Room*, 1876–1877. Detail: panels, south wall. Oil paint and gold leaf on wood. Freer Gallery of Art, Smithsonian Institution, Washington, DC: Gift of Charles Lang Freer, F1904.61.

intricate framework of carved walnut shelves to hold Leyland's porcelain collection.[22]

In the spring of 1876, the dining room nearly complete, Jeckyll consulted Leyland and Whistler about the color to paint the shutters and doors. Whistler suggested gilding them, to which Leyland assented. Jeckyll's faltering health soon forced him to stop work, and Whistler took up the gilding himself. He grew concerned that the roses on the leather wall hangings clashed with *La Princesse* and offered to touch up the leather with yellow and blue accents, and to paint the dado and cornice panels with a pattern based on a Japonesque ironwork pattern decorating one of Jeckyll's doors.[23] The ironwork motif can be described as resembling scales or waves, but Whistler came to link it with peacock feathers. Whistler later added motifs based on the ocelli, or eyespots, of peacocks' train feathers, and a pattern of overlapping scales filled in with more delicate brushstrokes, resembling softer and fluffier plumage (fig. 8.3). Artist and patron appear to have reached a verbal agreement

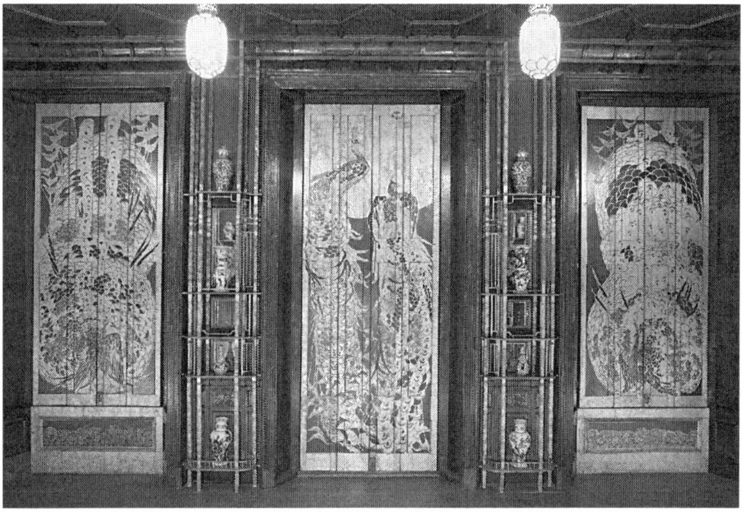

8.4 Whistler painted three full-length shutters with golden peacocks. James McNeill Whistler. *Harmony in Blue and Gold: The Peacock Room*, 1876–77. Detail: shutters, east wall. Oil paint and gold leaf on wood. Freer Gallery of Art, Smithsonian Institution, Washington, DC: Gift of Charles Lang Freer, F1904.61.

for a payment of five hundred pounds for the addition of these patterns to dado and cornice, a project expected to last only a few days.[24] Satisfied that his dining room was virtually complete, Leyland returned to Liverpool.

As days of work stretched into months through late 1876, Whistler continued to correspond with the Leyland family, reassuring them that his continuing improvements would make the Leylands the envy of artistic London. Whistler's feather motifs spread to the remaining dado panels and all across the walnut ceiling, and he gilded Jeckyll's walnut shelving. In a more daring intervention, the artist painted four towering peacocks in Prussian blue on the west wall's three floor-to-ceiling gilt shutters (fig. 8.4). This was the first addition of figural art to the decoration of the room, and made far more explicit the room's growing message of the superiority of Art to Nature. But Whistler did not inform Leyland of the addition, and when the patron made an impromptu visit in October 1876, Leyland was shocked and displeased by the artist's presumption. Despite his dissatisfaction with Whistler's changes, Leyland requested that Whistler finish the work and send him the bill.

Rather than simply putting the final touches on the room, however, Whistler redoubled his efforts. He painted over Jeckyll's floral-patterned leather wall coverings in coats of Prussian blue, then used the same blue to cover much of the gilt dado paneling. He deepened the room's hues still further by applying green glazes to the ceiling and wall panels. By this point all of Jeckyll's contributions were either gilded or painted over. By the close of 1876, the room was fully transformed from an airy garden pavilion to a lacquered, gilded chamber.

In early 1877, judging his work complete, Whistler led artists and journalists on tours of the Leyland's home. This imposition infuriated Leyland, as did Whistler's final bill: the artist requested the large sum of two thousand guineas, the old-fashioned coinage that denoted an exchange between gentlemen. Leyland retaliated by paying Whistler only one thousand pounds. Not only did this cut the artist's pay to under half, as pounds were worth five percent less than guineas, but pounds were the currency in which craftsmen, not artists, were paid.

Whistler was infuriated by what he perceived as Leyland's greed and a slight to his own status. Leyland, Whistler decided, had failed to appreciate his artistry, and failed to understand how the Peacock Room served as a tribute to Leyland as a patron of the arts. Whistler accepted Leyland's thousand-pound check, but he embarked on additional changes that would undermine the original, laudatory message of the room. Artist and patron continued to squabble, and in March 1877 Leyland finally barred Whistler from entering Leyland's home.[25] Whistler never saw the Peacock Room again.

Why Peacocks?

In a self-published leaflet distributed to viewers touring the room in February 1877, Whistler declared that "the Peacock is taken as a means of carrying out" the "arrangement" of the room.[26] However, Whistler's contemporaries and later critics recognized a range of artistic precedents for the Peacock Room, including Chinese blue-and-white porcelain, Japanese art, rococo and modern aesthetic decorating practices, and paintings by Whistler's fellow artists.[27] Peacocks and peacock-feather fans were familiar to British audiences as signifiers of Orientalist exoticism and luxury, and were frequently used as props in Orientalist fantasies of harems and odalisques.[28] Before the 1850s, British visual culture linked peacocks and their feathers most closely with India and the Near East, but the proliferation of Japanese prints and objects d'art in the West after midcentury established a link between peacocks and Japan in British culture.[29]

Whistler was familiar with a range of Japanese peacock imagery in the art that he collected or viewed in his friends' collections. He owned books of Japanese prints as well as hanging wall scrolls and standing screens, including the bird-patterned screen in the background of *La*

Princesse and at least one scroll with a single peacock.[30] Whistler would inevitably also have seen peacock-painted Chinese porcelain, possibly even in Leyland's own collection. Peacock feathers also proliferated as decorative elements in aesthetic interiors, including as popular "filler" in vases. Whistler could also look to any number of peacocks and peacock feathers in the art of his direct contemporaries.[31]

Alongside their popularity in artistic circles, peacocks were a primary symbol of an area of debate that had recently aroused extensive argument among men of science and expanded into popular consciousness: Charles Darwin's theory of evolution by sexual selection. Briefly proposed in *On the Origin of Species* in 1859, Charles Darwin expanded on the theory in *The Descent of Man, and Selection in Relation to Sex* (1871).[32] This theory had aroused extensive opposition due to Darwin's suggestion that all animals' predilection for the beautiful, including humankind's own aesthetic sense, was a mere byproduct of the heterosexual reproductive drive.

In a few paragraphs in the first edition of *On the Origin of Species*, Darwin briefly introduced sexual selection as a "less rigorous" force than natural selection, relying "not on a struggle for existence, but on a struggle between the males for possession of the females."[33] Violent competition for mates might lead to males evolving weaponry and armor to give them an edge, such as stags' antlers and cocks' spurs. Males could also develop features that encouraged females to choose them, like male birds' "gorgeous plumage," melodious songs, and elaborate dances.[34] The peahen's choosiness was among the few concrete examples of sexual selection presented in *Origin*. Darwin pointed out that bird owners "well know that [birds] often take individual preferences and dislikes," citing an acquaintance's pied peacock that "was eminently attractive to all his hen birds."[35] The naturalist concluded by likening sexual selection to the artificial selection that shaped domestic fowls: "If man can in a short time give elegant carriage and beauty to his bantams, according to his standard of beauty, I can see no good reason to doubt that female birds, by selecting, during thousands of generations, the most melodious or beautiful males, according to their standard of beauty, might produce a marked effect."[36]

This fleeting introduction proved unsatisfactory to critics, and the male peacock's elaborate train quickly became shorthand for the failings of Darwinian theory. As critics pointed out, the peacock's bright, heavy train makes the bird both visible and cumbersome, an easy target for predators. This feature is also "costly" to maintain in terms of the extra nourishment required, for what seems like so little potential advantage. Even if females found this sort of plumage attractive, how could it have arisen in the first place? Darwin's assertion that female animals could evaluate males' qualities and select mates based on their aesthetic preferences also ruffled Victorian feathers, rousing anxieties over the theory's implications for modern human gender relations. Reviewing *On the Origin of Species*, Henry Charles Fleeming Jenkin of the University of Edinburgh invoked the uneasy relationship between avian sexual selection and human courtship when he protested "how impossible it is to conceive" that a peacock's tail "could have been elaborated by the minute and careful inspection of rival gallants or admiring wives."[37]

Other critics argued that the beauties of nature were so improbable that they could only have come about through divine intervention. George Campbell, Duke of Argyll, for example, used the peacock to argue against Darwinian evolution in *The Reign of Law* (1867), the first full-length published riposte to *Origin*. *The Reign of Law* took a natural theological viewpoint, arguing that beauty arose through natural laws established by God. Argyll relied upon the anthropocentric argument that beauty was created for human appreciation and human benefit, a viewpoint that he shared with Ruskin. Campbell's conclusion that peacock feathers "are wonderful examples of ornament, but they do not represent anything except their own harmonies of colour," however, skirts rather close to the aesthetic argument for nonpurposive beauty.[38]

With this deluge of criticism, Darwin recognized that his explanation for the peacock's hyperbolic tail would have to be revisited. The problem of the peacock was grating; in a letter of 1860, Darwin confided to fellow naturalist Asa Gray, "The sight of a feather in a peacock's tail, whenever I gaze at it, makes me sick!"[39] With the publication of *The Descent of Man* in 1871, Darwin must have felt that he had developed his theory of sexual selection well enough to fend off critics. *Descent of Man* piles example upon example of sexually dimorphic features that might be explained by reference to sexual selection. Visual evidence was part of this strategy, and *Descent of Man* features several illustrations of male birds displaying for the benefit of potential mates. The second edition of the book added an image of a close relative of the peacock, the argus pheasant, presenting the eyespots on its wing feathers in an attempt to impress a mate.[40]

The potential implications of Darwinian sexual selection for the human aesthetic sense were widely recognized. As Larson has also noted in her essay in this volume, Ruskin was particularly repelled, and was a vociferous critic of both Darwin and Whistler in the 1870s and 1880s.[41] In lectures given in 1873 while serving as Slade Professor of Art at Oxford University, Ruskin reiterated his belief that birds were gifts from the Creator, signs of God's love and symbols exhorting humankind to behave morally.[42] Ruskin wrote a mocking parody of the "Darwinian system," in which "if you fasten a hair-brush to a mill-wheel . . . within continual hearing of a steam-whistle," the hairbrush will eventually "fall in love with the whistle; they will marry, lay an egg, and the product will be a nightingale."[43] Ruskin understood perfectly well that this was an outrageous exaggeration, but he was lashing out against what he saw as Darwinism's threat. As scholars have long noted, Ruskin objected to Darwin's willingness to muddle the boundaries that had set humans apart from nonhuman animals as uniquely conscious, ethical beings.[44] Decades earlier, in *Modern Painters II* (1846), Ruskin proposed that human were unique in possessing *theoria*, or the "theoretic faculty," the ability to distill morality from the beauties of the natural world. Darwin's argument that humanity's aesthetic sense evolved from the reproductive impulses of "lower" creatures threatened to undermine claims to a unique theoretic faculty. Moreover, to deny that animals' beauty was a gift from God, intended as a conduit toward divinity, was to court moral, cultural, and social degeneration.[45]

In its claim that beauty needed no moral purpose, Whistler's aestheticism was similarly threatening. Ruskin's famous statement in *The Stones of Venice* (1851) that "the most beautiful things in the world are the most useless; peacocks and lilies, for instance," was adopted by champions of l'art pour l'art, for whom peacocks and lilies became celebrated symbols of "useless" beauty.[46] The critic was outraged by this, and in an appendix to *The Stones of Venice* he explained that he meant "useless" only as "inapplicable to the service of the body," that is, having little utilitarian purpose to humans.[47] "Roast peacocks" may indeed be "unwholesome," Ruskin explained, and "dried lilies made bad hay: but I do not think peacocks superfluous birds, nor that the world could get on well without its lilies. Or, to look closer, I suppose the peacock's blue eyes to be very useless to him . . . yet I do not think there is a superfluous eye in all his tail."[48] The peacock's beauty, in Ruskin's argument, affords no benefit to the bird, but its ability to provide moral nourishment to humans secures it from superfluity.

The potential origin of human aesthetic feelings in Darwinian evolution was taken up vigorously by popular science writer Grant Allen, who rejected Ruskin's argument that the human aesthetic sense could never have arisen through naturalistic mechanisms. In the late 1870s Allen published several articles and two books on "physiological aesthetics," a concept he coined as a naturalistic explanation for why humans react positively or negatively to certain sights, smells, sounds, textures, and flavors. All aspects of the human aesthetic sense, Allen argued, from a general taste for bright colors to a specific preference for one artist over another, could be traced directly back to the effects of sexual selection.[49] In his first full-length book on the subject, *Physiological Aesthetics* (1877), Allen boldly took aim at John Ruskin and natural theology in his first sentence: while "Professor Ruskin" dismisses questions of "'why we receive pleasure from some forms and colours and not from others,'" these are the questions "which this little book asks, and, I hope, answers."[50] Allen straddled divisions between a man of science and a man of letters, publishing articles in the eminent scholarly journal *Nature* as well as in popular middle-class monthlies like the *Cornhill Magazine*.[51] It was a close friend of Allen, the author and anthropologist Andrew Lang, who linked Whistler's peacocks with Darwinian evolution—describing the golden birds as "such things as peacocks may come to be in the course of natural and sexual selection"—in 1877, while both men worked on the staff of the literary miscellany *London*.[52]

That Allen should open his book on evolutionary aesthetics by quoting Ruskin, and that Lang should casually invoke "natural and sexual selection" when reviewing Whistler's Peacock Room for a literary journal, indicate the degree to which competing ideas about aesthetics had saturated Victorian print culture. Nor were Ruskin and Darwin, art and science, siloed off for different audiences. The same journals that published poetry and fiction, literary criticism, and art reviews, such as the *Athenaeum*, *Cornhill Magazine*, *Macmillan's*, and the *Fortnightly Review*, among others, frequently included articles on contemporary science. *Punch*, *Fun*, and lesser-known periodicals also published topical cartoons satirizing Darwin's theories. As such, Darwin's and Ruskin's arguments would have been known to Whistler and his audiences, and perhaps even to Leyland. Considering Whistler's engagement with contemporary theories of physiological aesthetics, Carolyn Arscott has pointed out that the prominence of these theories "in the cultural nexus that Whistler occupied is easily established," and given Whistler's compulsive reading of art reviews in major journals, "looking for slights in his paranoid manner," he would have been aware of new theories.[53] Even without a letter or journal entry to show that Whistler had read Darwin, it is similarly reasonable to believe that the artist was aware of the peacock's multiple meanings, so prominent and contested, in the sophisticated London networks in which he moved.

A closer look at Whistler's peacocks demonstrates the degree to which his design dispensed with both Darwinian and Ruskinian prerogatives and departed from artistic precedents. The room's vivid colors, shining metal, and stylized motifs refer to natural forms, but the artist's hand and the patron's largesse have transmuted them into a higher art. Far from the priorities that Darwin ascribed to them, the four shutter

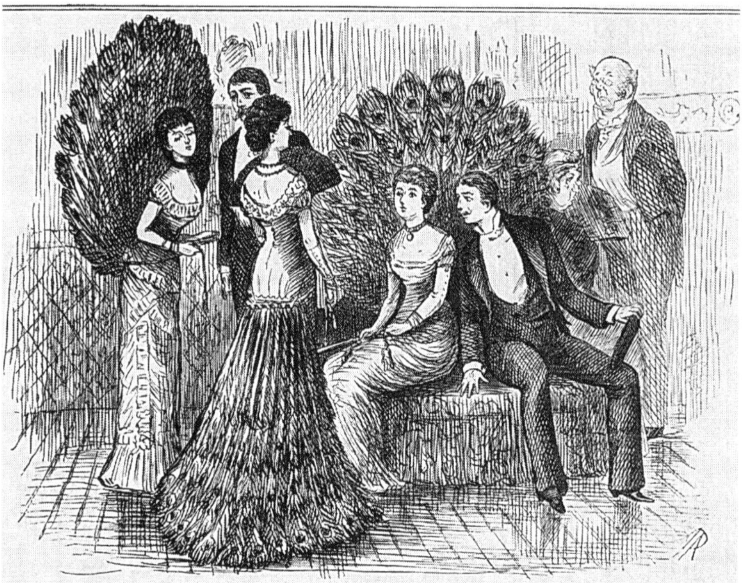

8.5 Cartoons compared female fashions to the attractive plumage of Darwin's birds. "The Peacock Train—'You just pull a String, and there you are!'" *Punch*, January 18, 1879, 15.

peacocks—a pair roosting together on the central shutter, and individual birds on the flanking shutters—behave as human aesthetes. Two birds display their full trains, a behavior that Darwin attributed to the drive to attract a peahen, yet there are no female birds available. Rather, the shutter peacocks display for themselves, for their fellow male birds, and for the aesthetic appreciation of the Leylands and their guests. On the middle set of shutters, two more birds roost companionably, silhouetted against an immense moon.[54] They both gaze to the right, toward the same point in the distance, or perhaps toward the exuberant display of the peacock on the righthand shutter. Gazing and inviting gazes, displaying and appreciating each other's displays, the shutter peacocks exhibit the delectation that Whistler and like-minded artists anticipated would take place in their lavishly decorated interiors. This behavior is completely at odds with the reproductive drive seen in other Victorian depictions of peacocks. It is open to a homosocial, even a homoerotic interpretation. As Andrew Stephenson has argued, self-consciously aesthetic spaces in the London art world of the late nineteenth century encouraged "modern modes of looking underpinned by homoerotic pleasures and desires that recast the male body as a living work of art."[55] Whistler's peacocks provide an avian example of this network of eroticized gazes between males.

This evacuation of a heterosexual drive in the presence of behaviors associated with male-female mating rituals is distinctive and unusual. Whistler had access to a number of Japanese prints and reproductions of Japanese artworks featuring peacocks, which he studied closely and drew from often for motifs and compositions.[56] As Merrill has observed, however, Japanese artists portrayed peafowl as solo males or in male-female pairs, but males were rarely pictured together without a hen.[57] And in the immediate context of British visual culture, peacock imagery was closely associated with women, not men. Whistler could hardly have missed the deluge of Victorian cartoons that pictured women as peacocks. As Susan Bernstein and other scholars have shown, the late Victorian flowering of imagery equating women with peacocks' trains and other animal fashions is linked to the publication of Darwin's *Descent of Man*.[58] These images based their humor on the lengths to which women will go to attract male attention. In the *Punch* cartoon called "The Peacock Train," for instance, three women show off a new sartorial innovation: a peacock-feather dress train that flips up with the pull of a string (fig. 8.5). A young man in the foreground is captivated, open-mouthed, pressing a phallic object (a fan or a flattened hat?) into his thigh.

The room further contravenes "nature" in picturing potential rivals living in mutual harmony under the reign of art and beauty. Darwin reminded his readers in *The Descent of Man* that "the peacock with his long train appears more like a dandy than a warrior, but he sometimes engages in fierce contests" over hens.[59] In Whistler's shutters, however, the peacocks are happily enclosed by art and beauty in a web that Melody Barnett Deusner has likened to a "social network" and which the poet and literary critic Arthur Symons, visiting at the turn of the twentieth century, described as "woven together into a web or network of almost alarming loveliness."[60] Symons recognized the essential artifice of the space, as he went on to describe it as "a room into which nature, sunlight, or any mortal compromise could never enter."[61]

The best-known species of peacock is *Pavo cristatus*, the blue peacock native to India, but almost all of Whistler's birds are painted to far more closely resemble a different species: *Pavo muticus*, the Java or green peacock.[62] This was the breed depicted in the Japanese artworks from which Whistler drew inspiration. The artist's attention to the breed's distinctive features attests to his study of Asian sources, but his choice of green peacocks also underscores the "unnatural" state of his painted harmony.

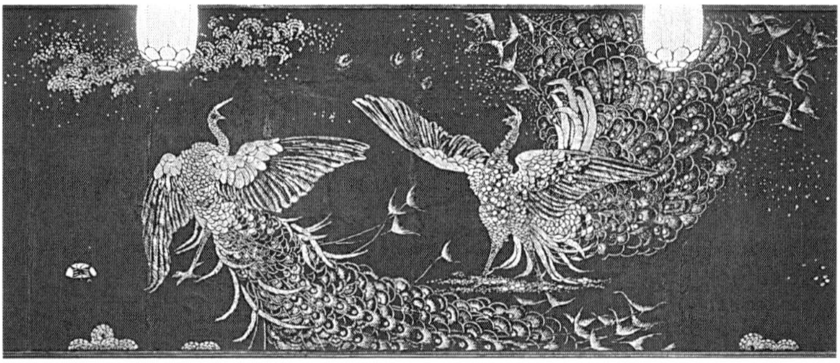

8.6 In the space intended for *The Three Girls*, Whistler painted a caricature of his patron. James McNeill Whistler, *Harmony in Blue and Gold: The Peacock Room*, 1876–1877. Detail: south wall mural, *Art and Money* (also called *L'Art et L'Argent, or the Story of the Room*), 1877. Oil paint, gold leaf, glass, and plaster over painted leather. Freer Gallery of Art, Smithsonian Institution, Washington, DC: Gift of Charles Lang Freer, F1904.61.

As Victorian naturalists observed, green peacocks were antisocial and combative, more difficult to keep in gardens or zoos than the more pliant blues, and adapted poorly to Britain's chilly climate. Like the plants in *The Three Girls*, green peacocks were beautiful rarities that required a hothouse.

Whistler's Revenge

After the relationship between Whistler and Leyland broke down, Whistler painted a number of images that reverse the harmony of the Peacock Room, depicting Leyland as a perverter of Art and Nature. The pointed criticism of these images is bolstered and heightened through its relationship to the Darwinian associations that equated peacocks with fertility, beauty, and reproduction.

As previously noted, in October 1876 Leyland informed Whistler that he would pay one thousand pounds rather than two thousand guineas for the artist's labor in the Peacock Room. Infuriated by what he perceived as a miserly, disrespectful rejection of his aesthetic accomplishment, Whistler initiated a process of artistic revenge against Leyland that would unfold over many months. Whistler swiftly resolved that he would never deliver *The Three Girls*, the magnum opus commissioned by Leyland a decade earlier and intended to occupy the south wall of the room. Instead, Whistler planned to fill the south wall with a mural. Working from a full-sized preparatory cartoon, Whistler painted over the south wall with a golden mural of a confrontation between two Java peacocks.[63] In a letter to Leyland, drafted in July 1877 but never sent,

Whistler named the mural "*L'Art et L'Argent*, or *The Story of the Room*."[64] Today, it is most often identified by the English translation of the title, *Art and Money* (fig. 8.6).

"Art" and "Money" identify the two birds, who also represent artist and patron. Leyland is Money, the righthand peacock, its eye a faux diamond, its body spangled with gold and silver coins. A ruff of silver bristles down its neck mimics the cambric frill that Leyland famously wore in place of a necktie. The bird's frill bristles, his wings are spread wide, and his train is unfurled; he appears to shed feathers as he shakes with rage. The Leyland peacock is defending the hoard of silver coins heaped at feet against the other, affronted bird, representing Art and Whistler himself. This lefthand bird is recognizable as a portrait of the artist by its crown of silver feathers, a reference to Whistler's distinctive forelock of prematurely silver hair.

With *Art and Money*, Whistler introduced a jarring disharmony into a previously harmonious space. Where the *Three Girls* was a peaceable encounter of young women gathering to tend a plant, and the Peacock Room shutters pictured an exchange of display and companionship, this is a clash. The scene becomes especially ironic in light of Darwin's evolutionist explanation for clashes between male peacocks. Darwinian conflict between males was productive; the victor was destined to mate with a female, and bring forth a new generation. But the conflict in *Art and Money* will bear no artistic fruit or flower. In hoarding his gold and silver, the Leyland peacock stifles the procreative power of art.[65]

Following his break with Leyland, Whistler's financial situation was desperate. He had once planned on making a tidy profit at the May 1877 opening of the Grosvenor Gallery, a new venue friendly to avant-garde artists.[66] But the room had so completely occupied Whistler's time that when the Grosvenor opened he had only one painting available to sell: *Nocturne in Black and Gold—The Falling Rocket* (ca. 1875).[67] Compounding his problems, Whistler had recently commissioned one of the aesthetic movement's leading architects to design an expensive new home in Chelsea.

In July 1877 John Ruskin published a caustic review of Whistler's paintings at the Grosvenor. In his self-published journal *Fors Clavigera*, Ruskin excoriated the "coxcomb" Whistler's "Cockney impudence" in trying to pass off unfinished sketches as valuable art. By offering *Nocturne in Black and Gold* for sale, Ruskin wrote, Whistler was "flinging a pot of paint in the public's face."[68]

Ever alert to the slightest criticism, Whistler responded by bringing a libel action against Ruskin, claiming a thousand guineas in damages. This sum was no coincidence: a thousand guineas was the difference

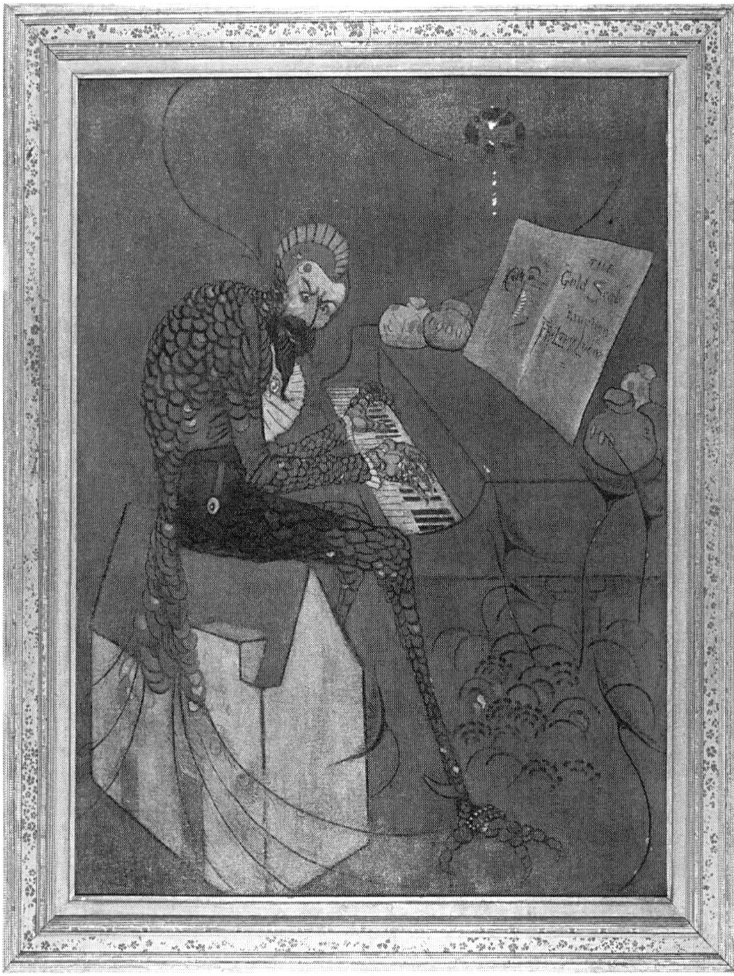

8.7 Creditors taking inventory of the bankrupt Whistler's possessions found *The Gold Scab* on display in his studio. James McNeill Whistler. *The Gold Scab: Eruption in Frilthy Lucre (The Creditor)*, 1879. Oil on canvas. 73 1/2 × 55 in (186.7 × 139.7 cm). Fine Arts Museums of San Francisco. Gift of Mrs. Alma de Bretteville Spreckels through the Patrons of Art and Music. 1977.11.

between Whistler's bill for the Peacock Room and Leyland's eventual payment. The amount of the libel suit was not only a practical consideration, recouping the lost fee for the Peacock Room, but perhaps also a symbolic one, suggesting that Whistler saw the nation's foremost critic as the appropriate source of restitution for Leyland's failings. Unfortunately for Whistler, while the court found in his favor, he was awarded only a farthing in damages, and the cost of the trial pushed Whistler into bankruptcy. The artist was forced to auction off his new home and most of its contents in order to satisfy his many creditors.[69]

Painfully for Whistler, Frederick Leyland was one of these creditors. For a decade, Leyland had been paying Whistler in advance for artworks that he never completed, most notably *The Three Girls*. In 1878, after the disastrous end of the Whistler-Ruskin trial, Leyland's agents descended upon Whistler's home to inventory his moveable goods for sale at auction. In his studio, they found further tokens of the artist's spite in the form of cruel caricatures of Leyland. The most significant, and the only one that survives today, was *The Gold Scab: Eruption in Frilthy Lucre* (fig. 8.7). The painting depicts Leyland as a scrawny, sickly peacock/human hybrid. He sits upon a model of Whistler's new home and plays a single, monotonous note on a piano.

If *Art and Money* had proved too subtle (and too beautiful) to convey Whistler's contempt for Leyland, *The Gold Scab* makes clear Whistler's condemnation of Leyland's sterile, sterilizing greed. The title phrase "Frilthy Lucre" puns on Leyland's frilly shirts and his ill-gotten gains. Leyland's lucre is represented not only by the bags of gold piled on the piano, each labeled "£1000," but by Whistler's new home, the White House, here serving as a piano bench. True Art, represented by the White House, is held hostage and forced to serve the vulgar purpose of accommodating Leyland's rear end. To drive home his contempt, Whistler added to his prominent butterfly signature a vast, prehensile stinger-tail that arcs across the picture to bury its point in Leyland's neck. If, as Syme has suggested, Whistler's butterfly signature drew upon "the polyvalence the act of floral fertilization allowed," exploring the queer possibility of artist as "both a pollinator and a flower," this iteration of Whistler's butterfly, with its hypertrophied, vengeful phallus, is a creator turned to destruction by righteous anger.[70]

The Gold Scab employs the Peacock Room's distinctive color scheme, but the glow of gold is transmuted to sickly yellow. The Leyland peacock is nightmarish. His face is sallow, his back hunched, his hands gnarled, his feathered train almost bald. Leyland's famous frill, in a stiffly pleated form resembling a reptile's neck frill, proliferates across his chest and cuffs and sprouts from a bald skullcap. His body is coated in the overlapping disc-shaped feathers that were the Peacock Room's main decorative motif. As the skin of the monstrous Leyland, though, the feathers resemble scabs. Some are painted gold, suggesting coins, as if Leyland has absorbed his wealth into his very body, and it has done him no good. Ill health and corruption have become the logical end of hoarding money.

The Gold Scab was made even more spiteful because Whistler placed it within the frame originally intended for *The Three Girls*. As *The Gold*

Scab has a vertical orientation, the frame had to be turned on its side, rendering the frame's musical passage illegible. The musical notes adorning *The Three Girls* were only one of that painting's references to strange, supernatural, non-Darwinian acts of creation. In *The Three Girls*, human artistry brought hothouse plants to full flower; men reproduced beautiful music through the interface of their hands and minds with the keys of a piano; patron and artist conceived a beautiful painting. But Whistler and Leyland's feud turned *The Three Girls* into the sick and twisted *Gold Scab*.[71]

Whistler was one among many artists, critics, and scientists "thinking with peacocks" in late Victorian Britain, but his contributions to the discourse are among the most fascinating, the most complex, and the least understood. As scholars continue to assess the impact of scientific theories on nineteenth-century art and visual culture, it is time to incorporate Whistler's Peacock Room into the growing body of artworks recognizably engaged in Victorian renegotiations of concepts as varied as gender, aesthetics, and evolution that played out in light of Darwin's theories of natural and sexual selection. In this analysis, I have begun to suggest some ways in which the Peacock Room might be productively viewed through a Darwinian lens. Whistler drew upon the peacock's Darwinian association with beauty in the service of (re)production, fertility, and sexuality, but presented new combinations of participants, mechanism, and outcome. In its early stages, before Whistler was derailed by his fury at Leyland, the Peacock Room and *The Three Girls* modeled an alternative to Darwinian and Ruskinian theories of beauty, presenting a vision of a queer new system in which "unnatural" combinations generate the most beautiful of all forms. At its best, then, the Peacock Room offered a beginning rather than an end.

Notes

Introduction. Victorian Science and Imagery

1. William Alexander Ross, *Pyrology, or, Fire Chemistry* (London: E. & F. N. Spon, 1875), vi. Similarly, biologist T. H. Huxley declared to the Royal Academy of Art in 1871: "Our purpose is the same as yours. . . . We both seek truth and we both seek beauty." "Banquet at the Royal Academy," *Times* (London), May 1, 1871, 6; quoted in Anne Helmreich, *Nature's Truth: Photography, Painting, and Science in Victorian Britain* (University Park: Pennsylvania State University Press, 2016), 2.

2. Charles Percy Snow, *The Two Cultures and the Scientific Revolution* (Cambridge: Cambridge University Press, 1959). For more on this, see Matthew Rampley, *The Seductions of Darwin: Art, Evolution, Neuroscience* (University Park: Pennsylvania State University Press, 2017), 1.

3. Stephen Wilson notes that "'scientific' questions . . . are also major cultural questions and demand widespread attention . . . [considering] the nature of life, about how brains and bodies work." Wilson, "Introduction: Art, Science + Technology," *Art + Science Now: How Scientific Research and Technological Innovation are Becoming Key to 21st-Century Aesthetics* (New York, NY: Thames and Hudson, 2010), 6.

4. Gowan Dawson and Sally Shuttleworth, "Science and Victorian Poetry," *Victorian Poetry* 41, no. 1 (Spring 2003): 2.

5. Caroline A. Jones and Peter Galison, *Picturing Science, Producing Art* (London: Routledge, 1998), 2–3.

6. Charles Darwin to Benjamin Waterhouse Hawkins [1848–51], George P. Merrill Collection box 4, file H., *Zoology of the Voyage of HMS Beagle*, pts. 4 and 5, 1838–43, Library of Congress.

7. Valerie Bramwell and Robert McCracken Peck, *All in the Bones: A Biography of Benjamin Waterhouse Hawkins* (Philadelphia: Academy of Natural Sciences of Philadelphia, 2008), 106–7.

8. *Gutta Percha: Its Discovery, History, and Manifold Uses* (London: Benjamin L. Green, 1851). Helen Godfrey, *Submarine Telegraphy and the Hunt for Gutta Percha: Challenge and Opportunity in a Global Trade* (Leiden: Brill, 2018), 18.

9. Nancy Rose Marshall, *City of Gold and Mud: Painting Victorian London* (New Haven, CT: Yale University Press, 2012), 233–49.

10. Bramwell and Peck, *All in the Bones*, 35–39.

11. "Joseph Leidy," Penn University Archives and Records, accessed June 20, 2019, https://archives.upenn.edu/exhibits/penn-people/biography/joseph-mellick-leidy. Nicolaas Rupke, *Richard Owen: Biology without Darwin* (Chicago: University of Chicago Press, 2009).

12. Benjamin Waterhouse Hawkins, *Comparative Anatomy as Applied to the Purposes of the Artist* (London: Winsor & Newton, 1860), 25. Regarding the plate, Hawkins said he was trying to suggest a handshake of brotherhood between a white man, such as abolitionist William Wilberforce, and an understandably wary Black man. Hawkins's views on evolution are discussed in Bramwell and Peck, *All in the Bones*, 82–85.

13. Kirk Savage, *Standing Soldiers, Kneeling Slaves: Race, War, and Monument in Nineteenth-Century America* (Princeton, NJ: Princeton University Press, 2018).

14. A work submitted to the Royal Academy's permanent collection upon his elevation to full membership. "Art Chronicle," *Portfolio* (1879): 106.

15. F. G. F., "Science is Measurement," *Leisure Hour* 31 (1882): 665.

16. The aphorism "science is measurement" was widely cited in the Victorian press, without any single author. It was sometimes attributed to the positivist Auguste Comte, who posited that the only truth was that gleaned through scientific assessment and logical analysis of empirical evidence. James Ward, "Mechanism and Morals: The World of Science and the World of History," *Hibbert Journal* (1906): 98.

17. See Bernard Lightman and Bennett Zon, eds., *Evolution and Victorian Culture* (Cambridge:: Cambridge University Press, 2014); Bernard Lightman, ed., *Victorian Science in Context* (Chicago: University of Chicago Press, 2009); Bernard Lightman, ed., *Victorian Popularizers of Science: Designing Nature for New Audiences* (Chicago: University of Chicago Press, 2009); Aileen Fyfe and Bernard Lightman, eds., *Science in the Marketplace: Nineteenth-Century Sites and Experiences* (Chicago: University of Chicago Press, 2007); and Carin Berkowitz and Bernard Lightman, eds., *Science Museums in Transition: Cultures of Display in Nineteenth-Century Britain and America* (Pittsburgh: University of Pittsburgh Press, 2017).

18. James Secord, *Visions of Science: Books and Readers at the Dawn of the Victorian Age* (Chicago: University of Chicago Press, 2014), 17.

19. Lightman, *Victorian Popularizers of Science*, 1–2, 15, 17.

20. Bernard Lightman, "Victorian Science and Popular Visual Culture," *Early Popular Visual Culture* 10, no. 1 (February 2012): 1. See also Joe Kember, John Plunkett, and Jill Sullivan, *Popular Exhibitions, Science and Showmanship 1840–1910* (London: Pickering and Chatto, 2013).

21. Sally Gregory Kohlstedt, "Place and Museum Space: The Smithsonian Institution, National Identity, and the American West, 1846–1896," in *Geographies of Nineteenth-Century Science*, ed. David N. Livingstone and Charles W. J. Withers (Chicago: University of Chicago Press, 2011), 399–437.

22. Gary D. Rosenberg and Renee M. Clary, eds., *Museums at the Forefront of the History and Philosophy of Geology: History Made, History in the Making*, special paper 535 (Boulder, CO: Geological Society of America, 2018), 11. See also Steven Conn, *Museums and American Intellectual Life, 1876–1926* (Chicago: University of Chicago Press, 1998). For a foundational study of the natural history museum in Britain, see Carla Yanni, *Nature's Museums: Victorian Science and the Architecture of Display* (New York: Princeton Architectural Press, 2006).

23. Arnold Thackray, "Reflections on the Decline Science in America and on Some of Its Causes," *Science* (Jul. 1971), cited in David Cahan, "Institutions and Communities," in *From Natural Philosophy to the Sciences: Writing the History of Nineteenth-Century Science*, ed. David Cahan (Chicago: University of Chicago Press, 2003), 321.

24. Geoffrey J. Martin, *American Geography and Geographers: Toward Geographical Science* (Oxford: Oxford University Press, 2015), 124.

25. Lois Marie Fink and Joshua C. Taylor, *Academy: The Academic Tradition in American Art* (Washington, DC: Smithsonian Institution Press, 1975); Mark Hain, Michael J. Lewis, and Stephen May, *Pennsylvania Academy of the Fine Arts, 1805–2005: 200 Years of Excellence* (Philadelphia: Pennsylvania Academy of the Fine Arts, 2005), and David Dearinger, *Paintings and Sculpture in the Collection of the National Academy of Design, 1826–1925* (New York: Hudson Hills, 2004).

26. Carole Paul, *The First Modern Museums of Art: The Birth of an Institution in Eighteenth and Early Nineteenth-Century Europe* (Los Angeles: J. Paul Getty Museum, 2012); Alan Wallach, *Exhibiting Contradiction: Essays on the Art Museum in the United States* (Amherst: University of Massachusetts Press, 1998). See also Giles Waterfield, *The People's Galleries: Art Museums and Exhibitions in Britain, 1800–1914* (New Haven, CT: Yale University Press, 2015) and Conn, *Museums and American Intellectual Life.*

27. For more on the history of museums, see John E. Simmons, *Museums: A History* (Lanham, MD: Rowman & Littlefield, 2016); Oliver Impey and Arthur MacGregor, eds., *The Origins of Museums: The Cabinet of Curiosities in Sixteenth- and Seventeenth-Century Europe* (Oxford: Ashmolean Museum, 2017); and Eileen Hooper-Greenhill, *Museums and the Shaping of Knowledge* (London: Routledge, 1992).

28. Jones and Galison, *Picturing Science, Producing Art*, 2. Jones and Galison remind us that "art, science, and the hermeneutical concepts that we bring to them are historically and culturally embedded" (3). A. I. Tauber similarly examines the "deeply shared philosophical foundations grounding science and

aesthetics." Tauber, *Elusive Synthesis: Aesthetics and Science* (Boston: Springer, 1997), 1. Donald R. Benson distinguishes "an integrated culture in which science and art are complementary enterprises, both of them significantly shaped by and at the same time giving shape to common cultural fictions and assumptions." Benson, "'Catching Light': Physics and Art in Walter Pater's Cultural Context," in *One Culture: Essays in Science and Literature*, ed. George Levine with Alan Rauch (Madison: University of Wisconsin Press, 1987).

29. Krzysztof Pomian, *Collectors and Curiosities: Paris and Venice, 1500–1800* (Oxford: Polity, 1990), cited in Tony Bennett, *The Birth of the Museum: History, Theory, Politics* (London: Routledge, 2013), 40–41.

30. Philip Fisher, *Making and Effacing Art: Modern American Art in a Culture of Museums* (New York: Oxford University Press, 1991) in Bennett, *Birth of the Museum*, 44. Stephen Greenblatt and Barbara Kirshenblatt-Gimblett observe that in art museums, unlike science collections, the unique object continued to inspire wonder. Stephen Greenblatt, "Resonance and Wonder," 42–56, and Barbara Kirshenblatt-Gimblett, "Objects of Ethnography," 386–443, in Ivan Karp and Stephen Levine, eds., *Exhibiting Cultures: The Poetics and Politics of Museum Display* (Washington, DC: Smithsonian Institution Press, 1991).

31. Lorraine Daston and Peter Galison, "The Image of Objectivity," in "Seeing Science," special issue, *Representations* 40 (Fall 1992): 81–128.

32. Sally Gregory Kohlstedt, "Nature by Design: Masculinity and Animal Display in Nineteenth-Century America," in *Figuring It Out: Science, Gender, and Visual Culture*, ed. Ann B. Shteir and Bernard Lightman (Hanover, NH: University Press of New England, 2006), 110–11.

33. Gowan Dawson and Bernard Lightman, introduction to *Victorian Science and Literature*, vol. 1 (London: Pickering and Chatto, 2011–12), 20.

34. Bernard Pullman, *The Atom in the History of Human Thought* (Oxford: Oxford University Press, 2001).

35. Robert D. Whitaker, "An Historical Note on the Conservation of Mass," *Journal of Chemical Education* 52, no. 10 (1975): 658–59. Ted Underwood, "How Did the Conservation of Energy Become 'The Highest Law in All Science'?" in *Repositioning Victorian Sciences: Shifting Centers in Nineteenth-Century Scientific Thinking*, ed. David Clifford, Elisabeth Wadge, Alex Warwick, and Martin Willis (London: Anthem, 2006), 119–30.

36. William F. Bynum, *Science and the Practice of Medicine in the Nineteenth Century* (Cambridge: Cambridge University Press, 1994).

37. Andrew Wilson, "Human Resemblances to Lower Life," *Longman's Magazine* (1883): 44.

38. Thomas Aldous Huxley, "On the Physical Basis of Life" (1868), in *Select Works of Thomas H. Huxley* (New York: J. B. Alden, 1886), 457. See also Robert Michael Brain, "*Protoplasmania*: Huxley, Haeckel, and the Vibratory Organism in Late Nineteenth-Century," in Larson and Brauer, *Art of Evolution*, 92–122.

39. Andrew Wilson, *Studies in Life and Sense* (London: Chatto & Windus, 1887), 315.

40. Trevor H. Levere, *Chemists and Chemistry in Nature and Society, 1770–1878* (New York: Routledge, 1994), 193–94; Martin Wallen, *City of Health, Fields of Disease: Revolutions in the Poetry, Medicine, and Philosophy of Romanticism* (Aldershot, UK: Ashgate, 2004), 14.

41. David Harvey, *Justice, Nature and the Geography of Difference* (London: Blackwell, 1996), 121.

42. Julia Voss, *Darwin's Pictures: Views of Evolutionary Theory, 1837–1874* (New Haven, CT: Yale University Press, 2010), 9.

43. Martin Kemp, *The Science of Art: Optical Themes in Western Art from Brunelleschi to Seurat* (New Haven, CT: Yale University Press, 1990), 1; Helmreich, *Nature's Truth*, 3.

44. John Ruskin, "The Laws of Fésole," in *The Works of John Ruskin*, vol. 15, ed. E. T. Cook and Alexander Wedderburn (London: George Allen, 1904), 360.

45. Among others, see Daniela Bleichmar, *Visible Empire: Botanical Expeditions and Visual Culture in the Hispanic Enlightenment* (Chicago: University of Chicago Press, 2012); Helmreich, *Nature's Truth*; Jonathan Crary, *Techniques of the Observer: On Vision and Modernity in the Nineteenth Century* (Cambridge: MIT Press, 1990); and Hal Foster, ed., *Vision and Visuality: Discussions in Contemporary Culture* (New York: Dia Art Foundation/New Press, 1999).

46. Ann B. Shteir and Bernard Lightman, introduction to Shteir and Lightman, *Figuring It Out*, xv–xvii.

47. Dennis Cosgrove, *Geography and Vision: Seeing, Imagining, and Representing the World* (London: I. B. Tauris, 2008), 4. Cosgrove cites Gillian Rose, "Making Space for the Female Subject of Feminism: The Spatial Subversions of Holzer, Kruger and Sherman," in *Mapping the Subject*, ed. Steve Pile and Nigel Thrift (London: Routledge, 1995), 332–54; and Derek Gregory, *Geographical Imaginations* (Oxford: Blackwell, 2000).

48. Klaus Hentschel, *Visual Cultures in Science and Technology: A Comparative History* (Oxford: Oxford University Press, 2014), 18–24. Foster's foundational *Vision and Visuality* is an essential source guide to these ideas in the format of a field-changing talks and responses by Foster himself, Martin Jay, Rosalind Krauss, Norman Bryson, and Jacqueline Rose.

49. See, for instance, Michael Baxandall, *Painting and Experience in Fifteenth Century Italy: A Primer in the Social History of Pictorial Style* (Oxford: Oxford University Press, 1972), 29–40; or David Lindberg, *Theories of Vision from al-Kindi to Kepler* (Chicago: University of Chicago Press, 1976).

50. Crary, *Techniques of the Observer*; Chris Otter, *The Victorian Eye: A Political History of Light and Vision in Britain, 1800–1910* (Chicago: University of Chicago Press, 2008); David Cahan, *Hermann von Helmholtz and the Foundations of Nineteenth-Century Science* (Berkeley: University of California Press, 1993). See

also Helmreich, *Nature's Truth*. Gillian Beer has noted a countercurrent arguing for the primacy of touch. She cites, for example, Charles Bell's Bridgewater Treatise on the hand. Beer, *Open Fields: Science in Cultural Encounter* (Oxford: Clarendon, 1999), 16–17.

51. Jonathan Crary, *Suspensions of Perception: Attention, Spectacle, and Modern Culture* (Cambridge: MIT Press, 1999), 3.

52. Barbara Maria Stafford, *Artful Science: Enlightenment Entertainment and the Eclipse of Visual Education* (Cambridge: MIT Press, 1996), 3.

53. Svetlana Alpers, *The Art of Describing: Dutch Art in the Seventeenth Century* (Chicago: University of Chicago Press, 1984) in John V. Pickstone, *Ways of Knowing: A New History of Science, Technology and Medicine* (Chicago: University of Chicago Press, 2001), 64–65. Ann Shelby Blum, *Picturing Nature: Nineteenth-Century Zoological Illustration* (Princeton, NJ: Princeton University Press, 1993), 10.

54. Blum, *Picturing Nature*, 10.

55. Bleichmar, *Visible Empire*, 7, 9.

56. Pamela Gilbert notes Foucault's insight that around 1800, the basis of knowledge became classification and an understanding of an "internal principle" that was "alien to the domain of the visible." Gilbert, *Cholera and Nation: Doctoring the Social Body in Victorian England* (Albany: State University of New York Press, 2008), 11. Michel Foucault, *The Order of Things: An Archaeology of the Human Sciences* (London: Routledge, 1989). Sylvia Pamboukian, "'Looking Radiant': Science, Photography and the X-Ray Craze of 1896," *Victorian Review* 27, no. 2 (2001): 56–74.

57. Daston and Galison, "Image of Objectivity," 123.

58. Sarah Alexander finds parallels in literature of the period: "Writers sought to theorize and describe what could not be detected empirically, untenable paradoxes, undetectable realities, unseen forces—to make sense of new experiences of modernity, same language as that of the imponderable." Alexander, *Victorian Literature and the Science of the Imponderable* (New York: Routledge, 2015), 6–8.

59. Alice Jenkins, *Space and the 'March of Mind': Literature and the Physical Sciences in Britain, 1815–1850* (Oxford: Oxford University Press, 2007), 177–78.

60. John Lucas Tupper, "Progress of the Species," in *Poems*, ed. William Michael Rossetti (London: Longmans, Green, 1897), 70.

61. Charles Darwin, letter to Leonard Jenyns, October 14, 1839, cited in Daniel Pauly, *Darwin's Fishes: An Encyclopedia of Ichthyology, Ecology, and Evolution* (Cambridge: Cambridge University Press, 2007), 107.

62. Martin J. S. Rudwick, *Scenes from Deep Time: Early Pictorial Representations of the Pre-historic World* (Chicago: University of Chicago Press, 1992), viii.

63. Henry Maudsley, *Body and Mind* (London: Macmillan, 1873), 111.

64. John Tyndall, *The Scientific Use of the Imagination: A Discourse* (London: Longmans Green, 1870), 6. This is an echo of Francisco Goya's famous epigram

in his print set *Los Caprichos* (1799), "Imagination abandoned by reason produces impossible monsters; united with her, she is the mother of the arts and source of their wonders." See also Lorraine Daston, "Fear and Loathing of the Imagination in Science," *Daedalus* 127, no. 1 (1998): 73–95; and George Levine, *Dying to Know: Scientific Epistomology and Narrative in Victorian England* (Chicago: University of Chicago Press, 2002), 79–100.

65. See, for instance, Nicola Bown, Carolyn Burdett, and Pamela Thurschwell, eds., *The Victorian Supernatural* (Cambridge: Cambridge University Press, 2004); or scholarship on spiritualism, such as Srdjan Smajić, *Ghost-Seers, Detectives, and Spiritualists: Theories of Vision in Victorian Literature and Science* (Cambridge University Press, 2010) or Anna Neill, *Primitive Minds: Evolution and Spiritual Experience in the Victorian Novel* (Columbus: Ohio State University Press, 2013); or work on Victorian fairy painting, such as Carole G. Silver, *Strange and Secret Peoples: Fairies and Victorian Consciousness* (New York: Oxford University Press, 1999); Nicola Bown, *Fairies in Nineteenth-Century Art and Literature* (New York: Cambridge University Press, 2001); and Melanie Keen, *Science in Wonderland: Scientific Fairy Tales of Victorian Britain* (Oxford: Oxford University Press, 2015).

66. William Carpenter, "Mesmerism, Odylism, Table-Turning, and Spiritualism," *Fraser's Monthly Magazine*, March 1877, 386–87.

67. Stafford, *Artful Science*, 73–103.

68. Jenkins, *Space and the 'March of Mind'*, 18, 98; Barri Gold, *Thermopoetics: Energy in Victorian Literature and Science* (Cambridge: MIT Press, 2010), 30–31.

69. Blake, "Auguries of Innocence," 1803, 1–2.

70. Humphry Davy, *Consolations in Travel, or, The Last Days of a Philosopher* (London: John Murray, 1830), 245–46. See also Trevor H. Levere, *Chemists and Chemistry in Nature and Society*, 191.

71. David Aubin, Charlotte Bigg, and Heinz Otto Sibum, eds., *The Heavens on Earth: Observatories and Astronomy in Nineteenth-Century Science and Culture* (Durham, NC: Duke University Press, 2010); Jutta Schickore, *The Microscope and the Eye: A History of Reflections, 1740–1870* (Chicago: University of Chicago Press, 2007).

72. Trevor H. Levere, "The Rich Economy of Nature: Chemistry in the Nineteenth Century," in *Nature and the Victorian Imagination*, ed. U. C. Knoepflmacher and G. B. Tennyson (Berkeley: University of California Press, 1977), 191–93.

73. Joseph Maclise, *Surgical Anatomy* (Philadelphia: Blanchard and Lea, 1850), vii.

74. John Ruskin, *Of Leaf Beauty*, vol. 5 of *Modern Painters*, in *The Works of John Ruskin*, vol. 7, ed. E. T. Cook and Alexander Wedderburn (London: George Allen, 1905), 207.

75. William Walker, *Handbook of Drawing* (London: Seeley, Jackson & Halliday, 1880), 118. Walker added: "And thus it is in contemplating any object, we do

so with the conviction that the designer of this one object was also the designer of the whole visible universe."

76. T. H. Huxley, "On the Physical Basis of Life," 462. Huxley resisted categorization as a materialist, however.

77. Philip Gilbert Hamerton, "Notes on Aesthetics (1879–80)," *Portfolio Papers* (London: Seeley, 1889), 172.

78. Philip Gilbert Hamerton, *The Graphic Arts: A Treatise on Varieties of Drawing, Painting, and Engraving in Comparison with Each Other and with Nature* (London: Seeley, Jackson, and Halliday, 1882), 1–2.

79. Gowan Dawson, "Intrinsic Earthliness: Science, Materialism, and the Fleshly School of Poetry," *Victorian Poetry* 41, no. 1 (2003): 113–29.

80. Barbara Larson, introduction to *Darwin and Theories of Aesthetics and Cultural History*, ed. Barbara Larson and Sabine Flach (Aldershot, UK: Ashgate, 2013), 9.

81. D. A. Miller, *The Novel and the Police* (Berkeley: University of California Press, 1989); Nicholas Dames, *The Physiology of the Novel: Reading, Neural Science, and the Form of Victorian Fiction* (Oxford: Oxford University Press, 2007).

82. Katharina Boehm, introduction to *Bodies and Things in Nineteenth-Century Literature and Culture*, ed. Katharina Boehm (New York: Palgrave Macmillan, 2012), 5. Quote from Maurice Merleau-Ponty, *The Visible and the Invisible*, ed. Claude Lefort, trans. Alphonso Lingis (Evanston, IL: Northwestern University Press, 1968), 137.

83. Martin Wallen, *City of Health, Fields of Disease: Revolutions in the Poetry, Medicine, and Philosophy of Romanticism* (Aldershot: Ashgate, 2004), 74.

84. Kemp, *Science of Art*, 1.

85. Phillip Prodger, *Darwin's Camera: Art and Photography in the Theory of Evolution* (Oxford: Oxford University Press, 2009), xxiv–xxv.

86. See for instance, Pamela Kort and Max Hollein, eds., *Darwin: Art and the Search for Origins* (Cologne: Wienand, 2009); Diana Donald and Jane Munro, eds., *Endless Forms: Charles Darwin, Natural Science and the Visual Arts* (Cambridge: Fitzwilliam Museum, 2009). Other prominent works on Darwinian visual culture include Jonathan Smith, *Charles Darwin and Victorian Visual Culture* (Cambridge: Cambridge University Press, 2006); Kathleen Pyne, *Art and the Higher Life: Painting and Evolutionary Thought in Late Nineteenth-Century America* (Austin: University of Texas Press, 1996); and Larson and Brauer, *Art of Evolution*.

87. Kate Flint, *Flash! Photography, Writing, and Surprising Illumination* (Oxford: Oxford University Press, 2017), 18–24.

88. Philip Ball, *Bright Earth: Art and the Invention of Color* (Chicago: University of Chicago Press, 2003), 162; Joyce Townsend, *Pre-Raphaelite Painting Techniques* (London: Tate, 2004); Helmreich, *Nature's Truth*, 59–62.

89. Wassily Kandinsky drew on Ernst Haeckel. Charissa N. Terranova and Meredith Tromble, introduction to *The Routledge Companion to Biology in Art*

and Architecture, ed. Charissa N. Terranova and Meredith Tromble (New York: Routledge, 2017), 1.

90. Goethe published his *Theory of Colors* in 1810. Ball, *Bright Earth*, 166.

91. Holmes, *Pre-Raphaelites and Science*, 9.

92. Holmes, *Pre-Raphaelites and Science*, 7.

93. Helmreich, *Nature's Truth*, 16. My italics.

94. Matthew C. Hunter, *Wicked Intelligence: Visual Art and the Science of Experiment in Restoration London* (Chicago: University of Chicago Press, 2013), 38, 116.

95. Martin Meisel, *Chaos Imagined: Literature, Art, Science* (New York: Columbia University Press, 2016); Michel Serres, "Turner Translates Carnot," in *Hermes: Literature, Science, Philosophy*, ed. Josué V. Harari and David F. Bell (Baltimore: Johns Hopkins University Press, 1982), 52–64.

96. Peter Galison places this shift in the 1970s. "Images Scatter into Data, Data Gather into Images," in *Iconoclash: Beyond the Image Wars in Science, Religion and Art*, ed. Bruno Latour and Peter Weibel (Cambridge, MA: MIT Press, 2002), 300.

97. Luc Pauwels, "The Role of Visual Representation in the Production of Scientific Reality," in *Visual Cultures of Science: Rethinking Representational Practices in the Production of a Scientific Reality*, ed. Luc Pauwels (Lebanon, NH: University Press of New England, 2006), vii. Pauwels is paraphrasing from Steve Woolgar, *Science: The Very Idea* (London: Tavistock, 1988), 30–31.

98. Blum, *Picturing Nature*, 10

99. Pauwels, "Introduction," viii.

100. Pauwels, "Introduction," vii.

101. Bruno Latour, "What Is Iconoclash?," in Latour and Weibel, *Iconoclash*, 19.

102. Pauwels, "Introduction," vii.

103. Daniela Bleichmar remarks that "historians of science have not considered images as loci of knowledge production and art histoirans have disregarded scientific illustrations." Bleichmar, *Visible Empire*, 5.

104. Alex Soojung-Kim Pang, "Visual Representation and Post-constructivist History of Science," *Historical Studies in the Physical and Biological Sciences* 28, no. 1 (1997): 148. Other publications of note in Pang's estimation are Martin Rudwick, "The Emergence of a Visual Language for Geological Science, 1760–1830," *History of Science* 14 (1976): 149–95; and Robert Scott Root-Bernstein, "On Paradigms and Revolutions in Art and Science: The Challenge of Interpretation," *Art Journal* 44, no. 2 (1984): 109–18.

105. Daston and Galison, "Image of Objectivity," 81–128.

106. Pang, "Visual Representation,"171.

107. Lightman and Shteir, "Introduction," xv–xx.

108. Lightman and Shteir, "Introduction," xxii.

109. Lightman, "Depicting Nature, Defining Roles: The Gender Politics of Victorian Illustration," in Shteir and Lightman, *Figuring It Out*, 220.

110. Lightman, "Depicting Nature," 234.

111. Gold, *Thermopoetics*, 16.

112. Claire Pettitt, *Patent Inventions: Intellectual Property and the Victorian Novel* (Oxford: Oxford University Press, 2004), 272–74. See also Jenkins, *Space and the 'March of Mind'*, 177.

113. Gillian Beer, *Darwin's Plots: Evolutionary Narrative in Darwin, George Eliot and Nineteenth-Century Fiction* (Cambridge: Cambridge University Press, 1983); Larson, "Darwin, Burke and the Biological Sublime," in Larson and Flach, *Darwin and Theories of Aesthetics*, 17–36. Smith considers the picturesque and the grotesque in light of Darwin's geology in *Charles Darwin and Victorian Visual Culture*, 245–85.

114. Voss, *Darwin's Pictures*, 10–11.

115. Blum, *Picturing Nature*, 3. Historians of botany were some of the earliest figures to argue for the work performed by images. See also Gill Saunders, *Picturing Plants: An Analytical History of Botanical Illustration* (Berkeley: University of California Press, 1995).

116. I am thinking of Caroline Levine's "strategic formalism" as well as other calls for neoformalist analysis. Levine, *Forms: Whole, Rhythm, Hierarchy, Network* (Princeton, NJ: Princeton University Press, 2017). See also Marjorie Levinson, "What Is New Formalism?" *PMLA* 122, no. 2 (2007): 558–69.

117. Smith, *Charles Darwin and Victorian Visual Culture*, 1–2.

118. Ann Shteir, for instance, considers how representations of Flora, the goddess of flowers, "reflect several traditional images of women" that might in turn have influenced how female scientists were received. Shteir, "Iconographies of Flora: The Goddess of Flowers in the Cultural History of Botany," in Shteir and Lightman, *Figuring It Out*, 5, 24.

119. Diana Donald, *Picturing Animals in Britain, c. 1750–1850* (New Haven, CT: Yale University Press, 2008), viii. Donald notes that "style and epistemology were inseparable" in natural history drawing.

120. Gold, *Thermopoetics*, 14. Gold observes that "literature and science draw on same root metaphors or myths" (18–19).

121. Larson, "Evolution and Victorian Art," in Lightman and Zon, *Evolution and Victorian Culture*, 123.

122. George Levine, *Darwin and the Novelists: Patterns of Science in Victorian Fiction* (Cambridge, MA: Harvard University Press, 1988), 3.

123. Daniel M. Fox and Christopher Lawrence, *Photographing Medicine: Images and Power in Britain and America since 1840* (New York: Greenwood, 1988), 4.

124. Benjamin Waterhouse Hawkins, *The Science of Drawing Simplified* (London: Smith, Elder, 1843), 5–6.

125. Hawkins, *Science of Drawing Simplified*, 7. Hawkins quotes artist John Flaxman's reminder that "art could not be perfected without Arithmetic and Geometry," 20.

Chapter 1. Measuring Native America

1. J. S. Rudwick, *Earth's Deep History: How It Was Discovered and Why It Matters* (Chicago: University of Chicago Press, 2014), 2–6.

2. For science and the visual, see Lynn Gamwell, *Exploring the Invisible: Art, Science, and the Spiritual* (Princeton, NJ: Princeton University Press, 2002); Martin Kemp, *Seen and Unseen: Art, Science, and Intuition from Leonardo to the Hubble Telescope* (New York: Oxford, 2006); M. Norton Wise, "Making Visible," *Isis* 97, no. 1 (March 2006): 75–82; Susana Maria Halpine, "Real Scientists Do It with Models: The Art of Science Visualization," *Teaching Artist Journal* 6, no. 1 (2008): 5–19.

3. David L. Browman and Douglas R. Givens, "Stratigraphic Excavation: The First 'New Archaeology,'" *American Anthropologist*, n.s., 98, no. 1 (March 1996): 80–81; Matthew R. Goodrum and Cora Olson, "The Quest for an Absolute Chronology in Human Prehistory: Anthropologists, Chemists, and the Fluorine Dating Method in Palaeoanthropology," *British Society for the History of Science* 42, no. 1 (March 2009): 95–114.

4. Terry A. Barnhart, *American Antiquities: Revisiting the Origins of American Archaeology* (Lincoln: University of Nebraska Press, 2015), 176–77; Isaiah Thomas, "Meeting of September 29, 1813. An Account of the American Antiquarian Society," in *Proceedings of the American Antiquarian Society, 1812–1849* (Worcester, MA: American Antiquarian Society, 1912), 15–16. See also Philip F. Gura, *The American Antiquarian Society, 1812–2012: A Bicentennial History* (Worcester, MA: American Antiquarian Society, 2012).

5. For the history of archaeology in Europe, including its relationship to nationalism, see Brian M. Fagan and Nadia Durrani, *A Brief History of Archaeology: Classical Times to the Twenty-First Century*, 2nd ed. (New York: Routledge, 2016), chs. 1, 3; Alain Schnapp, "French Archaeology: Between National Identity and Cultural Identity," in *Nationalism and Archaeology in Europe*, ed. Margarita Díaz-Andreu and Timothy Champion (New York: Routledge, 1996), 48–67; Margarita Díaz-Andreu, *A World History of Nineteenth-Century Archaeology: Nationalism, Colonialism, and the Past* (Oxford: Oxford University Press, 2007), chs. 3–4, 11–12; and Göran Blix, *From Paris to Pompeii: French Romanticism and the Cultural Politics of Archaeology* (Philadelphia: University of Pennsylvania Press, 2009).

6. Abiel Holmes, "Meeting of October 24, 1814. Address," in *Proceedings of the American Antiquarian Society, 1812–1849*, 58–59.

7. Holmes, "Meeting of October 24, 1814," 61–62.

8. For the history of archaeology in America, see Barnhart, *American*

Antiquities; Gordon R. Willey and Jeremy A. Sabloff, *A History of American Archaeology* (London: Thames and Hudson, 1974); Alice Beck Kehoe, *The Land of Prehistory: A Critical History of American Archaeology* (New York: Routledge, 1998); Thomas C. Patterson, *Toward the Social History of Archaeology in the United States* (Belmont, CA: Wadsworth/Thomson Learning, 2002); and Philip L. Kohl, Irina Podgorny, and Stefanie Gänger, eds., *Nature and Antiquities: The Making of Archaeology in the Americas* (Tuscon: University of Arizona Press, 2014). For geology as a form of national history, see Rebecca Bailey Bedell, *The Anatomy of Nature: Geology and American Landscape Painting, 1825–1875* (Princeton, NJ: Princeton University Press, 2001), 5–6, 94–107.

9. Barnhart, *American Antiquities*, 13–38, 253–55; Matthew N. Johnston, *Narrating the Landscape: Print Culture and American Expansion in the Nineteenth Century* (Norman: University of Oklahoma Press, 2016), esp. ch. 3.

10. The biographical information included here and throughout comes largely from the following sources: Montroville Wilson Dickeson, "Exploration of an Ancient Burial Ground, on the Estate of Major General Quitman," *Lotus* 1, no. 8 (May 20, 1848): 113–15; Bertha L. Heilbron, "A Mississippi Panorama," *Minnesota History* 23, no. 4 (December 1942): 349–54; Richard Veit, "A Case of Archaeological Amnesia: A Contextual Biography of Montroville Wilson Dickeson (1810–1882), Early American Archaeologist," *Anthropology of Eastern North America* 25 (1997): 97–123; Stephan Oettermann, *The Panorama: History of a Mass Medium*, trans. Deborah Lucas Schneider (New York: Zone Books, 1997), 337–40; Nanette Luarca-Shoaf, "Excavating a Nineteenth-Century Mass Medium," *American Art* 27, no. 2 (Summer 2013): 15–20; Luarca-Shoaf, "Currents of Time on the Lower Mississippi: M.W. Dickeson and *Monumental Grandeur of the Mississippi Valley*," chapter in "The Mississippi River in Antebellum Visual Culture" (PhD diss., University of Delaware, 2012), 170–239; and Barnhart, *American Antiquities*, 257–70. Luarca-Shoaf offers the most substantial interpretation of the panorama to date, including the panorama's representation of indigeneity and temporality and its scientific contexts, matters I treat at length as well.

11. Barnhart, *American Antiquities*, 253–57. See also Thomas C. Patterson, "Who Did Archaeology in the United States before There Were Archaeologists and Why? Preprofessional Archaeologies in the Nineteenth Century," in *Processual and Postprocessual Archaeologies: Multiple Ways of Knowing the Past*, ed. Robert W. Preucel (Carbondale: Center for Archaeological Investigations, Southern Illinois University at Carbondale, 1991), 242–50.

12. "Stated Meeting, October 20, 1846," *Proceedings of the Academy of Natural Sciences of Philadelphia* 3, no. 5 (September–October 1846): 109; Veit, "Case of Archaeological Amnesia," 115.

13. City Museum and John Ashton & Co., "City Museum. Callowhill Street, below Fifth. Another Great Bill," Broadsides Collection, American Philosophical

Society Library, Philadelphia, Pennsylvania; Veit, "Case of Archaeological Amnesia," 108, 110–13; Barnhart, *American Antiquities*, 262–70.

14. Gordon M. Sayre, "The Mound Builders and the Imagination of American Antiquity in Jefferson, Bartram, and Chateaubriand," *Early American Literature* 33, no. 3 (1998): 239. For detailed discussion of mound mania in the period, see also Robert Silverberg, *Mound Builders of Ancient America: The Archaeology of a Myth* (Greenwich, CT: New York Graphic Society, 1968), esp. ch. 2; and Roger Kennedy, "Jefferson and the Indians," *Winterthur Portfolio* 27, no. 2–3 (Summer–Autumn 1992): 105–21.

15. Silverberg, *Mound Builders of Ancient America*, 225–27, 233–42; Barnhart, *American Antiquities*, 44–45, 63–66; Jay Miller, *Ancestral Mounds: Vitality and Volatility of Native America* (Lincoln: University of Nebraska Press, 2015), esp. ch. 4.

16. Sayre, "Mound Builders," 239, 227–41; Bruce G. Trigger, "Anglo-American Archaeology," *World Archaeology* 13, no. 2 (October 1981): 143–44; Kehoe, *Land of Prehistory*, 64–65; Annette Kolodny, "Fictions of American Prehistory: Indians, Archeology, and National Origin Myths," *American Literature* 75, no. 4 (2004): 693–721; Scott Tribble, "Mounds, Myths, and Grave Mistakes: Wills De Hass and the Growing Pains of Nineteenth-Century Archaeology," *West Virginia History: A Journal of Regional Studies* 7, no. 1 (Spring 2013): 23–37; Jamie M. Arjona, "Sublime Perversions: Capturing the Uncanny Affects of Queer Temporalities in Mississippian Ruins," *Journal of Social Archaeology* 6, no. 2 (2016): 196–98. For late eighteenth-century commentary, see, for example, David Jones, "To the Editor of the Royal American Magazine," *Royal American Magazine, or, The Universal Repository of Instruction and Amusement* 2, no. 1 (January 1775): 29; William Bartram, *Travels through North and South Carolina, Georgia, East and West Florida, the Cherokee Country, the Extensive Territories of the Muscogulges or Creek Confederacy, and the Country of the Chactaws* (Dublin: J. Moore, W. Jones, R. McAllister, and J. Rice, 1793), 6, 136, 152, 178, 330, 336, 344, 370, 379, 388, 432, 460.

17. Jonathan Heart, "A letter from Major Jonathan Heart, to Benjamin Smith Barton, M.D. . . . containing observations on the Ancient Works of Art, the Native Inhabitants, &c. of the Western-Country, Fort Harmar, 5th January, 1791," in *Transactions of the American Philosophical Society*, vol. 3 (Philadelphia: American Philosophical Society, 1793): 217; Winthrop Sargent to William Maxwell, Cincinnati, September 2, 1794, "Territory of the United States North-west of the Ohio," *Centinel of the North-Western Territory*, 1, no. 43 (September 6, 1794): 2. See Barnhart, *American Antiquities*, 109–10, for a discussion of Sargent's letter.

18. Ephraim George Squier and Edwin Hamilton Davis, *Ancient Monuments of the Mississippi Valley*, ed. David J. Meltzer (Washington, DC: Smithsonian Institution Press, 1998), 301–2. See also Barnhart, *American Antiquities*, ch. 1.

19. Dickeson, "Exploration of an Ancient Burial Ground," 114, 115. For eighteenth-century archaeological interest in Native America, see Stephen

Williams, *Fantastic Archaeology: The Wild Side of North American Prehistory* (Philadelphia: University of Pennsylvania Press, 1991), ch. 2.

20. Montroville Wilson Dickeson, "Ancient Fortification, and Burial Place of the Beluxies, Pascagoula, Mississippi," *Lotus* 1 no. 11 (June 10, 1848): 162.

21. For moving panoramas, see Oettermann, *Panorama*; Erkki Huhtamo, *Illusions in Motion: Media Archaeology of the Moving Panorama and Related Spectacles* (Cambridge: MIT Press, 2013), esp. ch. 6, which considers the moving panorama craze in the 1840s in the United States; Angela Miller, "The Panorama, the Cinema, and the Emergence of the Spectacular," *Wide Angle* 18, no. 2 (April 1996): 34–69; Alison Griffiths, "'Shivers Down Your Spine': Panoramas and the Origins of the Cinematic Reenactment," *Screen* 44, no. 1 (Spring 2003): 1–37; and Vance Byrd, *A Pedagogy of Observation: Nineteenth-Century Panoramas, German Literature, and Reading Culture* (Lewisburg, PA: Bucknell University Press, 2017).

22. Montroville Wilson Dickeson, "Monumental Grandeur of the Mississippi River Valley! Now Exhibiting for a Short Time Only, With Scientific Lectures on American Aerchiology [*sic*] . . ." (Newark, NJ: Mercury Office, ca. 1851), printed broadside, 15 × 19 cm., Western Americana Collection, Rare Books and Special Collections, Firestone Library, Princeton University.

23. Dickeson, "Monumental Grandeur of the Mississippi River Valley."

24. In 1827 the property encompassed four thousand acres and by 1860, the Feriday family counted 149 enslaved people and 30 slave cabins among their holdings. Robert Dabney Calhoun, "A History of Concordia Parish, Louisiana: Seventh Installment," *Louisiana Historical Quarterly* 16, no. 3 (July 1933): 476; "Ferriday Started as a 3600 Acre Wedding Present," *Natchez Democrat*, November 22, 2006, https://www.natchezdemocrat.com/2006/11/22/ferriday-started-as-3600-acre-wedding-present/.

25. Montroville Wilson Dickeson, "Opening of Mounds upon the Plantation of William Feriday, Concordia Parish, Louisiana," *Lotus* 1, no. 10 (June 3, 1848): 145–47; Jeffrey L. Hantman and Gary Dunham, "The Enlightened Archaeologist," *Archaeology* 46, no. 3 (May/June, 1993): 45.

26. Dickeson, "Opening of Mounds upon the Plantation of William Feriday, Louisiana," 146, including Dickeson's drawing of the seated skeleton.

27. Arjona discusses at length the picturesque associations of the mounds in "Sublime Perversions," 199–203.

28. For discussion of European archaeology's revivifying mission, see Blix, *From Paris to Pompeii*.

29. Dickeson, "Monumental Grandeur of the Mississippi River Valley."

30. Montroville Wilson Dickeson, "Aboriginal Structures, upon the Premises of T. Benard, Esq.," *Lotus*1, no. 13 (June 24, 1848): 198; Dickeson, "Ancient Fortification," 162; Dickeson, "White Apple Village," *Lotus* 1, no. 24 (September 9, 1848): 372.

31. Barnhart, *American Antiquities*, 39.

32. For the link between racial prejudice and the development of archaeology in the United States, see Bruce G. Trigger, "Archaeology and the Image of the American Indian," *American Antiquity* 45, no. 4 (October 1980): 662–76. An extensive, multidisciplinary scholarly literature exists on representations of Native America in the nineteenth century, including Robert F. Berkhofer, *The White Man's Indian: Images of the American Indian from Columbus to the Present* (New York: Knopf, 1978); Julie Schimmel, "Inventing the 'Indian,'" in *The West as America: Reinterpreting Images of the Frontier, 1820–1920*, ed. William H. Truettner (Washington, DC: Smithsonian Institution Press, 1991), 172–74; Jean Fisher, "In Search of the 'Inauthentic': Disturbing Signs in Contemporary Native American Art," *Art Journal* 51, no. 3 (Autumn 1992): 44–50; T. J. Ferguson, "Native Americans and the Practice of Archaeology," *Annual Review of Anthropology* 25 (1996): 63–79; Susan Scheckel, *The Insistence of the Indian: Race and Nationalism in Nineteenth-Century American Culture* (Princeton, NJ: Princeton University Press, 1998); Chip Colwell-Chanthaphonh, "Reconciling American Archaeology and Native America," *Daedalus* 138, no. 2 (Spring 2009): 94–104; Elizabeth Hutchinson, "From Pantheon to Indian Gallery: Art and Sovereignty on the Early Nineteenth-Century Cultural Frontier," *Journal of American Studies* 47, no. 2 (May 2013): 313–37; and Cathy Covell Waegner, ed., *Mediating Indianness* (East Lansing: Michigan State University Press, 2015).

33. John Banvard, *Description of Banvard's Panorama of the Mississippi River, with the Adventures of the Artist* (Boston: John Putnam, 1847), 19; John R. Smith, *Descriptive Pamphlet of Smith's Leviathan Panorama of the Mississippi River!* (Philadelphia: Wm. S. Young, 1848), 13–14; Montroville Wilson Dickeson, "Indian Antiquities. A Course of Popular and Highly Interesting Lectures on American Archeology," Broadsides Collection, American Philosophical Society Library. For Jefferson and science, see Keith Thomson, *Jefferson's Shadow: The Story of His Science* (New Haven, CT: Yale University Press, 2012).

34. See Thomas Hariot, *A Briefe and True Report of the Newfoundland of Virginia* (Frankfurt-am-Main: Ioannis Wecheli and Theodori de Bry, 1590).

35. Pamela Regis, *Describing Early America: Bartram, Jefferson, Crèvecoeur, and the Influence of Natural History* (Philadelphia: University of Pennsylvania Press, 1999), 100–5; Robert J. Miller, *Native America, Discovered and Conquered: Thomas Jefferson, Lewis and Clark, and Manifest Destiny* (Lincoln: University of Nebraska Press, 2008), ch. 4; Sarah Rivett, *Unscripted America: Indigenous Languages and the Origins of a Literary Nation* (New York: Oxford University Press, 2017), ch. 7; Kennedy, "Jefferson and the Indians."

36. For more on Jefferson's excavation, see Hantman and Dunham, "Enlightened Archaeologist," 44–49; Silverberg, *Mound Builders of Ancient America*, 42–47; Barnhart, *American Antiquities*, 79–87; Willey and Sabloff, *History of American Archaeology*, 36–38; and Ronald Hatzenbuehler, "Questioning

Whether Thomas Jefferson Was the 'Father' of American Archaeology," *History and Anthropology* 22, no. 1 (2011): 121–29.

37. Thomas Jefferson, *Notes on the State of Virginia* (1787; repr., Richmond, VA: J. W. Randolph, 1853), 104. See Kennedy, "Jefferson and the Indians," for a discussion of Jefferson's changing views on Native American architecture.

38. Jefferson, *Notes on the State of Virginia*, 105–6.

39. Henry Rowe Schoolcraft, *An Address, Delivered before the Was-Ah Ho-De-No-Son-Ne, or New Confederacy of the Iroquois, at its Third Annual Council* (Rochester, NY: Jerome and Brothers, 1846), quoted in Johnston, *Narrating the Landscape*, 107 (the New Confederacy of the Iroquois was the name of a New York-based antiquarian society); Loudon *Free Press*, January 23, 1854, 2, quoted in Arjona, "Sublime Perversions," 198; George Catlin, "Letter—No. 1," in Catlin, *Letters and Notes on the Manners, Customs, and Conditions of the North American Indians*, vol. 1 (New York: Wiley and Putnam, 1841), 3; Banvard, *Description of Banvard's Panorama*, 15. For Catlin, see William H. Truettner, *The Natural Man Observed: A Study of Catlin's Indian Gallery* (Washington, DC: Smithsonian Institution Press, 1979); George Gurney and Therese Thau Heyman, eds., *George Catlin and His Indian Gallery* (Washington, DC: Smithsonian American Art Museum, 2002); Johnston, *Narrating the Landscape*, ch. 3.

40. Jefferson, *Notes on the State of Virginia*, 109–10; Renato Rosaldo, "Imperialist Nostalgia," *Representations* 26 (Spring 1989): 107–22; J. Gray Sweeney, "Racism, Nationalism, and Nostalgia," in *Race-ing Art History: Critical Readings in Race and Art History*, ed. Kymberly N. Pinder (New York: Routledge, 2002), ch. 8.

41. For a more detailed account of Dickeson's methods, see Veit, "Case of Archaeological Amnesia," 104–8, and Barnhart, *American Antiquities*, 259–61. Ultimately, Dickeson was most interested in securing relics. On an 1846 dig in Mississippi, he recalled, "Our time being limited, we paid but little attention to the stratification, and simply endeavored to get as many relics as possible." Stewart Culin, "The Dickeson Collection of American Antiquities," *Bulletin of the Free Museum of Science and Art of the University of Pennsylvania* 2, no. 3 (January 1900): 122.

42. Dickeson, "White Apple Village," 372; Dickeson, "Ancient Fortification," 161, 162, 163; Dickeson, "Opening of Mounds on the Plantation of Col. A. L. Bingaman," *Lotus* 1, no. 9 (May 27, 1848): 131.

43. Ruth B. Phillips, *Trading Identities: The Souvenir in Native North American Art from the Northeast, 1700–1900* (Seattle: University of Washington Press, 1998), 10. My brief account of the souvenir trade relies on Phillips's study, especially chapters 1 and 2. See also Janet Catherine Berlo, ed., *The Early Years of Native American Art History: The Politics of Scholarship and Collecting* (Seattle: University of Washington Press, 1992), and Elizabeth Hutchinson, *The Indian Craze: Primitivism, Modernism, and Transculturation in American Art, 1890–1915*

(Durham, NC: Duke University Press, 2009). See also Kehoe, *Land of Prehistory*, ch. 4, for the disconnect between archaeological evidence and settler colonial ideology.

44. George Catlin, "Letter—No. 2. Mouth of the Yellow Stone, Upper Missouri, 1832," in Catlin, *Letters and Notes*, 16.

45. Ajrona, "Sublime Perversions," 207–8; Johnston, *Narrating the Landscape*, 137.

46. Isaiah Thomas, "Meeting of September 29, 1813," 17–18; Holmes, "Meeting of October 24, 1814," 62–63.

47. Isaiah Thomas, "Communication from the President," in *Proceedings of the American Antiquarian Society, 1812–1849*, 51, 52n1; "Meeting of June 1, 1814," in *Proceedings of the American Antiquarian Society, 1812–1849*, 43–44n2; "Meeting of October 23, 1816," in *Proceedings of the American Antiquarian Society, 1812–1849*, 99–100; "Meeting of October 23, 1817," in *Proceedings of the American Antiquarian Society, 1812–1849*, 126–27; "Meeting of June 2, 1818," in *Proceedings of the American Antiquarian Society, 1812–1849*, 133.

48. Barnhart, *American Antiquities*, 29, 178; Rudwick, *Earth's Deep History*, 33–45; Trigger, "Anglo-American Archaeology," 139. For a thorough (and decolonizing) account of the collecting practices of the American Antiquarian Society, in particular Native American material culture, see Christine DeLucia, "Antiquarian Collecting and the Transits of Indigenous Material Culture: Rethinking 'Indian Relics' and Tribal Histories," *Commonplace* 17, no. 2, accessed January 29, 2020, http://commonplace.online/article/antiquarian-collecting-and-the-transits-of-indigenous-material-culture/; and DeLucia, "Fugitive Collections in New England Indian Country: Indigenous Material Culture and Early American History Making at Ezra Stiles's Yale Museum," *William and Mary Quarterly* 75, no. 1 (January 2018): 109–50.

49. See, for example, *Revolutionary Relics, or, Clinton Correspondence; Comprising the Celebrated Papers Found in André's Boots; Private and Confidential Letters from Washington, Hamilton, Lafayette, and Other Distinguished Officers and Statesmen of the American Revolution* (New York: New York Herald, 1842); *William Lanier Washington's Collection of Relics and Memorabilia of George Washington, to be sold at unrestricted public sale by order of Mr. Washington . . .* (New York: American Art Association, 1920).

50. Alexander von Humboldt, *Cosmos: A Sketch of a Physical Description of the Universe*, vol. 1, trans. E. C. Otté (London: Henry G. Bohn, 1864), 213; see also 199–206, 212. See also Humboldt, *Views of Nature*, ed. Stephen T. Jackson and Laura Dassow Walls, trans. Mark W. Person (Chicago: University of Chicago Press, 2014), 144–47, 189–90, 247–55.

51. W. Stump Forwood, *An Historical and Descriptive Narrative of the Mammoth Cave of Kentucky* (Philadelphia: J. B. Lippincott, 1870), 84–102; Alexander

von Humboldt, *Florae Fribergensis specimen: plantas cryptogamicas praesertim subterraneas exhibens* (Berolini: H. A. Rottmann, 1793); Humboldt, "Account of the Great Cavern of the Guacharo," *Edinburgh Philosophical Journal* 3 (1820): 83–92; Humboldt, *Views of Nature*, 248; Burton Saxe, *The Cave Secret; or, The Mystery of Night Island* (New York: American News, 1865); Edward Bulwer-Lytton, *The Coming Race* (Santa Barbara, CA: Woodbridge, 1989); Jules Verne, *Journey to the Centre of the Earth* (Mineola, NY: Dover, 2005); Allen A. Debus, "Re-Framing the Science in Jules Verne's *Journey to the Center of the Earth*," *Science Fiction Studies* 33, no. 3 (2006): 408–14; Rosalind Williams, *Notes on the Underground: An Essay on Technology, Society, and the Imagination*, new ed. (Cambridge, MA: MIT Press, 2008), 31–38; Rudwick, *Earth's Deep History*, chs. 5–6.

52. Saxe, *Cave Secret*, 15.

53. Browman and Givens, "Stratigraphic Excavation," 80–81.

54. Dickeson, "Indian Antiquities."

55. James Madison, "A Letter on the Supposed Fortifications of the Western Country, from Bishop Madison of Virginia to Dr. Barton," *Transactions of the American Philosophical Society* 6, no. 1 (1804): 140.

56. Huhtamo, *Illusions in Motion*, 180.

57. Miller, "Panorama, the Cinema"; Griffiths, "'Shivers Down Your Spine.'"

58. John Banvard, *Description of Banvard's Geographical Painting of the Mississippi, with Historical Additions of the War* (New York: L. H. Biglow, 1862), 19; Smith, *Descriptive Pamphlet*, 10–11, 32; John R. Smith, *Descriptive Book of the Tour of Europe, the Largest Moving Panorama in the World* (New York: Pettiner and Gray, 1855), 7–33, 44.

59. Both Byrd, in *Pedagogy of Observation*, and Huhtamo, in *Illusions in Motion*, discuss at length the paratextual aspect of panoramas.

60. See, for example, Smith, *Descriptive Book of the Tour of Europe*, vol. 3, and George R. Gliddon, *Hand-Book to the American Panorama of the Nile* (London: James Madden, 1849), 8, 17.

Chapter 2: "All That Is Solid Melts into Air"

Parts of this material were presented at the 2016 Interdisciplinary Nineteenth-Century Studies conference; at Wolfson College, Cambridge, and Guelph University in 2017; and at the Victorian Studies Association of Ontario annual conference in 2018. I thank my INCS copanelists Deb Gettelman and Anne Leonard, and members of the audience at each venue, including Tina Choi, Anne Colley, Meredith Hale, Lorraine Janzen Kooistra, Christina Smylitopoulous, and Lindsay Wells for their insightful questions and comments. I am also very grateful to Nancy Rose Marshall and the two peer reviewers of the manuscript for their excellent suggestions.

1. Letter from Burne-Jones to Charles Eliot Norton, dated December 20, 1894, quoted in Duncan Robinson, *William Morris, Edward Burne-Jones and the Kelmscott Chaucer* (Mount Kisco, NY: Moyer Bell, 1986), 36.

2. Helen Phillips, "Chaucer and the Nineteenth-Century City," in *Chaucer and the City*, ed. (Cambridge: D. S. Brewer, 2006), 196. Steve Ellis discusses Burne-Jones's illustrations as primarily interested in "presenting Chaucer as the poet of Nature." Ellis, *Chaucer at Large: The Poet in the Modern Imagination* (Minneapolis: University of Minnesota Press, 2000), 2.

3. Jennifer L. Roberts, "The Power of Patience: Teaching Students the Value of Deceleration and Immersive Attention," *Harvard Magazine*, November–December 2013, 40–43.

4. Elizabeth Carolyn Miller, *Slow Print: Literary Radicalism and Late Victorian Print Culture* (Stanford, CA: Stanford University Press, 2013), 2.

5. Bruce Clarke, *Energy Forms: Allegory and Science in the Era of Classical Thermodynamics* (Ann Arbor: University of Michigan Press, 2001), 18.

6. Geoffrey Chaucer, *The Works of Geoffrey Chaucer*, ed. F. S. Ellis (Hammersmith: Kelmscott, 1896), 459–60.

7. Kathy Cawsey, "'Alum de glas' or 'Alymed glass'? Manuscript Reading in Book III of *The House of Fame*," *University of Toronto Quarterly* 73, no. 4 (Fall 2004): 975, 976.

8. John Ruskin, *Of Mountain Beauty*, vol. 4 of *Modern Painters* (1856; New York: John Wiley and Sons, 1880), 139.

9. Ruskin, *Of Mountain Beauty*, 114, 95, 112, 114.

10. John Ruskin, *Deucalion: Collected Studies of the Lapse of Waves and Life of Stones*, in *The Works of John Ruskin*, vol. 26, ed. E. T. Cook and Alexander Wedderburn (London: George Allen, 1906), 126; Ruskin, "La Douce Dame" letter 34 (October 1873), in *Fors Clavigera: Letters to the Workmen and Labourers of Great Britain*, vol. 3, in *The Works of John Ruskin*, ed. E. T. Cook and Alexander Wedderburn, vol. 27 (London: George Allen, 1907), 639.

11. James D. Forbes, *Travels through the Alps of Savoy and Other Parts of the Pennine Chain with Observations on the Phenomena of Glaciers* (Edinburgh: Adam and Charles Black, 1843), 22. As Flint notes, glaciers "invited linguistic comparisons." Kate Flint, *The Victorians and the Visual Imagination* (Cambridge: Cambridge University Press, 2000), 129.

12. Marc-Théodore Bourrit, *A Relation of a Journey to the Glaciers, in the Dutchy of Savoy*, trans. C. and F. Davy (Norwich: Richard Beatniffe, 1775), 112. Bourrit is describing the Glacier Montanvert, as the Mer de Glace was formerly known.

13. Fiona MacCarthy, *The Last Pre-Raphaelite: Edward Burne-Jones and the Victorian Imagination* (London: Faber and Faber, 2011), 108, 140. Ruskin rejected the sliding and dilation theories of glacier movement; generally agreed with

James Forbes that the glacier is a "semifluid," its motion that of a "viscous fluid"; and heaped contempt upon John Tyndall's theories of fracture and regelation. Forbes, *Travels through the Alps*, 366, 368. E. T. Cook's introduction to volume 26 of *The Works of John Ruskin* offers a summary of the controversy over glacier movement. See also Flint, *Victorians and the Visual Imagination*, 117–38; Bruce Hevly, "The Heroic Science of Glacier Motion," *Osiris* 11 (1996): 66–86; and Paul L. Sawyer, "Ruskin and Tyndall: The Poetry of Matter and the Poetry of Spirit," *Annals of the New York Academy of Sciences* 360, no. 1 (April 1981): 217–46. Nanna Kaalund has argued that the glacier controversy precipitated scientists' use of the periodical press to challenge and establish cultural authority. Nanna Katrine Lüders Kaalund, "A Frosty Disagreement: John Tyndall, James David Forbes, and the Early Formation of the X-Club," *Annals of Science* 74, no. 4 (2017): 282–98.

14. "Grosvenor Gallery Winter Exhibition," *Academy* 454 (January 15, 1881): 50.

15. Allison Smith, entry for cat. no. 86 in Alien Staley and Christopher Newall, *The Pre-Raphaelite Vision: Truth to Nature* (London: Tate, 2004), 153.

16. Christopher Newall, "Understanding the Landscape," in Staley and Newall, *Pre-Raphaelite Vision*, 139; Kenneth Bendiner, *An Introduction to Victorian Painting* (New Haven, CT: Yale University Press, 1985), 47.

17. Bendiner, *Introduction to Victorian Painting*, 48–49; Christiana Payne, *John Brett: Pre-Raphaelite Landscape Painter* (New Haven, CT: Yale University Press, 2010), 32–33. Agassiz praised the painting when it was exhibited in Boston in 1858. Payne, *John Brett*, 35.

18. Bendiner, *Introduction to Victorian Painting*, 47, 56, 50, 47, 50.

19. Smith, entry in *Pre-Raphaelite Vision*, 153.

20. Payne, *John Brett*, 32.

21. Entries for July 25, July 19, August 20, July 22, July 21, and August 6, 1871 in William Morris, *Icelandic Journals* (London: Mare's Nest, 1996), 53, 30, 31, 121, 42, 62, 61, 14, 84. One of Burne-Jones's many caricatures of Morris depicts him climbing a mountain in Iceland, his prominent rear end stressing the undignified nature of his ascent (ca. 1871, collection of Peter and Renate Nahum). The rock formations in this drawing informed Burne-Jones's early sketches for the names of the famous engraved in ice.

22. Chaucer, *Works of Geoffrey Chaucer*, 460.

23. Ruskin, *Of Mountain Beauty*, 165; Ruskin, "La Douce Dame," 635. His photographs and careful sketches enabled this monitoring and measurement.

24. Ian Warrell, entry for cat. no. 82 in Staley and Newall, *Pre-Raphaelite Vision*, 149; Ruskin, "La Douce Dame," 635–36.

25. Bert Hansen, "The Early History of Glacial Theory in British Geology," *Journal of Glaciology* 9, no. 55 (1970): 135.

26. Glaciers were not accurately measured until the end of the nineteenth century, but glaciologists have used nineteenth-century texts, drawings, paintings,

and photographs to reconstruct the movements of several well-known Swiss glaciers, including the Mer de Glace, revealing the extent of loss since 1850. See, for example, Samuel U. Nussbaumer and Heinz J. Zumbühl, "The Little Ice Age History of the Glacier des Bossons (Mont Blanc Massif, France): A New High-Resolution Glacier Length Curve Based on Historical Documents," *Climatic Change* 111, no. 2 (2012): 302. In 2014 several papers in the "Ice Roundtables" organized by Jeffrey Jerome Cohen at the New Chaucer Society congress in Reykjavík considered the mountain on which Fame's palace is built in terms of glacial ice. For the abstracts, see Cohen, "Ice at #NCS14," *In the Middle* (blog), June 20, 2014, http://www.inthemedievalmiddle.com/2014/06/ice-at-ncs14.html.

27. Chaucer, *Works of Geoffrey Chaucer*, 460.

28. Laurel Amtower, "Authorizing the Reader in Chaucer's *House of Fame*," *Philological Quarterly* 79, no. 3 (Summer 2000): 275, 283.

29. Although the classics still formed an essential part of British schooling, advocacy for scientific education increased in the second half of the nineteenth century, and the values attached to traditional cultural capital seemed fragile if not already fading in the wake of rapid cultural shifts. "Who now reads the ancient writers?" asked Frederic Harrison in *The Choice of Books and Other Literary Pieces* (London: Macmillan, 1886), 13. For an overview of educational debates and Tyndall's role in them, see Ursula DeYoung, *A Vision of Modern Science: John Tyndall and the Role of the Scientist in Victorian Culture* (New York: Palgrave Macmillan, 2011), ch. 4.

30. See Gregg Hecimovich, *Puzzling the Reader: Riddles in Nineteenth-Century British Literature* (New York: Peter Lang, 2008).

31. David K. Coley, "'Withyn a temple ymad of glas': Glazing, Glossing, and Patronage in Chaucer's *House of Fame*," *Chaucer Review* 45, no. 1 (2010): 63.

32. Ruskin, *Of Mountain Beauty*, 59.

33. Isobel Armstrong, *Victorian Glassworlds: Glass Culture and the Imagination, 1830–1880* (Oxford: Oxford University Press, 2008), 295.

34. As Nicholas Daly has shown, walking through the nineteenth-century city entailed "negotiating a hasty path through a baffling array of *actual* texts" in the form of posters, billboards, sandwichmen, window cards, handbills, newspapers, etc.; "adults and children alike . . . experienced the actual three-dimensional city as a printed surface." Daly, *The Demographic Imagination and the Nineteenth-Century City: Paris, London, New York* (Cambridge: Cambridge University Press, 2015), 108.

35. William Morris, *Stories in Prose, Stories in Verse, Shorter Poems, Lectures and Essays*, ed. G. D. H. Cole (London: Nonesuch Press, 1934), 524.

36. A. L. Morton, "Morris, Marx, and Engels," *Journal of William Morris Studies* 7, no. 1 (Autumn 1986): 47.

37. Karl Marx and Frederick Engels, *Manifesto of the Communist Party*, trans. Samuel Moore with Frederick Engels (London: William Reeves, 1888), 10, 9.

Marshall Berman and Zygmunt Bauman explore the image of "all that is solid melt[ing] into air" in modernity; Bauman, focusing on the imagery of melting solids, argues that modernity was "a process of 'liquefaction' from the start," essentially "fluid." Bauman, *Liquid Modernity* (Cambridge: Polity, 2000), 2, 3; Berman, *All That Is Solid Melts into Air: The Experience of Modernity* (New York: Simon and Schuster, 1982).

38. William Morris, *The Ideal Book: Essays and Lectures on the Arts of the Book*, ed. William S. Peterson (Berkeley: University of California Press, 1982), 1. Miller considers the later nineteenth-century print marketplace, with its emphasis on speed and profit, "a synecdoche for capitalism, an automatic machine for reproducing the logic of mass production." *Slow Print*, 2.

39. Harrison, *Choice of Books*, 16, 2. Harrison and Morris shared a concern for the protection of ancient buildings. Florence S. Boos, ed., "William Morris's 'Commercial War': A Critical Edition," *Journal of Pre-Raphaelite Studies* 19 (Fall 2010): 64n4.

40. Morris, *Ideal Book*, 1.

41. Morris, *Ideal Book*, 75–76.

42. While the torrent of print in the Victorian period might be imagined as resulting in a greater likelihood of individual works' survival, Morris himself believed the poor quality of paper, ink, and printing would damage their longevity. He never tired of praising Richard Jefferies's 1885 novel *After London* and may have particularly appreciated its discussion of book survival: after a catastrophe that destroys London and much of the population of England, most printed books have "mouldered away," leading to a reversion to manuscript copies of the classics. Richard Jefferies, *After London; or Wild England*, ed. Mark Frost (Edinburgh: Edinburgh University Press, 2017), 93.

43. William S. Peterson, *The Kelmscott Press: A History of William Morris's Typographical Adventure* (Berkeley: University of California Press, 1991), 151; Robinson, *William Morris*, 33–34.

44. I thank Lorraine Janzen Kooistra for pointing this out. On the significance of wood engraving as a visual technology of Victorian modernity, see Kooistra's *Poetry, Pictures, and Popular Publishing: The Illustrated Gift Book and Victorian Visual Culture, 1855–1875* (Athens: Ohio University Press, 2011).

45. Armstrong, *Glassworlds*, 151. Glass is now considered an amorphous solid.

46. Ruskin, *Deucalion*, 127.

47. Coley, "'Withyn a temple ymad of glas,'" 63. Burne-Jones's preliminary studies draw out the initial uncertainty presented by Chaucer's text; one is labeled "the glass mountain full of names," another, "foundations of ice."

48. Armstrong, *Glassworlds*, 1. Jasmine Allen has suggested the importance of contextualizing Burne-Jones's stained-glass designs within the culture of popular spectacle, mass consumption, and optical devices. Allen, "Stained

Glass and the Culture of Spectacle, 1780–1862," *Visual Culture in Britain* 13, no. 1 (2012): 1–23.

49. Armstrong, *Glassworlds*, 5, 92.

50. Viewing Ruskin's daguerreotype of the Mer de Glace would entail similar optical effects.

51. Armstrong, *Glassworlds*, 141. Some of the thicker black lines may also evoke the leadwork of stained-glass windows, a reading not incompatible with the illustration's suggestion of commodification. See Allen, "Stained Glass."

52. London's coal-fueled smogs, which were at their worst in winter, reached a peak in the 1880s. They left "greasy, oily deposits on surfaces such as windows and pavements." Christine L. Corton, *London Fog: The Biography* (Cambridge, MA: Belknap Press of Harvard University Press, 2015), 85, 48.

53. Commodification was not solely a feature of the urban landscape, and it is also possible to see in the peering figure of Chaucer's traveling narrator a tourist. Alpine tourism grew significantly in the mid- to late nineteenth century, partially in response to midcentury investigations of glacier movement. Travelers in search of the mountain sublime and solitude in the majestic peaks, however, were increasingly disappointed as the Alps became crowded and permeated by "the commercial spirit," as Anne Colley has shown. Colley, *Victorians in the Mountains: Sinking the Sublime* (Farnham, UK: Ashgate, 2010), 29. The conflation of Alpine and urban experience in Burne-Jones's illustration suggests modernity's collapsing of the distances and differences between places. Flint, *Victorians and the Visual Imagination*, 135.

54. Chaucer, *Works of Geoffrey Chaucer*, 461, 460, 461. William Thackeray, "Mr. Molony's Account of the Crystal Palace, *Punch*, April 26, 1851, 171.

55. Greenland's Crystal Palace Glacier took the building's name, and authors invoked the structure to describe icy phenomena: an article in *St. Nicholas*, for example, described an iceberg as "a floating crystal palace with transparent walls." John R. Coryell, "The Seals' Crystal Palace," *St. Nicholas*, November 1891–April 1892, 366–67. On the frosted glass glaciers in the Sydenham Crystal Palace, see "The Morning Chronicle," *Morning Chronicle*, June 12, 1854, 6. I am grateful to Nancy Rose Marshall for this reference.

56. "The Crystal Palace," *Times* (London), January 15, 1851, 5.

57. Diary of Emily Hall of Ravenswood, West Wickham, June 5, 1854 and June 13, 1854, Bromley Library, quoted in Nancy Rose Marshall, *City of Gold and Mud: Painting Victorian London* (New Haven, CT: Yale University Press, 2012), 233, 281n3.

58. Elizabeth Eastlake, "The Crystal Palace," *Quarterly Review* 96, no.192 (March 1855): 304, 305.

59. Armstrong, *Glassworlds*, 96.

60. Armstrong, *Glassworlds*, 161; Eileen Gillooly, "Rhetorical Remedies for Taxonomic Troubles," in *Victorian Prism: Refractions of the Crystal Palace*, ed.

James Buzard, Joseph W. Childers, and Eileen Gillooly (Charlottesville: University of Virginia Press, 2007), 36.

61. Gillooly, "Rhetorical Remedies," 35; Peter Gurney, "'A Palace for the People'? The Crystal Palace and Consumer Culture in Victorian England," in Buzard, Childers, and Gillooly, *Victorian Prism*, 138.

62. "The Morning Chronicle," *Morning Chronicle*, 6. See Marshall on the Sydenham Crystal Palace and its grounds as "capitalism made manifest, a visual testimony to . . . free enterprise," yet one in which elements like the ethnology exhibit and the dinosaurs suggested that "progress could lead to extinction rather than higher achievement." Marshall, *City of Gold and Mud*, 237, 245.

63. "The Morning Chronicle," *The Morning Chronicle*, 6.

64. Eastlake, "Crystal Palace," 322.

65. Georgiana Burne-Jones, *Memorials of Edward Burne-Jones* (London: Macmillan, 1904), 1:101.

66. Jesse Oak Taylor, "The Novel as Climate Model: Realism and the Greenhouse Effect in *Bleak House*," *Novel: A Forum on Fiction* 46, no. 1 (Spring 2013): 6, 7, 4, 8.

67. William Shakespeare, *The Tempest*, in *The Norton Shakespeare*, 3rd ed., ed. Stephen Greenblatt et al. (New York: W. W. Norton, 2016), 4.1.150, 152–54. References are to act, scene, and line.

68. Ruskin, *Of Mountain Beauty*, 69.

69. John Tyndall, *Heat Considered as a Mode of Motion; Being a Course of Twelve Lectures Delivered at the Royal Institution of Great Britain in the Season of 1862* (London: Longman, Green, Longman, Roberts, & Green, 1863), 434. Tyndall "wrote extensively about the conservation of energy" but not entropy. Roland Jackson, *The Ascent of John Tyndall: Victorian Scientist, Mountaineer, and Public Intellectual* (Oxford: Oxford University Press, 2018), 178.

70. Tobias Meneley, "Anthropocene Air," *Minnesota Review* 83 (2014): 97.

71. William A. Quinn, "Chaucer's Recital of Presence in the *House of Fame* and the Embodiment of Authority," *Chaucer Review* 43, no. 2 (2008): 185.

72. Nina Dubin, *Futures and Ruins: Eighteenth-Century Paris and the Art of Hubert Robert* (Los Angeles: Getty Research Institute, 2010).

73. Burne-Jones, *Memorials*, 2:282. I thank Nancy Rose Marshall for pointing out the resemblance to the Doré illustration.

74. T. B. Macaulay, review *The Ecclesiastical and Political History of the Popes, during the Sixteenth and Seventeenth Centuries*, by Leopold van Ranke, trans. Sarah Austin, in *Edinburgh Review* 72, no. 145 (October 1840): 228.

75. Robert Dingley, "The Ruins of the Future: Macaulay's New Zealander and the Spirit of the Age," in *Histories of the Future: Studies in Fact, Fantasy and Science Fiction*, ed. Alan Sandison and Robert Dingley (Basingstoke, UK: Palgrave, 2000), 17, 25.

76. Blanchard Jerrold and Gustave Doré, *London: A Pilgrimage* (London: Grant, 1872), 5, 190.

77. David John Skilton, "Tourists at the Ruins of London: The Metropolis and the Struggle for Empire," *Cercles: revue pluridisciplinaire du monde anglophone* 17 (2007): 96.

78. Dingley, "Ruins of the Future," 25.

79. Velma Bourgeois Richmond, "Edward Burne-Jones's Chaucer Portraits in the Kelmscott 'Chaucer,'" *Chaucer Review* 40, no. 1 (2005): 26.

80. Richmond notes the impact of the journeys to Italy both artists took on their development, their synthesis of the classical and the medieval, and the resemblance between Burne-Jones's Chaucer portraits and his self-portraits in caricature. "Edward Burne-Jones's Chaucer Portraits," 6. For the portrait, see Thomas Hoccleve, *The Regiment of Princes*, ca. 1411–1420, British Library, MS Harley 4866, f. 88r.

81. Richmond, "Edward Burne-Jones's Chaucer Portraits," 28.

82. Morris, *Stories in Prose*, 630. For a recent overview of the literature on Morris and ecology, see Patrick O'Sullivan, "'Morris the Red, Morris the Green'—A Partial Review," *Journal of the William Morris Society* 19, no. 3 (Winter 2011): 22–38.

83. John Ruskin, *Queen of the Air, Being a Study of the Greek Myths of Storm and Cloud* (1869; London: George Allen, 1906), viii–ix.

84. Richmond, "Edward Burne-Jones's Chaucer Portraits," 27.

85. Chaucer, *Works of Geoffrey Chaucer*, 464.

86. Diana C. Archibald, "Beauty, Unity, and the Ideal: Wholeness and Heterogeneity in the Kelmscott *Chaucer*," *Studies in Medievalism* 7 (1995): 176.

87. The verdant environment created by Morris's borders offers a striking contrast to the landscapes of the House of Fame. Caroline Arscott argues that Morris's borders offered Burne-Jones's fragile figures a "protective wrapping," one that enabled the expression of the artist's anxieties. Arscott, *William Morris and Edward Burne-Jones: Interlacings* (New Haven, CT: Yale University Press, 2008), 57.

88. Ruskin, "La Douce Dame," 636.

89. The American Eunice Foote had made the discovery three years earlier, unbeknownst to Tyndall. Roland Jackson suggests that if Ruskin and Tyndall had had a more amicable relationship, Ruskin's observations of glacial retreat and awareness of the threat industrial activity posed to the planet "could have benefited from Tyndall's rigorous science and challenged Tyndall to explore the consequences of burning fossil fuels." Jackson, *Ascent of John Tyndall*, 158, 161, 255.

90. John Tyndall, *Forms of Water in Clouds and Rivers, Ice and Glaciers* (New York: D. Appleton, 1872), xi.

91. Burne-Jones's work, as John Holmes points out, is not modeled on science. But, like some of the Rossetti poems that, Holmes argues, combine "the

investigative gaze of science with the humane and open-minded sympathy of poetry," Burne-Jones's fusion of literary, artistic, and scientific sources may help viewers "to reach a more comprehensive understanding" than can be "achieved through the reductive procedures of science on their own." Holmes, *The Pre-Raphaelites and Science* (New Haven, CT: Yale University Press, 2018), 7, 92.

92. I thank Tina Choi for pointing out the resemblance to Humboldt's illustration.

93. Burne-Jones, *Memorials*, 1:51.

94. John Tresch, *The Romantic Machine: Utopian Science and Technology after Napoleon* (Chicago: University of Chicago Press, 2012), 86.

95. Michael Dettelbach, "The Face of Nature: Precise Measurement, Mapping, and Sensibility in the Work of Alexander von Humboldt," in *Studies in the History and Philosophy of Biological and Biomedical Sciences* 30, no. 4 (1999): 503.

96. Nicholas Mirzoeff, "Visualizing the Anthropocene," *Public Culture* 26, no. 2(Spring 2014): 213.

97. Mirzoeff, "Visualizing the Anthropocene," 217.

98. Arscott, *William Morris and Edward Burne-Jones*, 27.

Chapter 3: Grasping the Elusive

1. George MacDonald, *At the Back of the North Wind* (London: Strahan, 1871). The story is featured in each monthly issue of *Good Words for the Young* from November 1868 to October 1870.

2. The fifth chapter from my PhD dissertation, titled "Waves, Wind, and the Body," forms the basis for this chapter. See Rebecca Carey Gibbons, "The Limits of the Body in Victorian Illustration: Arthur Hughes and Frederick Sandys" (Courtauld Institute of Art, University of London, 2016), 243–90.

3. In a letter from Arthur Hughes to Mary MacDonald, George MacDonald's daughter, on September 26, 1873, Hughes mentions the month he spent with her family in Hastings. Mary MacDonald was engaged to Hughes's nephew, the artist Edward Robert Hughes, for four years before her death from tuberculosis. On December 16, 1873, Arthur Hughes wrote to George MacDonald's wife, Louisa MacDonald, reluctantly turning down an invitation to the MacDonald house for the holidays. Mss 103, series I, box 2, folder 86, George MacDonald Collection, Beinecke Rare Book and Manuscript Library, Yale University. Additionally, in his biography of MacDonald, Joseph Johnson relates Hughes's fond memories of the theatrical performances at "The Hermitage," the MacDonald home in Hammersmith, and his admiration for the author. Johnson, *George MacDonald: A Biographical and Critical Appreciation* (London: Sir Isaac Pitman and Sons, 1906), 180, 59–60. Greville MacDonald writes of his friend Sir Johnston Forbes-Robertson's memories of a performance at the MacDonald home, described as a "charade representing the *Story of the Three Bears*," in which his father George covered himself with a skin rug and pretended to be the Big

Bear, while Hughes joined in "with great gusto." Greville MacDonald, *George MacDonald and His Wife* (London: G. Allen and Unwin, 1924), 312–13.

4. MacDonald, *MacDonald and His Wife*, 361.

5. "At the Back of the North Wind," *Art Journal* 33 (March 1871): 96.

6. Alexander Strahan to Lilia (Lily) MacDonald, September 22, 1870; December 30, 1870. Strahan also mentioned the volumes in a letter dated December 13, 1873, to Mrs. MacDonald, writing, "This is a debt that will be discharged some day—gold and ivory permitting." It is unclear whether the project was ever carried out; in his December 30 letter, Strahan blames Otto von Bismarck's siege of Paris for the delay in having them bound in France. Mss 103, series I, box 3, folder 134, George MacDonald Collection.

7. MacDonald, *At the Back of the North Wind*, 125, 75.

8. MacDonald, *At the Back of the North Wind*, 73.

9. MacDonald, *At the Back of the North Wind*, 11.

10. See Bernard Lightman, *Victorian Popularizers of Science: Designing Nature for New Audiences* (Chicago: University of Chicago Press, 2010) and Geoffrey Cantor et al., *Science in the Nineteenth-Century Periodical: Reading the Magazine of Nature* (Cambridge: Cambridge University Press, 2004). Hughes humorously suggests a familiarity with contemporary scientific debates in a letter referencing the Victorian biologist Thomas Huxley, writing, "I have just read this quotation, from Huxley I think: 'Man is a Parasite on the surface of a tiny Satellite revolving round a tenth-rate Sun in an obscure corner of a petty Constellation. . . . He can at most look forward or back-ward a few hundred years, a mere swing of the Cosmical Pendulum which marks the interminable history of Suns and Systems,' and I think now I wont be conceited ever any more . . ." Arthur Hughes to Alice Boyd, May 20, 1892, RBSC-ARC-1433-6-6, Penkill Collection, University of British Columbia Library Rare Books and Special Collections.

11. Bernard Lightman, ed., *Victorian Science in Context* (Chicago: University of Chicago Press, 1997), 192–94.

12. John George Wood, *The Common Objects of the Country* (London: G. Routledge, 1858), 1–2.

13. Wood writes, "Almost every object that is described by the pen will be figured by the pencil." Wood, *Common Objects of the Country*, 2.

14. Useful sources for information on Victorian meteorology are Katharine Anderson, *Predicting the Weather: Victorians and the Science of Meteorology* (Chicago: University of Chicago Press, 2005) and Mark Monmonier, *Air Apparent: How Meteorologists Learned to Map, Predict, and Dramatize Weather* (Chicago: University of Chicago Press, 1999).

15. Monmonier provides a helpful overview of the history of weather reports, charts, and maps. See *Air Apparent*, 42–46, 154–57. See also Robert H. Scott, "Weather Charts in Newspapers," *Journal of the Society of Arts* 23, no. 1183 (July 23, 1875): 776–82; and R. G. K. Lempfert, "British Weather

Forecasts: Past and Present," *Quarterly Journal of the Royal Meteorological Society* 39, no. 167 (1913): 173–84. The National Meteorological Archive in Exeter, England, holds a full set of Daily Weather Reports from 1860 to the present day. They have been scanned and are available via their digital library and archive: https://digital.nmla.metoffice.gov.uk/SO_86058de1-8d55-4bc5-8305-5698d0bd7e13/. The archive also holds Daily Synoptic Charts and Quarterly Weather Reports.

16. Beginning in 1871, the *Shipping Gazette*, a London trade newspaper, published a less comprehensive daily wind chart from the Meteorological Office.

17. Alexander Buchan, *Handy Book of Meteorology*, 2nd ed. (Edinburgh: William Blackwood and Sons, 1868), viii.

18. See Bruce Clarke, *Energy Forms: Allegory and Science in the Era of Classical Thermodynamics* (Ann Arbor: University of Michigan Press, 2001), 163–92; Daniel M. Siegel, "Thomson, Maxwell, and the Universal Ether in Victorian Physics," in G. N. Cantor and M. J. S. Hodge, *Conceptions of Ether: Studies in the History of Ether Theories, 1740–1900* (Cambridge: Cambridge University Press, 1981), 239–66; and Steven Connor, *The Matter of Air: Science and Art of the Ethereal* (London: Reaktion Books, 2010), 148–72.

19. Gillian Beer, *Open Fields: Science in Cultural Encounter* (Oxford: Clarendon, 1996), 298.

20. Beer, *Open Fields*, 299.

21. John F. W. Herschel, "The Weather and Weather Prophets," *Good Words* 5 (January 1864): 59.

22. John F. W. Herschel, *Meteorology from the Encyclopaedia Britannica* (Edinburgh: Adam and Charles Black, 1861), 14.

23. Anderson, *Predicting the Weather*, 89–90. See also Vladimir Jankovic, "Ideological Crests versus Empirical Troughs: John Herschel's and William Radcliffe Birt's Research on Atmospheric Waves, 1843–50," *British Journal for the History of Science* 31, no. 1 (March 1998): 21–40; Aitor Anduaga, *Politics, Statistics and Weather Forecasting, 1840–1910: Taming the Weather* (London: Routledge, 2020), 48–53.

24. John Herschel to William Radcliffe Birt, July 28, 1843, RS:HS 19.106, Herschel Papers 19.106, Archives, Royal Society of London. Quoted in Jankovic, "Ideological Crests versus Empirical Troughs," 29, note 34. See also Gregory A. Good, "A Shift of View: Meteorology in John Herschel's Terrestrial Physics," in James Rodger Fleming, Vladimir Jankovic, and Deborah R. Coen, eds., *Intimate Universality: Local and Global Themes in the History of Weather and Climate* (Sagamore Beach, MA: Science History Publications, 2006), 50–54.

25. William Radcliffe Birt, "A Contribution to Our Knowledge of Atmospheric Waves," *London, Edinburgh, and Dublin Philosophical Magazine and Journal of Science* 44, series 4, no. 291 (August 1872): 138, and Birt, *Handbook of the Law of Storms: being a Digest of the Principal Facts of Revolving Storms for the Use*

of Commanders in Her Majesty's Navy and Mercantile Marine (London: George Philip and Son, 1856), 120.

26. Robert Fitzroy, *The Weather Book: A Manual of Practical Meteorology* (London: Longman, Green, Longman, Roberts and Green, 1863), 3, 75.

27. Fitzroy, *Weather Book*, 70. Fitzroy cites the prominent nineteenth-century German meteorologist Heinrich Dove as the proponent of the law of gyration. Gyration, or wind veering, was the result of equatorial (warm, moist) and polar (cold, dry) air currents alternatively displacing each other. See also Anderson, *Predicting the Weather*, 89; H. W. Dove, *The Law of Storms Considered in Connection with the Ordinary Movements of the Atmosphere*, 2nd ed., trans. Robert H. Scott (London: Longman, Roberts, Green, 1862).

28. Herschel, *Meteorology from the Encyclopaedia Britannica*, 69. Fitzroy, *Weather Book*, 79, 95, 105. Fitzroy also disagrees with the "wave theory" assertion that the lowest part of the wave, or the trough, occurs with the lowest barometer reading and the crest, or highest point, occurs with the highest reading. He argues that instead the lightest portions of winds should rise the highest and the heaviest portions should be lowest.

29. Fitzroy, *Weather Book*, 104. James Pollard Espy, *Report on Meteorology* (Washington DC: Government Printing Office, 1850). Espy's book includes charts of comparative barometric readings from his many weather correspondents.

30. "Admiral Fitzroy's Weather-Book," *London Review of Politics, Society, Literature, Art, and Science* 6, no. 147 (April 25, 1863): 444.

31. Fitzroy, *Weather Book*, 103, foldout in back of book.

32. Similar charts tracking barometric pressure and featuring wavy lines can be found in the copies of the *Quarterly Weather Report of the Meteorological Office* for the years 1869–1880 (London: E. Stanford), X31.F-B, National Meteorological Library and Archive, Met Office, Exeter, UK. Although they did not incorporate wavy forms, wind star charts also sought to provide an ideal grasp or summary of the behavior of wind. They consist of a central circle, marked with latitude and longitude, surrounded by "rays" spreading out in triangular shapes. The longest, broadest rays correspond to the greatest number of observations of wind in that direction. Wind Charts (London: Board of Trade, 1855–1867), W05.P3, National Meteorological Library and Archive.

33. Scott, "Weather Charts," 777.

34. B.S. [Balfour Stewart], *Encyclopaedia Britannica*, 9th ed., vol. 16 (1883), s.v. "meteorology." *Isobar* is derived from two Greek words meaning "equal pressure." The map is constructed from mean values calculated for eleven years (1870–1880 inclusive), with pressure of 30 inches and upward indicated with solid lines and 29.9 inches and under by dotted lines. Arrows show the direction of the prevailing winds at the localities indicated by the respective arrow points. Winds are understood to be strongest where the isobars are closest together.

35. The strong appeal of weather charts and maps within meteorology should be considered against the larger trend of proliferating forms of visual representation in an era of new scientific instruments, observation methods, and printing techniques. See Jonathan Crary, *Techniques of the Observer: On Vision and Modernity in the Nineteenth Century* (Cambridge: MIT Press, 1992); Caroline A. Jones and Peter Galison, eds., *Picturing Science, Producing Art* (London: Routledge, 1998); Larry J. Schaaf, *Out of the Shadows: Herschel, Talbot, and the Invention of Photography* (New Haven, CT: Yale University Press, 1992); Klaus Hentschel, *Mapping the Spectrum: Techniques of Visual Representation in Research and Teaching* (Oxford: Oxford University Press, 2009).

36. Francis Galton, *Meteorographica, or Methods of Mapping the Weather* (London: Macmillan, 1863), 5.

37. Francis Galton, "A Development of the Theory of Cyclones," *Proceedings of the Royal Society of London* 12, no. 54 (January 8, 1863): 385–86.

38. Robert H. Scott, "A Word about Weather," *Chambers's Journal of Popular Literature, Science, and Art* 51, no. 554 (August 8, 1874): 511.

39. Fitzroy, *Weather Book*, 170.

40. MacDonald, *At the Back of the North Wind*, 67, 358. The magical women featured in MacDonald's works are often maternal. Robert Lee Wolff writes, "In his works of fancy and imagination again and again he can imagine the mother-goddess, all-loving and all-knowing." Wolff, *The Golden Key: A Study of the Fiction of George MacDonald* (New Haven, CT: Yale University Press, 1961), 317.

41. Ronald Hutton, *The Triumph of the Moon: A History of Modern Pagan Witchcraft* (Oxford: Oxford University Press, 1999), 32–42. Additionally, in *On the Origin of Species* (1859), Darwin consistently personifies Nature as "She." Stanley Edgar Hyman has written on the centrality of Darwin's female Nature to his creation myth. See *The Tangled Bank: Darwin, Marx, Frazer and Freud as Imaginative Writers* (New York: Atheneum, 1962), 37–38.

42. Bonnie Gaarden, *The Christian Goddess: Archtype and Theology in the Fantasies of George MacDonald* (Madison, NJ: Fairleigh Dickinson University Press, 2011). MacDonald had served as the minister of a Congregationalist chapel in Arundel but was forced to resign, apparently due to the inability of a narrow-minded church community to accept his theological sophistication. Although Greville MacDonald devotes a significant portion of his father's biography to his time at Arundel, leading many scholars to regard the resignation as a tragic turning point, George MacDonald seems to have always possessed a greater interest in engaging with theology through literature. See Muriel Hutton, "The George MacDonald Collection," *Yale University Library Gazette* 51, no. 2 (October 1976): 76–78. MacDonald, *George MacDonald and His Wife*, 135–87.

43. MacDonald, *At the Back of the North Wind*, 38.

44. George MacDonald, "The Child in the Midst," in *Epea Aptera: Unspoken Sermons*, First Series (London: Alexander Strahan, 1867), 5, 12.

45. MacDonald, *At the Back of the North Wind*, 59.

46. MacDonald, *At the Back of the North Wind*, 65.

47. Auerbach compares MacDonald's description to Freud's iconography of female hysteria in which portions of patients' bodies become dissociated and take on a strange and powerful life of their own. Nina Auerbach, *Woman and the Demon: The Life of a Victorian Myth* (Cambridge, MA: Harvard University Press, 1982), 48.

48. See Freud's essay "Fetishism" (1927), in which Freud's focus on the fetishist is bound to the disappointment and anxiety of the male child. The essay appears in Freud, *On Sexuality: Three Essays on the Theory of Sexuality and Other Works*, edited by Angela Richards, translated by James Strachey (Harmondsworth: Penguin, 1977), 345–57.

49. Henry Frith, *How to Read Character in Features, Forms, and Faces: A Guide to the General Outlines of Physiognomy* (London: Ward, Lock, Bowden, 1891), 14, 16, 18–22; Mary Cowling, *The Artist as Anthropologist: The Representation of Type and Character in Victorian Art* (Cambridge: Cambridge University Press, 1989), 111–15; Galia Ofek, *Representations of Hair in Victorian Literature and Culture* (Farnham, UK: Ashgate, 2009), 60–63.

50. The North Wind's long, sensuous hair references Victorian representations of the Christian figure Mary Magdalene, following a tradition in Renaissance painting in which her hair often appears in place of clothing, becoming "a locus of both sexual sin and redemption." Galia Ofek observes the sexual intimacy of women classified by Victorians as "Magdalene types" (in contrast to respectable "Madonna types"), demonstrated by the undressed state of the woman's hair, inviting viewers to "undress" her in their imagination. Ofek, *Representations of Hair*, 66.

51. Virginia M. Allen, "'One Strangling Golden Hair': Dante Gabriel Rossetti's *Lady Lilith*," *Art Bulletin* 66, no. 2 (June 1984): 285–94. See also Elisabeth G. Gitter, "The Power of Women's Hair in the Victorian Imagination," *PMLA* 99, no. 5 (October 1984): 936–54. Gitter explores the Victorian fascination with the problem of "reading" women's hair.

52. MacDonald, *At the Back of the North Wind*, 73.

53. Medusa was undergoing a transformation in the nineteenth century through the work of Algernon Swinburne and Dante Gabriel Rossetti, whose Medusas eliminated the response of delight or sympathy and created instead a fear of female sexuality. See Jerome J. McGann, "The Beauty of Medusa: A Study in Romantic Literary Iconology," *Studies in Romanticism* 11, no. 1 (Winter 1972): 3–25.

54. MacDonald, *At the Back of the North Wind*, 69–70.

55. MacDonald, *At the Back of the North Wind*, 70.

56. MacDonald, *At the Back of the North Wind*, 71.

57. I follow Lucetta Johnson in arguing that hair is not exclusively connected to uncontrolled sexuality and should be interpreted in terms of the overall

structure of the body and the picture plane. Johnson, "Beyond the Hair-Line: The Representation of Hair in the Work of Dante Gabriel Rossetti" (PhD diss., Courtauld Institute of Art, 2010).

58. Susan Casteras, *Images of Victorian Womanhood in English Art* (London: Associated University Presses, 1987), 177.

59. Laurence Housman, "The Illustrations of Arthur Hughes," *Bibliophile* 1, no. 5 (July 1908): 235.

60. MacDonald, *At the Back of the North Wind*, 109, 55.

61. MacDonald, *At the Back of the North Wind*, 321.

62. MacDonald, *At the Back of the North Wind*, 294.

63. Sally Shuttleworth, *The Mind of the Child: Child Development in Literature, Science, and Medicine, 1840–1900* (Oxford: Oxford University Press, 2010), 75–76.

64. Stephanie L. Schatz, "Lewis Carroll's Dream-Child and Victorian Child Psychopathology," *Journal of the History of Ideas* 76, no. 1 (January 2015): 93–114.

65. Henry Maudsley, *The Physiology and Pathology of Mind* (London: Macmillan, 1867), 268–70.

66. Schatz, "Lewis Carroll's Dream-Child," 113.

67. R. B. Shaberman, "George MacDonald and Lewis Carroll," *North Wind: A Journal of George MacDonald Studies* 1 (1982): 10–30, https://digitalcommons.snc.edu/northwind/vol1/iss1/1.

68. See Jenny Bourne Taylor and Sally Shuttleworth, eds., *Embodied Selves: An Anthology of Psychological Texts, 1830–1890* (Oxford: Clarendon, 1998), 67–161, for writings on the unconscious mind and the workings of memory during the Victorian period.

69. William Benjamin Carpenter, *Principles of Human Physiology*, 5th ed. (London: John Churchill, 1855). Thomas Laycock argued that he actually developed the concept first through his reference to the "automatic action of the cerebrum," in his articles on hysteria in the 1840s. See Thomas Laycock, *Mind and Brain: Or the Correlations of Consciousness and and Organisation*, vol. 2 (Edinburgh: Sutherland and Knox, 1860), 479–80.

70. Carpenter, *Principles of Human Physiology*, 643–44.

71. William Benjamin Carpenter, *Principles of Mental Physiology*, 2nd ed. (London: H. S. King, 1874), 436, 519.

72. George MacDonald, *A Dish of Orts: Chiefly Papers on the Imagination, and on Shakespeare* (London: Sampson, Low, Marston, 1893), 24.

73. MacDonald, *Dish of Orts*, 25.

74. MacDonald, *At the Back of the North Wind*, 143.

75. MacDonald, *At the Back of the North Wind*, 138.

76. MacDonald, *At the Back of the North Wind*, 154–55.

77. MacDonald, *At the Back of the North Wind*, 165, 212.

78. Diamond temporarily visits the back of the North Wind before the end of the novel. While he is there, he lies ill at home, indicating the presence of a threshold that he does not fully cross until his death.

79. Examples include Irene's great-great-grandmother in *The Princess and the Goblin* (1872) and the Wise Woman in *The Wise Woman or The Lost Princess: A Double Story* (1875).

80. Eneas Sweetland Dallas, *The Gay Science*, vol. 1 (London: Chapman and Hall, 1866), 199.

81. Dallas, *Gay Science*, 201.

82. Dallas, *Gay Science*, 200–201.

83. Carpenter, *Principles of Mental Physiology*, 526. See also Frances Power Cobbe, *Hours of Work and Play* (London: N. Trubner, 1867), 87–113.

84. Cobbe also remarks on the "vivid sense of the separation" between "the phenomena of Unconscious Thought on one side, and over against them our conscious personality," despite their "mutual interdependence." See Cobbe, "Unconscious Cerebration: A Psychological Study," *Macmillan's Magazine* 23, no. 133 (November 1870): 25, 37. Furthermore, Cobbe and Carpenter were both interested in the suspension of volition that occurs during unconscious cerebration and the potential of the Will, associated with the conscious mind, to control reflex or involuntary responses, comparing the relationship between the unconscious and conscious mind to that of an excitable horse controlled by a skilled, attentive rider. Cobbe, "Unconscious Cerebration," 36. Carpenter, *Principles of Mental Physiology*, 24.

85. Dallas, *Gay Science*, 249.

86. Dallas, *Gay Science*, 207.

87. Hughes may have had the illustration in mind when writing to the poet and artist William Bell Scott years later, "This is a most unusual year—and here, altho' it is almost hot in the sun now, a continual North & East wind comes thro' one at most unexpected moments." Arthur Hughes to William Bell Scott, April 26, 1886, RBSC-ARC-1433, Penkill Collection.

88. Dallas, *Gay Science*, 207.

89. Dallas, *Gay Science*, 250. It is also worth noting that Alexander Bain, a prominent figure within nineteenth-century psychology, developed a theory of diffuse emotional waves of nervous influence linking the mind and body. He associates the unconscious mind with bodily processes, and observes that the performance of actions, and their accompanying emotions, become more unconscious with repetition, producing a contraction of the portion of the wave within the brain that signals less involvement of the conscious mind. Alexander Bain, *The Emotions and the Will* (London: John W. Parker and Son, 1859), 11–12, 620.

90. "Literature," *Aberdeen Weekly Journal*, December 9, 1886, 2.

91. Buchan, *Handy Book of Meteorology*, 231.

92. Buchan, *Handy Book of Meteorology*, 1–2.

93. Herschel, *Meteorology from the Encyclopaedia Britannica*, 5.

94. Galton, *Meteorographica*, A2. See also Derek William Forrest, *Francis Galton: The Life and Work of a Victorian Genius* (London: Paul Elek, 1974), 78–82.

95. Galton, *Meteorographica*, 5.

96. "Mapping the Weather," *London Review of Politics, Society, Literature, Art, and Science* 7, no. 174 (October 31, 1863): 469.

97. *Report of a Committee Appointed to Consider Certain Questions Relating to the Meteorological Department of the Board of Trade* [Galton Report], Parliamentary Papers, 1866 (3646) LXV.329, 20, quoted in Anderson, *Predicting the Weather*, 124. Fitzroy, *Weather Book*, 15.

98. Anderson, *Predicting the Weather*, 178. Hughes acknowledges these associations in an 1873 letter to MacDonald's wife, Louisa, in which he declines an invitation for the holidays and complains of being overwhelmed with work, writing, "Oh, that I had quicksilver for a spinal marrow, flowering into a corresponding kind of lump of lightning flashes in my head!" Arthur Hughes to Louisa MacDonald, December 16, 1873, mss 103, series I, box 2, folder 86, George MacDonald Collection.

99. Carpenter, *Principles of Mental Physiology*, 35, 37–38.

100. "The Winds," *Cornhill Magazine*, 11, no. 63 (March 1865): 353.

101. William Radcliffe Birt, "On Certain Atmospheric or Barometric Waves which Traversed Europe During November 1842," *London, Edinburgh, and Dublin Philosophical Magazine and Journal of Science* 30, series 3, no. 204 (1847): 485.

102. Napier Shaw, *Manual of Meteorology*, vol. 1, *Meteorology in History* (Cambridge: Cambridge University Press, 1926), 154–55.

Chapter 4: A Haunting Picture, in Light of Victorian Science

I am grateful to Chrissie Bradstreet, Bob Cozzolino, Domenica DiPiazza, Kate Flint, Elizabeth Franz, Thomas Hull, Deborah Kragten, and Kimberly James Overdevest for constructive conversations informing this argument; thanks also to Melissa Buron and Alison Hokanson for organizing the College Art Association session in 2017 at which this essay first took shape. The two anonymous readers for the University of Pittsburgh Press also offered invaluable suggestions.

1. Reviewers remarked: "He has limned no shadowy presence through whose filmy vagueness objects beyond can be traced, but rather has stopped just short of fully realizing a living human creature," and "the apparition is too tangible and realized—its rivalry with flesh and blood is too positive and confusing." "Collated Opinion upon the Exhibition of the Royal Academy: Sir John E. Millais, R.A.," supplement to *Artist*, June 1895, np. "The ghostly visitant, however, seems almost of 'too solid flesh compounded.'" "Notes on the Royal Academy Exhibition," *Bow Bells: A Magazine of General Literature and Art*, May 10, 1895, 466.

2. [David Masson], "The Academy II," *Spectator*, May 18, 1895, 684.

3. M. H. Spielmann, *Millais and His Works* (London: Blackwood, 1898), 119.

4. Anne Helmreich, *Nature's Truth: Photography, Painting, and Science in Victorian Britain* (University Park: Pennsylvania State University Press, 2016), 80.

5. See later discussion. Michael Faraday, *A Course of Six Lectures on the Chemical History of a Candle* (London: Griffin, Bohn, 1861); Francisco Vera, Rodrigo Rivera, and César Núñez, "Burning a Candle in a Vessel, a Simple Experiment with a Long History," *Science & Education* 20, no. 9 (September 2011): 884.

6. It was listed as having been in the works for twenty years for the *Royal Academy Illustrated*; and for forty in "A Septette of Favourite Paintings," *Strand*, May 1907, 489.

7. "A Night Terror in Africa," *Temple Bar: A London Magazine for Town and Country Readers*, 43 (December 1874): 72. The writer remarks that "by speaking you personify," meaning that in the process of addressing a ghost it becomes more than a hallucination to take on a human reality.

8. Malcolm Warner, "Millais in Reproduction," in *Writing the Pre-Raphaelites: "Text, Context, Subtext"*, ed. Michaela Giebelhausen and Tim Barringer (London: Routledge, 2017), 228.

9. A reviewer mentioned the placement of the pictures. "The Royal Academy of Arts 1895," *Art Journal* 57 (June 1895): 318. Martin Postle originally identified the model in "Leighton's Lost Model: The Rediscovery of Mary Lloyd," *Apollo*, February 1996, 27–29. Also in the same room was Leighton's well-known *Flaming June*. Jason Rosenfeld, *John Everett Millais* (London: Phaidon, 2012), 234.

10. "Septette of Favourite Paintings," 490.

11. Marion Spielmann, "The Royal Academy Exhibition 2," *Magazine of Art* 18 (1895): 282; similarly: "One perceives and admits . . . a force it is not easy to define," said the *Standard*; "he had again vindicated his right to the great reputation he enjoys," claimed the *Athenaeum* in "Collated Opinion," n.p.; it was the work in which "the renewal of [Millais's] style and the return to the old starting-point is most clearly to be traced," argued Claude Phillips, "Fine Art: The Royal Academy I," *Academy*, May 11, 1895, 408; and was clearly "one of the best of the artist's recent works," according to the *Morning Post* in "Collated Opinion," n.p..

12. This fund, which purchased works of art for the nation, acquired Millais's work for £2,000, a large amount at that time. Alison Smith, entry for *Speak! Speak!*, in Smith and Jason Rosenfeld, *Millais* (London: Tate, 2008), 168.

13. Others related to Millais's circle were more overtly linked to scientific investigators; William Michael Rossetti, for instance, belonged to the same club as naturalist and evolutionist Charles Darwin, biologist T. H. Huxley, scientist and philosopher William Whewell, and physicist John Tyndall. Julie Codell, "Empiricism, Naturalism and Science in Millais's Paintings," in *John Everett Millais: Beyond the Pre-Raphaelite Brotherhood*, ed. Deborah Mancoff (New

Haven, CT: Yale University Press, 2001), 138–39. Codell concludes that it is not really possible to know with any certainty Millais's precise level of knowledge or interest in any one area of science (137).

14. Helmreich, *Nature's Truth*, 89.

15. John Holmes, *The Pre-Raphaelites and Science* (New Haven, CT: Yale University Press, 2018), 73; 78–79; Julie Codell, "The Dilemma of the Artist in Millais's *Lorenzo and Isabella*: Phrenology, the Gaze and the Social Discourse," *Art History* 14, no. 1 (March 1991): 51–66; Stephanie Grilli, "Pre-Raphaelitism and Phrenology," in *Pre-Raphaelite Papers*, ed. Leslie Parris (London: Tate, 1984), 44–60.

16. Helmreich, *Nature's Truth*, 79–117.

17. Codell, "Empiricism, Naturalism and Science," 138–39. Codell cites Stephens, "Modern Giants," *Germ* 4 (May 1850): 171. Helmreich, *Nature's Truth*, 48–49, 58–62.

18. Laura Mulvey, "Visual Pleasure and Narrative Cinema," *Screen* 16, no. 3 (October 1975): 6–18.

19. Sally Shuttleworth, *Charlotte Brontë and Victorian Psychology* (Cambridge: Cambridge University Press, 1996), 4, 28. Millais had already represented women in white as beyond rational control in his *Somnambulist* of 1863 (held in private collection). One reviewer asked of *Speak! Speak!*, "Is it a somnambulist, one asks, instead of a wraith, that has taken up her station in the mysterious chamber?" "Collated Opinion," n.p. *The Somnambulist* was related to Millais's friend Wilkie Collins's famous novel *Woman in White* (1859–1860), in which the character Anne Catherick is of wandering mind. Collins claimed to have been inspired by an actual event in the 1850s in which he and Millais, walking in London, encountered a screaming woman who "seemed to float rather than run" through the night in "flowing white robes." John Guille Millais, *Life and Letters of John Everett Millais* (London: Methuen, 1899), 1:281.

20. Marion Spielmann described her as a "spectre-bride." Spielmann, "Royal Academy Exhibition 2," 282.

21. Margaret Cameron Coss Flower, *Victorian Jewellery* (North Chelmsford: Courier, 2002), 6.

22. Thank you to Kimberly James Overdevest for observing the similarity between the two pictures.

23. *Punch* humorously interpreted the image as a scene of a man who, having retired to the wrong hotel room, wakes to find the rightful occupant returning from a ball; in his disorientation, he thinks of "the ballad of 'Margaret's Grim Ghost,'" better known as "Fair Margaret and Sweet William," in Francis James Child, *English and Scottish Ballads* (Boston: Little, Brown, 1860), no. 74, in which a woman's spirit returns to bring her unfaithful lover with her into death. "First Impressions of the R.A.," *Punch*, May 11, 1895, 221.This subject, with a reversal of the sex of the dead partner, was popular among the Pre-Raphaelites, as

Elizabeth Siddall (1857) and Edward Coley Burne-Jones (1861) had both previously portrayed the ballad *Clerk Saunders* (Child, no. 69).

24. Suzanne Fagence Cooper, *Effie: The Passionate Lives of Effie Gray, John Ruskin and John Everett Millais* (New York: St. Martin's, 2010), 95. The stone walls and spiral staircase suggesting a fortified mansion that may have been painted from Murthly Castle near Millais's favorite Scottish painting retreat. Alison Smith sets the scene at Murthly near where, according to his biographer, Millais was staying with his brother-in-law George Gray at his home, Bowerswell, at the time the painting was begun. M. H. Spielmann, *Millais and his Works*, 120, in Alison Smith and Jason Rosenfeld, *Millais*, 168. J. G. Millais, however, says that it was painted at Bowerswell in a special room set up for the purpose with a four-poster bed scouted and secured for the purpose, before being finished in London. Millais, *Life and Letters*, 2:307. Others identify the staircase as the same one he painted in Scotland from Saint Mary's Tower, Birnam, that appears in the *Gray Lady* and the *Princes in the Tower*. A staircase at Murthly Castle does indeed resemble the one in the picture. With thanks to Thomas Steuart Fotheringham for the photo from Murthly (private communication, January 6, 2017).

25. "Septette of Favourite Paintings," 490.

26. With thanks to Dawn Hoskin, assistant curator of metalwork, ceramics and glass at the Victoria & Albert Museum, who located and photographed the lamp for me (private communication, August 18, 2016).

27. Alex Soo-jung Kim Pang, *Empire and the Sun: Victorian Solar Eclipse Expeditions* (Stanford, CA: Stanford University Press, 2002); Nancy Rose Marshall, "The Many Shades of Shakespeare: Representations of Othello and Desdmona in Victorian Visual Culture," in *Transculturation in British Art, 1770–1930*, ed. Julie F. Codell (New York: Routledge, 2017), 73–92.

28. The "grey dawn, forming, with the light of the flaring taper at the bedside, a harmonious discord. . . . Millais . . . used it to good effect in an earlier picture, 'The Rescue.'" "Septette of Favourite Paintings," 490.

29. Lynda Nead, *Victorian Babylon: People, Streets, and Images in Nineteenth-Century London* (New Haven, CT: Yale University Press, 2000), 83.

30. Stella Pratt-Smith, "The Poetic Science of Nineteenth-Century Electricity," *Nineteenth-Century Contexts* 34, no. 1 (February 2012): 51–62. Graeme Gooday notes that the "fear of death by intangible mysterious force of electricity was a significant social phenomenon." "Electricity and the Sociable Circulation of Fear and Fearlessness," in *Geographies of Nineteenth-Century Science*, ed. David N. Livingstone and Charles W. J. Withers (Chicago: University of Chicago Press, 2011), 204.

31. There is no evidence that Millais followed through with this commission. "Electricity at the World's Fair," *Electrical World* 20, no. 9 (1892): 133, 132.

32. "Artificial Light in Studios," *Lightning*, January 7, 1892, 268.

33. Frith stated, "My own opinion is that the extreme blackness of the shadows always thrown . . . makes Electric Light useless as an illuminant for an artist's studio." "Artificial Light in Studios," *Lightning*, January 14, 1892, 288.

34. "Notes," *Telegraphic Journal and Electrical Review*, January 6, 1883, 17.

35. Ernest Freeberg, private communication, July 27, 2016. Hélène Valance notes that American artist Henry Ossawa Tanner "erased the line between electric power and religious faith" by portraying the angel in a way that evokes Nicola Tesla's experiments of the same moment in his *Annunciation* (held at Philadelphia Museum of Art, 1898). *Nocturne: Night in American Art, 1890–1917*, trans. Jane Marie Todd (New Haven, CT: Yale University Press, 2018), 37–39.

36. Millais himself depicted a transparent ghost, *The Gray Lady*, in 1883 (held in private collection), which his daughter records was the only picture he painted by electric light. Mrs. Stuart-Wortley (Alice Millais), in Millais, *Life and Letters*, 2:162.

37. Thomas Griffiths, *Chemistry of the Four Ancient Elements, Fire, Air, Earth, and Water: An Essay Founded upon Lectures Delivered before Her Most Gracious Majesty the Queen* (London: Samuel Highley, 1842). By the end of the century, as thermodynamics and the study of energy emerged after the work of William Thomson, Lord Kelvin, physics increasingly claimed the study of combustion, a seemingly simple process that in fact has continued to yield new understandings of this apparently simple action on board space stations.

38. Robert E. Schofield, *The Enlightened Joseph Priestley: A Study of his Life and Work from 1773 to 1804* (University Park: Pennsylvania State University Press, 2010), 111–12.

39. Michael Faraday, *A Course of Six Lectures on the Chemical History of a Candle* (London: Griffin, Bohn, 1861), 167.

40. Joe Jackson, *A World on Fire: A Heretic, An Aristocrat, and the Race to Discover Oxygen* (New York: Viking, 2005).

41. Faraday, *Course of Six Lectures*; Vera, Rivera, and Núñez, "Burning a Candle in a Vessel," 884.

42. Barri Gold, *Thermopoetics: Energy in Victorian Literature and Science* (Cambridge: MIT Press, 2010), 7.

43. Henry Maudsley, *Body and Will* (London: Kegan, Paul, Trench, 1883), 213.

44. Thomas Wright Hall, *Correlation Theory of Chemical Action and Affinity* (London: Remington, 1888), 238; Matthew Moncrieff Pattison Muir, *Chemistry of Fire* (London: Methuen, 1893), 1–2.

45. Faraday, *Chemical History of a Candle*, 46–48.

46. Kate Flint, *The Victorians and the Visual Imagination* (Cambridge: Cambridge University Press, 2000), 284.

47. "The species of faith, or credulity, which leads to the belief in ghosts, belongs not to science . . . for by knowledge it is as once destroyed." Justus Liebig,

Familiar Letters on Chemistry (London: Walton and Maberly, 1859), 24–25.

48. "Address by the President on 'The Census of Hallucinations,'" *Proceedings of the Society for Psychical Research*, July 8, 1889, 9.

49. "This is presumably an apparition, but in spite of the somewhat ghostly light which surrounds it, the figure has so earthly a character as almost to preclude the notion." "The Art of the Year," *Month: A Catholic Magazine and Review*, June 1895, 159. Another wrote, "The realism that finds utterance in the rendering of the diamonds seems a little tawdry." "The Millais and Rossetti Exhibitions," *Fortnightly Review*, February 1, 1898, 192.

50. Louis Simonin, *Underground Life, or, Mines and Miners*, trans. H. W. Bristow (London: Chapman and Hall, 1869), 491.

51. Faraday, *Course of Six Lectures*, 21.

52. David J. Jones, *Gothic Machine: Textualities, Pre-cinematic Media and Film in Popular Visual Culture, 1670–1910* (Cardiff: University of Wales Press, 2011).

53. See John Henry Pepper, *True History of a Ghost and All about Metempsychosis* (London: Cassell, 1890).

54. Bence Jones's *Memoir of Faraday* is cited as a source for a letter to the editor of *Nature* concerning an anecdote by which the chemist called his niece to "see a spectre," by placing a candle behind the two of them and projecting "two gigantic shadowy beings" on the window glass. "Faraday Exhibiting Ghosts," *Nature* 22, no. 583 (October 21, 1880): 583.

55. Iwan Rhys Morus, "Illuminating Illusions, or the Victorian Art of Seeing Things," *Early Popular Visual Culture* 10, no. 1 (2012): 37–50.

56. Shuttleworth, *Charlotte Brontë and Victorian Psychology*, 12, 28–29.

57. Edina, "Some Recent Materializations: No. 3, Séance with Mrs. Davidson," *Light: A Journal of Psychical, Occult, and Mystical Research* 15 (June 15, 1895): 281.

58. "The Capture of a Spirit," *Graphic*, February 7, 1880, 1.

59. See J. Jeffrey Franklin, *Spirit Matters: Occult Beliefs, Alternative Religions, and the Crisis of Faith in Victorian Britain* (Ithaca, NY: Cornell University Press, 2018).

60. Richard Noakes, "Spiritualism, Science, and the Supernatural in Mid-Victorian Britain," in *The Victorian Supernatural*, ed. Nicola Bown, Carolyn Burdett, and Pamela Thurschwell (Cambridge: Cambridge University Press, 2004), 23–43; Tatiana Kontou, ed., *The Ashgate Research Companion to Nineteenth-Century Spiritualism and the Occult* (London: Routledge, 2016); Srdjan Smajić, *Ghost-Seers, Detectives, and Spiritualists: Theories of Vision in Victorian* (Cambridge University Press, 2010).

61. Claude Phillips believed the picture to be set in "the dim borderland between waking and sleeping." Phillips, "Fine Art," 408.

62. "The Looker On," *Blackwood's Magazine*, 1895, 919. The *Art Journal* described the figure as "a man in a high fever [who] has been tossing, and dreaming of his betrothed or wife." "Royal Academy of Arts 1895," 164.

63. Alan Collins, "England," in *The Oxford Handbook of the History of Psychology: Global Perspectives*, ed. David B. Baker (Oxford: Oxford University Press, 2012), 182. See also Rick Rylance, *Victorian Psychology and British Culture, 1850–1880* (Oxford: Oxford University Press, 2000).

64. Myers was well known to Millais as the husband of Eileen Tennant, whose portrait he painted, and as brother-in-law of artist Dorothy Tennant, who took lessons with Millais.

65. Collins, "England," 188.

66. "Report on the Census of Hallucinations," in *Proceedings of the Society for Psychical Research* (London: Kegan, Paul, Trench, Trübner, 1894), 36, 43, 46. In Shane McCorristine, *Spectres of the Self: Thinking about Ghosts and Ghost-Seeing in England, 1750–1920* (Cambridge: Cambridge University Press 2010), 198.

67. Frederic Myers, "On Recognised Apparitions Occuring More than a Year after Death," *Proceedings of the Society for Psychical Research* 6 (1889): 14–15; see also McCorristine, *Spectres of the Self*, 179.

68. Edmund Gurney, *Phantasms of the Living*, 2 vols. (London: Trübner, 1886).

69. William Benjamin Carpenter, *Principles of Mental Physiology*, 4th ed. (London: Henry S. King, 1876), 515–43.

70. Carpenter, *Principles of Mental Physiology*, 344, 455–56.

71. McCorristine, *Spectres of the Self*, 193.

72. Edmund Parish, *Hallucinations and Illusions: A Study of the Fallacies of Perception* (London: W. Scott, 1897), 14, 15; Edmund Gurney, "Hallucinations," *Proceedings of the Society for Psychical Research* 3 (1885): 155, in Parish, *Hallucinations*, 15.

73. G. H. Lewes, *Physiology of Common Life*, vol. 2 (Edinburgh: Blackwood, 1860). I am grateful to Anne Helmreich for pointing me to this source; see Helmreich, *Nature's Truth*, 90.

74. Charles Godlove Raue, *Psychology as a Natural Science Applied to the Solution of Occult Psychic Phenomena* (Philadelphia: Porter and Coates, 1889), 469.

75. Thomas Case, *Physical Realism: Being an Analytical Philosophy from the Physical Objects of Science to the Physical Data of Sense* (London: Longmans, Green, 1888), 20. For Case, materialism was a subset of realism holding both that physical material things existed and that they were the only realities, or at least that they were prior to psychical realities. The proliferating theories of thinkers in each school ultimately renders any attempt at clear categorization impossible, and in fact the two systems can overlap or bleed into each other. Among idealists ranged René Descartes, John Locke, David Hume, and Immanuel Kant.

76. William James, "The Sentiment of Rationality," in *Essays in Popular Philosophy* (London: Longmans, Green, 1896), 90–91.

77. William James, *Psychology* (New York: H. Holt, 1892), 330.

78. Bernard Lightman, "Science and Culture," in *The Cambridge Companion to Victorian Culture,* ed. Francis O'Gorman (Cambridge: Cambridge University Press, 2010), 12–42; Helmreich, *Nature's Truth,* 87.

79. Helmreich, *Nature's Truth,* 87–90, 89.

80. Phillips, "Fine Art," 408. Another remarked, "The literalness which still clings to the former PreRaphaelite is not . . . married to the true imaginative power which can convincingly present its subjects in the glamour of romance." "Collated Opinion," n.p.

81. "Collated Opinion," n.p.; Phillips, "Pictures of the Year," *Fortnightly Review,* June 1, 1895, 934.

82. Note that realism in art is not entirely mappable onto the philosophic category.

83. "Notes on the Royal Academy Exhibition," 466.

84. "The Supernatural in Art," *Borderland: A Quarterly Review and Index* (July 1895): 195.

85. Helmreich, *Nature's Truth,* 79ff.

86. [Tom Taylor], "Exhibition of the Royal Academy," *Times,* May 2, 1874, 12.

87. Helmreich, *Nature's Truth,* 99.

88. Cosmo Monkhouse, "Wright of Derby," *Magazine of Art* 6 (1883): 385. Monkhouse reproduces *Three Men Looking at the Gladiator,* 385.

89. Review of *The Life and Works of Joseph Wright, A.R.A., Commonly Called "Wright of Derby",* by William Bemrose, *Art Journal* 48 (1886): 95. Millais would have known Wright of Derby's best-known picture, the *Experiment on a Bird in an Airpump* (1768, held at Tate Britain), in which a wild-haired scientist-magician performs a popular educational demonstration of the fact that oxygen is needed to live: one turn of the stopcock at the top of the glass vessel will save the weakening bird. Also featuring a hidden taper, the painting ties together the same tropes as *Speak! Speak!*—of life, death, and burning candles—conflating science and magic in a way distinctive to the Enlightenment and the century afterward.

90. Wright became an associate of the academy under president Joshua Reynolds, only to later decline election as full academician. The *Gladiator* was by Agasias of Ephesis. Its fragments had been found in Rome by 1611 and restored, then looted by Napoleon and taken to the Louvre in 1808. Brigitte Bourgeois, "Secure for Eternity: Assembly Techniques for Large Statuary in the Sixteenth to the Nineteenth Century," in *History of Restoration of Ancient Stone Sculptures,* ed. Janet Burnett Grossman, Jerry Podany, and Marion True (Los Angeles: J. Paul Getty Museum, 2003), 153–58. The artist's "elaborate effects of artificial light" and this painting's "largeness of style and majesty of design" were admired. Monkhouse, "Wright of Derby," 388, 389.

91. "Septette of Favourite Paintings," 490.

92. Phillips, "Pictures of the Year," 933. See also "Collated Opinions," n.p.

93. Catherine Williams, *Ruskin's Philosophy* (Penzance: Triton, 1975), 22.

94. "The awkwardness of the attitude in which she holds the curtains asunder is suggestive of an amount of effort inconsistent with such a character." "Art of the Year," 159. *Ophelia* was described in 1895 as "a canvas of the very first importance and as having last changed hands for $15,000, "Sir John Everett Millais, Bart., R.A. A Character Sketch," *Review of Reviews* 11 (1895): 546.

95. "Sir John Everett Millais, Bart., R.A.," 551.

96. D. C. Thomson, "Realism in Painting," *Art Journal* 19 (1880): 283.

97. Martin Kemp notes the extent to which the "intellectual and observational concerns in the visual arts and the sciences of Europe" were founded on "direct study of nature through a faculty of vision." Kemp, *The Science of Art: Optical Themes in Western Art from Brunelleschi to Seurat* (New Haven, CT: Yale University Press, 1990), 1.

98. Thomson, "Realism in Painting," 283.

99. Unidentified Pierpont Morgan Library ms., cited in Gordon H. Fleming, *John Everett Millais, A Biography* (London: Constable, 1998), 47.

100. "Septette of Favourite Paintings," 490. "The 'bit' that most pleased the painter . . . was the rendering of the man's wonderful Italian throat." *Royal Academy Illustrated* (London: Cassell, 1895), vi.

101. Fleming, *John Everett Millais*, 287; Morell Mackenzie, *Diseases of the Pharynx, Larynx, and Trachea* (New York: William Wood, 1880), 1:244.

102. "At the Banquet, Saturday, May 4," *Punch*, May 11, 1895, 221. "Millais . . . was himself suffering from a serious cold, and had almost lost his voice. During all the first part of the evening, indeed, his remarks were inaudible to all but those in his immediate neighborhood; but later on, his voice returned to him, and he spoke to admirable effect." "London Letter," *Critic*, May 25, 1895, 387.

103. Spielmann, *Millais and His Works*, 129.

104. J. Eadie Reid, *Sir J. E. Millais, P.R.A.* (London: Walter Scott, 1909), 136. Millais invokes *Hamlet*'s "to be or not to be" monologue: "To sleep, perchance to dream." Thanks to the anonymous reader for this observation.

105. "The Case of Sir John Millais: Passages from the Notebooks of Sir Frederick Treves," *British Royal College Surgeons Annual* (1961): 284–388. Treves's somewhat gothic account describes his performing the procedure in a room with "neither gas nor electric light, while the lamps and candles threw prodigious shadows on the wall." He recounts Millais's "terrible breathing," a "tense whistling in the throat of a suffocating man," noting that "he was speechless" (385).

106. Barbara Maria Stafford, *Body Criticism: Imaging the Unseen in Enlightenment Art and Medicine* (Cambridge: MIT Press, 1993), 5.

107. Painting is without sound and without words, nonnarrative, as David Masson recounted with much vehemence in his review of the Royal Academy

exhibition that year (dismissing Millais's painting as "infected with this vice of narrative illustration"). [Masson], "Academy II," 684.

Chapter 5: Photographing Ether, Documenting Pain

I thank Nancy Rose Marshall for organizing this volume and corresponding 2015 College Art Association panel, and for the incisive editorial suggestions of Marshall and the anonymous peer reviewers, as well as Jack Eckert, Former Public Services Librarian, Center for the History of Medicine, Francis A. Countway Library of Medicine; Jeffrey Mifflin, former archivist, Massachusetts General Hospital; Dominic Hall, curator, Warren Anatomical Museum; Patricia Hills; Keith Morgan; Kim Sichel; Ross Barrett; and Samuel Alberti.

1. The commemoration of October 16 continues. In 2001, MGH commissioned an oil painting of a reenactment of the October 16, 1846, ether demonstration by Warren and Lucia Prosperi. It hangs in the MGH surgical amphitheater, known as the Ether Dome, which is a National Historic Site.

2. Fred S. Kleiner, *Gardner's Art through the Ages* (Boston: Cengage Learning, 2013), 793.

3. My analysis of these daguerreotypes is built upon the research of Bates and Isabel Lowry, who chart the production of these images, confirm authorship, establish an order of commission, and posit provenance. While they identify the surgeons in each image, I disagree with some identifications, as well as some of their other claims. Lowry and Lowry, "Simultaneous Developments: Documentary Photography and Painless Surgery," in *Young America: The Daguerreotypes of Southworth & Hawes*, ed. Grant B. Romer and Brian Wallis (New York: International Center of Photography, 2005), 75–88.

4. The history of the ether debate is treated in primary and secondary scholarship, including Nathaniel J. Bowditch, *The Ether Controversy: Vindication of the Hospital Report of 1848* (Boston: J. Wilson, 1848); Truman Smith, *An Examination of the Question of Anaesthesia, Arising on the Memorial of Charles Thomas Wells* (New York: John A. Gray, 1859); James Young Simpson, *History of Modern Anaesthetics: A Second Letter to Dr. J. Bigelow* (Edinburgh: Edmonston and Douglas, 1870); Julie M. Fenster, *Ether Day: The Strange Tale of America's Greatest Medical Discovery and the Haunted Men who Made It* (New York: HarperCollins, 2001); Richard J. Wolfe, *Tarnished Idol: William Thomas Green Morton and the Introduction of Surgical Anesthesia: A Chronicle of the Ether Controversy* (San Anselm, CA: Norman, 2001).

5. Stephanie Browner addresses the discovery of ether and ensuing controversy; see "Ideologies of the Anesthetic: Professionalism, Egalitarianism, and the Ether Controversy," *American Quarterly* 51, no. 1 (March 1999): 108–43.

6. René Fülop-Miller, "The Day Man Conquered Pain," *Rotarian*, September 1946, 41. This first appeared in 1901 as a republished letter from Dr. G. M.

Angell, a student at Harvard Medical College, who observed the 1846 demonstrations. Angell, "Recollections of the First Operation Performed Under Anesthesia," *International Journal of Surgery* 14 (August 1901): 249–50. Fülop-Miller's article marks the centenary of the discovery. The quotation is not grounded in fact but is relevant for the way in which it looks back at 1846, reshapes historical truth, and emphasizes certain aspects of anesthesia.

7. Canonical texts that address the mid-nineteenth-century transformation of American medicine include Charles E. Rosenberg, *No Other Gods: On Science and American Social Thought* (Baltimore, MD: Johns Hopkins University Press, 1976); Paul Starr, *The Social Transformation of American Medicine: The Rise of a Sovereign Profession and the Making of a Vast Industry* (New York: Basic Books, 1982); Kenneth M. Ludmerer, *Learning to Heal: The Development of American Medical Education* (New York: Basic Books, 1985); John Harley Warner, *Against the Spirit of the System: The French Impulse in Nineteenth-Century American Medicine* (Princeton, NJ: Princeton University Press, 1997); John Harley Warner, *The Therapeutic Perspective: Medical Perspective, Knowledge, and Identity in America, 1820–1885* (Princeton, NJ: Princeton University Press, 1997).

8. See Henry Jacob Bigelow, "Insensibility during Surgical Operations Produced by Inhalation," *Boston Medical and Surgical Journal* 35, no. 16 (November 18, 1846): http://doi.org/10.1056/NEJM184611180351601; John Collins Warren, *Etherization: with Surgical Remarks* (Boston: William D. Ticknor, 1848); Henry Jacob Bigelow, *Surgical Anesthesia: Addresses and Other Papers* (Boston: Little, Brown, 1900).

9. For a full account, see Henry Viets, "The Earliest Printed References in Newspapers and Journals to the First Public Demonstration of Ether Anesthesia in 1846," *Journal of the History of Medicine and Allied Sciences* 4, no. 2 (March 1949): 149–69.

10. See Webster Bull and Martha Bull, *Something in the Ether: A Bicentennial History of Massachusetts General Hospital, 1811–2011* (Beverly, MA: Memoirs Unlimited, 2011); John A. Kastor, *Mergers of Teaching Hospitals in Boston, New York, and Northern California* (Ann Arbor: University of Michigan Press, 2001).

11. John D. Thompson and Grace Goldin, *The Hospital: A Social and Architectural History* (New Haven, CT: Yale University Press, 1975), 102.

12. R. A. Strickland and J. F. Butterworth found that "by the late 1850s, the concept that ether and chloroform could produce sexually related dreams and ideations was well known in Europe, Britain, and America." Strickland and Butterworth, "Sexual Dreaming during Anesthesia: Early Case Histories (1849–1888) of the Phenomenon," *Anesthesiology* 106, no. 6 (June 2007): 1232–36.

13. John Ashhurst, *The Principles and Practice of Surgery* (Philadelphia: Henry C. Lea, 1871), 80.

14. Edward Warren, *Some Account of the Letheon; Or, Who Was the Discoverer?* (Boston: Dutton and Wentworth, 1847), 84–85.

15. John Mason Good, *A Physiological System of Nosology* (London: Printed by Bensley for E. Cox, 1817). See also Robert Hooper, *Lexicon Medicum: or, Medical Dictionary* (New York: J. & J. Harper, 1829), 102, 147.

16. René Descartes, *The Philosophical Writings of Descartes*, trans. John Cottingham, Robert Stoothoff, and Dugald Murdoch, 3 vols. (Cambridge: Cambridge University Press, 1984–1991), esp. *The Description of the Human Body*, originally published in 1647.

17. Robert Whytt, *An Essay on the Vital and Other Involuntary Motions of Animals* (Edinburgh: Hamilton, Balfour, and Neill, 1751), 266–326.

18. Charles Bell, *Idea of a New Anatomy of the Brain—Submitted for the Observations of His Friends* (London: Strahan and Preston, 1811); Johannes Müller, *Elements of Physiology*, trans. William Baly (London: Taylor and Walton, 1842).

19. *Oxford English Dictionary*, 3rd ed., s.v. "pain," accessed March 2015, http://www.oed.com.

20. Susan Sontag, *Regarding the Pain of Others* (New York: Farrar, Straus and Giroux, 2003).

21. Steven Bruhm, *Gothic Bodies: The Politics of Pain in Romantic Fiction* (Philadelphia: University of Pennsylvania Press, 1994), 8. See also Joanna Bourke, *The Story of Pain. From Prayer to Painkillers* (Oxford: Oxford University Press, 2014); Javier Moscoso, *Pain: A Cultural History* (New York: Palgrave Macmillan, 2012); Keith Wailoo, *Pain: A Political History* (Baltimore, MD: Johns Hopkins University Press, 2014); Lucy Bending, *The Representation of Bodily Pain in Late Nineteenth-Century English Culture* (Oxford: Clarendon, 2004).

22. Michel Foucault argues that by the mid-nineteenth century disciplinary punishment had transitioned from corporeal conditioning via pain to a system based on capital; i.e., incarceration to lessen work value and monetary accumulation. Foucault, *Discipline and Punish: The Birth of the Prison* (New York: Vintage Books, 1995).

23. "Gossip of the Month," *United States Magazine and Democratic Review*, vol 21 (New York: Publication Office, 1847), 182.

24. See Martin S. Pernick, *A Calculus of Suffering: Pain, Professionalism, and Anesthesia in Nineteenth-Century America* (New York: Columbia University Press, 1985). For more on pain in relation to slavery, see Elizabeth C. Clark, "'The Sacred Rights of the Weak': Pain, Sympathy, and the Culture of Individual Rights in Antebellum America," *Journal of American History* 82, no. 2 (September 1995): 463–93; and Saidiya V. Hartman, *Scenes of Subjection: Terror, Slavery, and Self-Making in Nineteenth-Century America* (Oxford: Oxford University Press, 1997).

25. The administration of ether and, later, chloroform for obstetrics was debated. In April 1853, chloroform was administered by John Snow to Queen Victoria during the birth of her eighth child, ushering in its more widespread obstetrical use. One report declared: "Thus a new era in the destiny of woman

was inaugurated: she was absolved from the curse which is said to have been pronounced on Eve, and upon all her daughters to the end of time; pain was subordinated to volition, and she can now look forward to the joys of maternity unmixed with sorrow." John Chapman, *Chloroform and Other Anaesthetics: Their History, and Use during Childbirth* (London: Williams and Norgate, 1859), 18. Today, gender and racial biases persist when it comes to pain. Female patients are still less likely to be taken seriously when it comes to medical complaints like chronic pain. Similarly, medical racism persists and patients who are Black, Indigenous, and people of color are more likely to be dismissed by doctors, especially when it comes to pain. See Janice A. Sabin, "How We Fail Black Patients in Pain," AAMC, January 6, 2020, https://www.aamc.org/news-insights/how-we-fail-black-patients-pain, and Jennifer Billock, "Pain Bias: The Health Inequality Rarely Discussed," *The Health Gap* (blog), BBC, May 22, 2018, https://www.bbc.com/future/article/20180518-the-inequality-in-how-women-are-treated-for-pain.

26. For more on vivisection see Sarah Burns, *Painting the Dark Side: Art and the Gothic Imagination in Nineteenth-Century America* (Berkeley: University of California Press, 2004); Robert Baker, *Before Bioethics: A History of American Medical Ethics from the Colonial Period to the Bioethics Revolution* (Oxford: Oxford University Press, 2013); Susan Lederer, *Subjected to Science: Human Experimentation in America before the Second World War* (Baltimore, MD: Johns Hopkins University Press, 1995). Henry Jacob Bigelow condemned vivisection in an 1871 address to the Massachusetts Medical Society. Bigelow, *Medical Education in America: Being the Annual Address Read before the Massachusetts Medical Society* (Cambridge: Welch, Bigelow, 1871).

27. Bigelow, "Insensibility during Surgical Operations Produced by Inhalation," 16.

28. Michel Foucault, *The Birth of the Clinic: An Archeology of Medical Knowledge* (New York: Vintage Books, 1994), xviii–xix, xi.

29. For pictures of pain, see Maria Pia Di Bella and James Elkins, eds., *Representations of Pain in Art and Visual Culture* (New York: Routledge, 2013).

30. On drawing manuals and human expression in art, see Martin Kemp, *The Human Animal in Western Art and Science* (Chicago: University of Chicago, 2007); Elliot Bostwick Davis, "Training the Eye and the Hand: Drawing Books in Nineteenth-Century America" (PhD diss., Columbia University, 1992).

31. Charles Le Brun, *Expressions des passions de l'âme* (Augsburg, 1732).

32. Michael Yonan, "The Man behind the Mask? Looking at Franz Xaver Messerschmidt," *Eighteenth-Century Studies* 42, no. 3 (Spring 2009): 431–51.

33. G.-B. Duchenne, *Mécanisme de la physionomie humaine ou analyse électro-physiologique de l'expression des passions* (Paris: Librairie J.-B. Bailliere et Fils, 1862). For more on how photography was adopted to aid the fields of physiognomy, public health, criminology, and eugenics, see Alan Sekula, "The Body and the Archive," *October* 39 (Winter 1986): 3–64.

34. Naomi Slipp, "For the Edification of All: Nineteenth-Century American Medicine, Art, and the Role of the Classical Cast in Cultural Life," *Athanor 32* (2014): 65–73.

35. Bates Lowry and Isabel Barrett Lowry, *Silver Canvas: Daguerreotype Masterpieces from the J. Paul Getty Museum* (Los Angeles: J. Paul Getty Museum, 1998), 182.

36. Photographic theory continues to grapple with this tension. Daston and Galison have identified how mid-nineteenth-century photography operated to separate the object of study from the interpreter, injecting "objectivity" into scientific studies, interpretations, and images. This fundamental shift in Western knowledge, revolutionized scientific epistemology and our relationship to scientific knowledge, aided by the camera. Lorraine Daston and Peter Galison, *Objectivity* (New York: Zone Books, 2007), 121.

37. Daston and Galison, *Objectivity*, 139.

38. Tanya Sheehan, *Doctored: The Medicine of Photography in Nineteenth-Century America* (University Park: Pennsylvania State University Press, 2011), 74.

39. Bigelow, "Insensibility of Surgical Operations Produced by Inhalation," 312, 314.

40. Sarah Kate Gillespie, *The Early American Daguerreotype: Cross-Currents in Art and Technology* (Cambridge: MIT Press, 2016).

41. The 1851 introduction of the wet collodion process, which used glass negatives to create paper prints, allowed for the reproduction of images and superseded daguerroetypy. A detailed technical account of daguerreotypy is found in M. Susan Barger and William B. White, *The Daguerreotype: Nineteenth-Century Technology and Modern Science* (Baltimore, MD: Johns Hopkins University Press, 2000), 44–46.

42. Trachtenberg expands on this when he writes: "Was it black magic? Something demonic?—that marked responses to the daguerreotype." Elsewhere he claims: "In the daguerreotype, the photographic image retains its association with magic, still excites an awe that follows not from subject matter but from its own inherently ephemeral character, floating just within a mirror-like surface of flickering light." Alan Trachtenberg, "The Daguerreotype and Antebellum America," in *American Photography: Local and Global Contexts*, edited by Bettina Gockel, with Patrizia Munforte (Berlin: Akademie, 2012), 7, 17.

43. Alan Trachtenberg, *Reading American Photographs: Images as History, Mathew Brady to Walker Evans* (New York: Hill & Wang, 1989), 3. Quoting James E. McClees, *Elements of Photography* (Philadelphia: J. H. Jones, 1855), 3.

44. A. Claudet, "Production of Natural Colours by Photography," *British Journal of Photography* 14, no. 351 (1867): 40.

45. Barger and White, *Daguerreotype*, 1, quoting Antoine Claudet, "The Progress and Present State of the Daguerreotype Art," *Journal of the Franklin Institute* 40, no. 1 (1845): 49.

46. James Wylde, *The Magic of Science: A Manual of Easy and Instructive Scientific Experiments* (London: Richard Griffin, 1861). For more on magic, alchemy, and early photography see Laurie Virginia Dahlberg, "The Material Ethereal: Photography and the Alchemical Ancestor," in *Art & Alchemy*, ed. Jacob Wamberg (Copenhagen: Museum Tusculanum Press, 2006), 83–100. For more on magic lanterns, see Jonathan Crary, *Suspensions of Perception: Attention, Spectacle, and Modern Culture* (Cambridge, MA: MIT Press, 1999); Linda M. Shires, *Perspectives: Modes of Viewing and Knowing in Nineteenth-Century England* (Columbus: Ohio State University Press, 2009); David J. Jones, *Gothic Machine: Textualities, Pre-cinematic Media and Film in Popular Visual Culture, 1670–1910* (Cardiff: University of Wales Press, 2011); Wendy Bellion, *Citizen Spectator: Art, Illusion, and Visual Perception in Early National America* (Chapel Hill: University of North Carolina Press, 2011).

47. Douglas Nickel, "Talbot's Natural Magic," *History of Photography* 26, no. 2 (Summer 2002): 139.

48. Nathaniel Hawthorne, *The House of Seven Gables* (New York: Charles Merrill, 1907); Alan Trachtenberg, "Seeing and Believing: Hawthorne's Reflections on the Daguerreotype in *The House of the Seven Gables*," *American Literary History* 9, no. 3 (Autumn 1997): 460–81.

49. "Discussion on Anaesthesia," *Boston Medical and Surgical Journal* 46, no. 11 (1852): 224–25.

50. Review of *Rudimentary Electricity*, by W. Snow Harris, *Regulations of the Electric Telegraph Company*, and *Traité de télégraphie electrique*, by François Moigno, *Edinburgh Review*, American ed., 90 (October 1849): 231.

51. Lunsford P. Yandell, "A Plea for Anaesthetics in Midwifery," *Transactions of the Indiana State Medical Society* 21 (1871): 44.

52. Margaret Fuller, *At Home and Abroad; or, Things and Thoughts in America and Europe* (Boston: Roberts Brothers, 1895), 203.

53. *Oxford English Dictionary*, 3rd ed., s.v. "ether," accessed May 2019, http://www.oed.com.

54. *Oxford English Dictionary*, 3rd ed., s.v. "lethe," accessed May 2019, http://www.oed.com.

55. "Ether produces a deep sleep, chloroform a deeper still; but too often it is the sleep of death." Thomas Green, "On Death from Chloroform," *British Medical Journal* 2 (1872): 629.

56. In March of 1847, chloroform, which was more dangerous but faster acting, was introduced as a competing anesthetic, complicating the widespread acceptance and adoption of ether. In January of 1848, British patient Hannah Greener was the first to die from chloroform anesthesia. Subsequently, nitrous oxide, cocaine, and heroin were also introduced as anesthetics. The subfield of anesthesiology did not professionalize in Great Britain until 1893. Barbara Duncum, *The Development of Inhalation Anaesthesia* (London: Oxford University

Press, 1947). Prior to the introduction of ether, mesmerism was one method by which surgeons "anesthetized" patients. In this practice—essentially hypnotism—medical professionals would lull patients into a deep sleep. See Alison Winter, *Mesmerized: Powers of Mind in Victorian Britain* (Chicago: University of Chicago Press, 1998). For a highly amusing period account scientifically disproving mesmerism in a medical context, see Ernest Abraham Hart, *Hypnotism, Mesmerism and the New Witchcraft* (New York: Appleton, 1893).

57. "Every large hospital in Europe and America, except our own Pennsylvania Hospital in Philadelphia, uses one of the other of them habitually before great operations." "Abstract of Discussion on Anaesthesia, before the Philadelphia Co. Medical Society, March 9, 1852," *New Jersey Medical Reporter* 6 (1852): 347.

58. On nineteenth-century vision and the camera, see Jonathan Crary, *Techniques of the Observer: On Vision and Modernity in the Nineteenth Century* (Cambridge: MIT Press, 1990); Corey Keller, *Brought to Light: Photography and the Invisible, 1840–1900* (New Haven, CT: Yale University Press, 2008); and Shawn Michelle Smith, *At the Edge of Sight: Photography and the Unseen* (Durham, NC: Duke University Press, 2013).

59. George Adams, *The Massachusetts Register: State Record for the Year 1852* (Boston: Office, No. 19, Washington Street, 1852), 328.

60. See Grant B. Romer, "'A High Reputation with All True Artists and Connoisseurs': The Daguerrian Careers of A. S. Southworth and J. J. Hawes," in Romer and Wallis, *Young America*, 20–55.

61. George Adams, *The Massachusetts Register: State Record for the Year 1853* (Boston: Office, No. 19, Washington Street, 1853), 326.

62. Robert A. Sobieszek, *The Spirit of Fact: The Daguerreotypes of Southworth & Hawes* (Boston: D. R. Godine, 1976), xn11.

63. *History of Wages in the United States from Colonial Times to 1928; Revision of Bulletin No. 499 with supplement, 1929–1933* (Washington: Government Printing Office, 1934), 160.

64. Lowry and Lowry, "Simultaneous Developments," 86n13.

65. Albert S. Southworth, "An Address to the National Photographic Association of the United States, Delivered at Cleveland, Ohio, June 1870," *Philadelphia Photographer* 8, no. 94 (October 1871): 320.

66. John D. Stoeckle and Guillermo C. Sanchez, "On Seeing Medicine's Science and Art: Cure and Care, Body and Patient," in *The Invention of Photography and Its Impact on Learning*, ed. Louise Todd Ambler and Melissa Banta (Cambridge, MA: Harvard University Press, 1989), 73.

67. The intellectual debate over the place of photography within cultural life, as a scientific tool or artistic medium, began with its introduction. Baudelaire's opinion that it would pollute the spheres of high art is articulated in "Salon de 1859," in Baudelaire, *Critique d'art suivi de critique musicale* (Paris: Gallimard,

1992): 268–342. Following the acceptance of photography as a fine art in the second half of the twentieth century, scholarship has tried to parse the ways in which practice and intent produced radically different photographic works. See Abigail Solomon-Godeau, *Photography at the Dock: Essays on Photographic History, Institutions, and Practices* (Minneapolis: University of Minnesota Press, 2009); Andrea Henderson, "Magic Mirrors: Formalist Realism in Victorian Physics and Photography," *Representation* 117, no. 1 (Winter 2012): 120–50; Jennifer Tucker, *Nature Exposed: Photography as Eyewitness in Victorian Science* (Baltimore, MD: Johns Hopkins University Press, 2005).

68. Romer, "High Reputation," 44.

69. George K. Behlmer, "Grave Doubts: Victorian Medicine, Moral Panic, and the Signs of Death," *Journal of British Studies* 42, no. 2 (2003): 206–35.

70. Julie Bresson, Ngai Liu, Marc Fischler, and Alain Bresson, "Anesthesia, Sleep and Death: From Mythology to the Operating Room," *Anesthesia & Analgesia* 117, no. 5 (November 2013): 1257–59.

71. Lowry and Lowry, "Simultaneous Developments," 75.

72. Lowry and Lowry, "Simultaneous Developments," 87n58.

73. Both hang in the Francis A. Countway Library of Medicine, Harvard Medical School. The authorship and dating have not been determined. Warren possibly saw Rembrandt's painting in 1801. Edward Warren, *The Life of John Collins Warren*, vol. 1 (Boston: Ticknor and Fields, 1860), 54.

74. Warren, *Life of John Collins Warren*, 399.

75. Oliver Wendell Holmes, *An Introductory Lecture, Delivered at the Massachusetts Medical College, November 3, 1847* (Boston: Ticknor, 1847), 21.

Chapter 6: Drawing Racial Comparisons in Nineteenth-Century British and American Anatomical Atlases

First and foremost I would like to thank William Shupbach, Curator of Iconographic Collections at the Wellcome Collection, for introducing me to the work of Joseph Maclise. I am grateful to William, along with the curators and librarians at the National Library of Medicine in Bethesda (Michael Sappol and Stephen Greenberg), Georgetown University Special Collections (Ethan Henderson), and the Royal Academy of Arts (Annette Wickham) for their generosity and patience. I presented this material at the College Art Association in 2015, the American Association for the History of Medicine in 2016, and the Objects in Motion workshop (supported by the Terra Foundation for American Art and the journal *British Art Studies*) in 2018. I am deeply grateful for the feedback I received from conference organizers, copanelists, and audience members. This paper has also benefited from the expert input of numerous generous readers: David Bronstein, Philippa Levine, Nancy Rose Marshall, Natasha Ruiz-Gómez, Michael Sappol, Tania Woloshyn, and the two anonymous readers. I have learned a great deal from these discussions and exchanges.

1. Nancy Stepan, *Picturing Tropical Nature* (Ithaca, NY: Cornell University Press, 2001), 94.

2. Joseph Maclise, preface to *Surgical Anatomy* (Philadelphia: Blanchard and Lea, 1851), vi.

3. Maclise, *Surgical Anatomy* (1851), vii.

4. Deanna Petherbridge and Ludmilla Jordanova, *The Quick and the Dead: Artists and Anatomy* (Berkeley: University of California Press, 1997), 63. See also Londa Schiebinger, "Skeletons in the Closet: The First Illustration of a Female Skeleton in Eighteenth-Century Anatomy," *Representations* 14 (1986): 42–82.

5. My examination of the relationship between art and medicine during the nineteenth century owes much to the pioneering interdisciplinary studies into the visual culture of medicine undertaken by historians of science and medicine, and historians of art: Fae Brauer, Anthea Callen, Mary Hunter, Ludmilla Jordanova, Martin Kemp, Natasha Ruiz-Gómez, Susan Sidlauskas, Tania Woloshyn, and others.

6. As was brought to my attention by medical historian Michael Sappol, Maclise changed the text of the preface to the second British edition of *Surgical Anatomy*. These textual changes, while significant in terms of the controversies that dogged Maclise's later career, do not relate to the question of race and its depiction. Hence, they lie beyond the scope of the current chapter.

7. Ludmilla Jordanova, *Sexual Visions: Images of Gender in Science and Medicine between the Eighteenth and Twentieth Centuries* (Madison: University of Wisconsin Press, 1989), 106.

8. My thanks to Tania Woloshyn for sharing these thoughts with me.

9. David Batchelor, *Chromophobia* (London: Reaktion Books, 2000), 13.

10. An interesting point of comparison is the controversy surrounding the nineteenth-century American surgeon and pioneer of modern gynecology J. Marion Sims. Sims performed experimental operations on enslaved Black women, recently igniting a heated scholarly debate about whether he should be celebrated as a surgical pioneer or condemned for unethical behavior. Durrenda Ojanuga, "The Medical Ethics of the 'Father of Gynaecology,' Dr J. Marion Sims," *Journal of Medical Ethics* 19, no. 1 (1993): 28–31; and L. L. Wall, "The Medical Ethics of Dr J Marion Sims: A Fresh Look at the Historical Record," *Medical Ethics* 32, no. 6 (2006): 346–50.

11. For a recent account of the career of Charles Bell, see Carin Berkowitz, *Charles Bell and the Anatomy of Reform* (Chicago: University of Chicago Press, 2015).

12. This sketch is reproduced in Michael Sappol, ed., *Hidden Treasure: The National Library of Medicine* (New York: Blast Books, 2012), 89.

13. Charles Bell, *Engravings of the Arteries; Illustrating the Second Volume of the Anatomy of the Human Body, and Serving as An Introduction to the Surgery of the Arteries*, 3rd ed. (London: Longman, 1811), vii.

14. Bell, *Engravings of the Arteries*, 11.

15. Richard Quain, *The Anatomy of the Arteries of the Human Body with Its Applications to Pathology and Operative Surgery in Lithographic Drawings with Practical Commentaries. The Drawings from Nature and on Stone by Joseph Maclise* (London: Taylor and Walton, 1844).

16. Quain, preface to *Anatomy of the Arteries*, n.p. Royal College of Surgeons of England, *Plarr's Lives of the Fellows Online*, s.v. "Quain, Richard," accessed June 2017, http://livesonline.rcseng.ac.uk/biogs/E000194b.htm.

17. This sculptural fragment is believed to be part of a first-century BCE marble copy of the original Greek bronze.

18. Joshua Reynolds, "Discourse VIII, Delivered to the Students on the Distribution of the Prizes, December 10th, 1778," in *The Discourses of Joshua Reynolds*, edited by John Burnet (London: James Carpenter, 1842), 142.

19. Reynolds, *Discourses of Joshua Reynolds*, 143.

20. Lorraine Daston and Peter Galison, *Objectivity* (New York: Zone Books, 2007), 22, 19–22.

21. Entry for Royal College of Surgeons of England, *Parr's Lives of the Fellows Online*, s.v. "Maclise, Joseph" accessed June 2017, http://livesonline.rcseng.ac.uk/biogs/E002613b.htm.

22. Writes Sappol: "Looking at Bidloo's *Anatomia*, we see disturbing echoes of sexual fetishism and political torture." *Dream Anatomy* (Bethesda, MD: National Library of Medicine, 2002), 34. Additionally, Natasha Ruiz-Gómez examines the association of nineteenth-century écorchés, especially those produced by Paul Richer, with the violence and torture of dissection, in "Shaking the Tyranny of the Cadaver: Doctor Paul Richer and the 'Living Écorché,'" in *Bodies beyond Borders: Moving Anatomies, 1750–1950*, ed. Kaat Wils, Raf De Bont, and Sokhieng Au (Leuven, Belgium: University of Leuven Press, 2017), 231–57.

23. Nancy Weston, *Daniel Maclise: Irish Artist in Victorian London* (Dublin: Four Courts, 2001), 229.

24. For more on the professional and social connections between surgeons and artists during the second half of the nineteenth century in Britain see Keren Rosa Hammerschlag, "The Gentleman Artist-Surgeon in Late Victorian Group Portraiture," *Visual Culture in Britain* 14, no. 2 (2013): 154–78.

25. Leon Litvack, "Continental Art and the 'Cockneyfied Corkonian': German and French Influences on Daniel Maclise," in *Daniel Maclise 1806–1870: Romancing the Past*, ed. Peter Murray (Cork: Crawford Art Gallery and Gandon Editions, 2008), 199.

26. My thanks to Philippa Levine for alerting me to this. John Turpin, "Maclise, Daniel," in *Oxford Dictionary of National Biography*, ed. H. C. G. Matthew and Brian Harrison (Oxford: Oxford University Press, 2004), accessed December 1, 2017, https://doi.org/10.1093/ref:odnb/17682. Dante Gabriel Rossetti, for instance, lived at 16 Cheyne Walk from 1862 to 1882.

27. Daniel Maclise won the commission in 1858.

28. Annette Wickham and Mark Murray-Flutter, eds., *Daniel Maclise: The Waterloo Cartoon* (London: Royal Academy Publications, 2015), 35. There is an example of a Petit-type tourniquet in the Science Museum in London.

29. Wickham and Murray-Flutter, *Daniel Maclise*, 32.

30. To cite but one striking example, in Daniel's painting of 1842, *The Origin of the Harp* (held at the Manchester Art Gallery), he depicted a sea nymph in a distinctive gesture of head and body angled to the right, and arm extended backward to the left, the same gesture assumed by the female figure in plate 7 of the 1851 edition of *Surgical Anatomy*. This observation was made by Erin DeSimone, a student in my Art and Medicine course at Georgetown University.

31. Nancy Weston writes about the inclusion of the Black figure in *Trafalgar* that "Maclise seems to be referring to John Singleton Copley's (1782–84) battle scene *Death of Major Peirson*." Weston, *Daniel Maclise*, 245. The figure may also reference the American Indian crouching in the foreground of Benjamin West's *Death of General Wolfe* (1771; held at the National Gallery of Canada).

32. It must be noted that, while there is a similarity in the roles played by the Black figures in these two "contemporary history paintings," racial understandings altered dramatically from the late eighteenth century, when Copley produced his work, to the mid-nineteenth century, when Maclise produced his. For a fuller discussion of conceptions of racial difference and beauty during the eighteenth century, see David Bindman, *Ape to Apollo: Aesthetics and the Idea of Race in the Eighteenth Century* (London: Reaktion Books, 2002).

33. Litvack, "Continental Art," 200.

34. This observation is also made by Weston, *Daniel Maclise*, 214.

35. Weston, *Daniel Maclise*, 245.

36. Catalogue entry for Daniel Maclise's *The Death of Nelson* in Jan Marsh, ed., *Black Victorians: Black People in British Art, 1800–1900* (Aldershot, UK: Lund Humphries, 2006), 143.

37. See Adrian Desmond, *The Politics of Evolution: Morphology, Medicine, and Reform in Radical London* (Chicago: University of Chicago Press, 1989), 368–69.

38. Nicolaas A. Rupke, "Richard Owen's Vertebrate Archetype," *Isis* 84 (1993): 234–35.

39. According to one reviewer, the main difference between Owen's *On the Archetype* and Maclise's *Comparative Osteology* was that Maclise focused his attention almost entirely on the higher forms of vertebrate animals, with the illustrations drawn almost exclusively from mammals and the highest orders, "man and the quadrumana." "Owen and Maclise on the *Archetype Skeleton*," *British and Foreign Medico-chirurgical Review or Quarterly Journal of Practical Medicine and Surgery* 2 (July 1848): 120. Adrian Desmond notes that Owen and Maclise disagreed over the number of skull vertebrae and the origin of arms, but that most commentators "considered Owen safer to follow." *Politics of Evolution*, 369.

Desmond also discusses the clash between Huxley and Owen over the concept of the archetype in *Archetypes and Ancestors: Palaeontology in Victorian London, 1850–1875* (London: Blond and Briggs, 1982), 19–55.

40. Joseph Maclise, "Introduction," in *Comparative Osteology: Being Morphological Studies to Demonstrate the Archetype Skeleton of Vertebrated Animals* (London: Taylor and Walton, 1847), 1–15.

41. Desmond, *Politics of Evolution*, 8.

42. Desmond, *Politics of Evolution*, 8.

43. Desmond, *Politics of Evolution*, 425.

44. Maclise, "Remarks on the Figures of Plate XXIX," in *Comparative Osteology*.

45. Maclise, "Remarks on the Figures of Plate XLVII," in *Comparative Osteology*.

46. Petherbridge and Jordanova, *Quick and the Dead*, 82.

47. Bernard Lightman and Bennett Zon, introduction to *Evolution and Victorian Culture*, ed. Bernard Lightman and Bennett Zon (Cambridge: Cambridge University Press, 2014), 1–16; Martin Kemp, *The Human Animal in Western Art and Science* (Chicago: University of Chicago Press, 2007), 221–24; and Fae Brauer, "The Transparent Body: Biocultures of Evolution, Eugenics, and Scientific Racism," in *A History of Visual Culture: Western Civilization from the 18th to the 21st Century*, ed. Jane Kromm and Susan Benforado Bakewell (Oxford: Berg, 2009), 89–103.

48. Maclise, "Remarks on the Figures of Plate XLVII," in *Comparative Osteology*.

49. Maclise, "Remarks on the Figures of Plate XLVIII," in *Comparative Osteology*. We can assume that Maclise was here referring to "native" New Zealanders, rather than white settlers.

50. Huxley argued that humans had descended from an apelike ancestor. Thomas H. Huxley, *Evidence as to Man's Place in Nature* (London: Williams and Norgate, 1863).

51. A photograph from ca. 1900 in the Wellcome Library of medical students dissecting includes illustrations, possibly by Maclise, hanging on the wall behind the dissecting group.

52. Martin Kemp and Marina Wallace, *Spectacular Bodies: The Art and Science of the Human Body from Leonardo to Now* (London: Hayward Gallery, 2000); and Sappol, *Dream Anatomy*. See, for example, the illustrations in Bernhard Siegfried Albinus's *Tabulae sceleti et musculorum corporis humani* (1747).

53. Maclise, *Surgical Anatomy* (1851), v.

54. Maclise, *Surgical Anatomy* (1851), viii.

55. William A. Alcott, *The House I Live In, or, The Human Body, for the Use of Families and Schools*, 13th ed. (Boston: Charles H. Peirce, 1847), 130.

56. Michael Sappol, *A Traffic of Dead Bodies: Anatomy and Embodied Social Identity in Nineteenth-Century America* (Princeton, NJ: Princeton University Press, 2002), 184.

57. Grigsby, "Food Chains: French Abolitionism and Human Consumption (1787–1819)," in *An Economy of Colour: Visual Culture and the Atlantic World, 1600–1830*, ed. Geoff Quilley and Kay Dian Kriz (Manchester: Manchester University Press, 2003), 170.

58. It appears as plate 4 in the second British edition.

59. J. C. Nott and Geo. R. Gliddon, *Types of Mankind, or, Ethnological Researches*, 6th ed. (Philadelphia: Lippincott, Grambo, 1854), 414.

60. Rebecca E. May, "'This Shattered Prison': Bodily Dissolution, *Wuthering Heights*, and Joseph Maclise's Dissection Manuals," *Nineteenth-Century Contexts* 33, no. 5 (2011): 431.

61. The first (1812), second (1816) and third (1833) American editions of Bell's *Engravings of the Arteries* were claimed to have been made from the third London edition (1811), and it was the third London edition that included the illustration of the dissected Black man. But, on close examination, it is apparent that the American editions were made from the second London edition, which did not include that particular illustration.

62. Stephen Jay Gould, *The Mismeasure of Man*, rev. ed. (London: Norton, 1996), 74.

63. In 1850 Reverend John Bachman published *The Doctrine of the Unity of the Human Race, Examined on the Principles of Science*, which attempted to integrate a literal acceptance of the biblical account of creation with the data concerning human biological diversity. C. Loring Brace, *"Race" Is a Four-letter Word: The Genesis of the Concept* (New York: Oxford University Press, 2005), 106 and 108.

64. John Harley Warner makes this point in "Witnessing Dissection: Photography, Medicine, and American Culture," in *Dissection: Photographs of a Rite of Passage in American Medicine, 1880–1930*, ed. J. H. Warner and James M. Edmonson (New York: Blast Books, 2009), 15–21.

65. The literature includes Nicholas Hudson, " The 'Hottentot Venus,' Sexuality, and the Changing Aesthetics of Race, 1650–1850," *Mosaic* 41, no. 1 (2008): 19–41; Deborah Willis, ed., *Black Venus 2010: They Called Her "Hottentot"* (Philadelphia: Temple University Press, 2010); and Kara Reilly, "Two Venuses: Historicizing the Anatomical Female Body," *Performance Research* 19, no. 4 (2014): 111–21.

Chapter 7: The Post-Darwinian Eye, Physiological Aesthetics, and the Early Years of Aestheticism, 1860–1876

1. The Grosvenor Gallery was founded in London in 1877 and became an immediate success. It was home to the aesthetic artists and it was here and in

that first year that critic John Ruskin saw an exhibition of Whistler's paintings that led to a vitriolic review. Whistler in turn sued Ruskin. For more on the gallery see Susan Casteras and Colleen Denner, *The Grosvenor Gallery: A Palace of Art in Victorian England* (New Haven, CT: Yale University Press, 1996).

2. Bain was an early figure in the establishment of psychology as a legitimate field of study. An active author in psychology from the mid-nineteenth century, he established the important journal *Mind* in 1876. On the early years of psychology in Great Britain see Rick Rylance, *Victorian Psychology and British Culture, 1850–1880* (Oxford: Oxford University Press, 2000).

3. Herbert Spencer, *The Principles of Psychology*, 2nd ed., vol. 1 (London: Williams and Norgate, 1870), v.

4. Herbert Spencer, *The Principles of Psychology*, 2nd ed., vol. 2, (1872; repr., New York: Appleton, 1897), 643–45.

5. Spencer, *Principles of Psychology*, vol. 2, 643–45. In his hierarchy of pleasures he begins with simple sensations related to single colors or tones, then moves to slightly higher harmonies of colors and tones. Above this is the greater complexity in combined forms and colors, but the highest aesthetic response is to be found in pictures where form and color are the agents of excitations of ideal sentiments remote from simple sensations, followed by ideal feelings that are suggested to the imagination through viewing a beautiful scene (such as a landscape). The greatest aesthetic feelings are due to both intellectual and complex emotional powers in advanced humans. As early as his essay "Use and Beauty" in 1852, Spencer had noted that the sense of the beautiful was untethered from the preservation of society. "Use and Beauty" (1852), in Herbert Spencer, *Essays: Scientific, Political and Speculative*, vol. 2 (London: Williams and Norgate, 1891), 373.

6. The phrase was used in France during the Romantic period by philosopher Victor Cousin and became a bohemian rallying cry, embodied notably in Théophile Gautier's 1835 *Mademoiselle de Maupin*.

7. Alexander Bain, *The Emotions and the Will* (London: John W. Parker and Son, 1859), 230. On the influence of Spencer on developing theories of aesthetic forms, see Benjamin Morgan, *The Outward Mind: Materialist Aesthetics in Victorian Science and Literature* (Chicago: University of Chicago Press, 2017).

8. Alexander Bain, *The Senses and the Intellect* (London: J. W. Parker, 1855), 246.

9. Bain, *Senses and the Intellect*, 246.

10. George Lewes, "Mr. Darwin's Hypotheses, Part I," *Fortnightly Review*, April 1, 1868, 353–73; Lewes, "Mr. Darwin's Hypotheses, Part II," *Fortnightly Review*, June 1, 1868, 611–28; Lewes, "Mr. Darwin's Hypotheses, Part III," *Fortnightly Review*, July 1, 1868, 61–80; Lewes, "Mr. Darwin's Hypotheses, Conclusion," *Fortnightly Review*, November 1, 1868, 492–509. On the *Fortnightly Review* see Edwin Mallard Everett, *The Party of Humanity: The* Fortnightly

Review *and Its Contributors, 1865–1874* (Chapel Hill: University of North Carolina Press, 1939).

11. Charles Darwin, *On the Origin of Species by Means of Natural Selection or the Preservation of Favoured Races in the Struggle for Life*, 1st ed. (London: John Murray, 1859; repr., New York: Avenel Books, 1979), 186–89.

12. Darwin, *On the Origin of Species*, 1st ed., 177–78.

13. Darwin, *On the Origin of Species*, 1st ed.,188.

14. Darwin, *On the Origin of Species*, 1st ed., 202.

15. The texts are Johannes Müller, *Zür vergleichenden Physiologie des Gesichtsinnes des Menschen und Thiere* (Leipzig: C. Cnobloch, 1826) and *Über die phantastischen Gesichterscheintungen: Eine physiologie Untersuchung mit einer physiologischen des Aristotles über des Traum* (Coblenz: J. Hölscher, 1826).

16. See David Cahan, "Helmholtz and the British Scientific Elite: From Force Conservation to Energy Conservation," *Notes and Records of the Royal Society of London* 66, no. 1 (2012): 55–68. The ophthalmoscope was introduced in Great Britain in 1862.

17. Kate Flint, *The Victorians and the Visual Imagination* (Cambridge: Cambridge University Press, 2000), 256.

18. Charles Darwin, *On the Origin of Species*, 6th ed. (London: John Murray, 1872), 163.

19. Charles Darwin, *The Descent of Man, and Selection in Relation to Sex*, 2nd ed. (London: John Murray, 1874), 441. Here he cites his copy of Helmholtz, *Popular Lectures on Scientific Subjects*, trans. E. Atkinson, introduction by John Tyndall (London: Longman, Green, 1873), 219, 227, 269, 390.

20. Charles Darwin, *The Descent of Man, and Selection in Relation to Sex*, 1st ed. (London: John Murray, 1871), 118.

21. Darwin, *Descent of Man*, 1st ed., 118.

22. Darwin, *Descent of Man*, 1st ed., 47.

23. Darwin, *On the Origin of Species*, 1st ed., 193.

24. Phillip Prodger, "Ugly Disagreements: Darwin and Ruskin Debate Sex and Beauty," in *The Art of Evolution: Darwin, Darwinisms, and Visual Culture*, ed. Barbara Larson and Fae Brauer (Hanover, NH: Dartmouth College Press), 40–58.

25. Evelleen Richards mapped Darwin's growing interest in the arbitrary nature of beauty by species in her book *Darwin and the Making of Sexual Selection* (Chicago: University of Chicago Press, 2017). She traces Darwin's interest in the relativism of beauty by race and even by individuals among humans back to, among other sources, Darwin's reading of Joshua Reynolds.

26. Darwin, *On the Origin of Species*, 5th ed. (London: Murray, 1869), 247.

27. Bain, *Senses and the Intellect*, 259.

28. Richard O. Prum, *The Evolution of Beauty: How Darwin's Forgotten Theory of Mate Choice Shapes the Animal World—and Us* (New York: Doubleday, 2017),

233. Prum notes that at first Darwin saw sexual selection and beauty as an aid in natural selection, but later uncoupled the two, concluding that sexual selection leading to sometimes less than pragmatic extravagant ornaments conflicted with survival (by calling attention to the organism). Aesthetic evolution by mate choice is "self-directed" and does not rely on external forces in nature.

29. Sidney Colvin, "English Painters and Painting in 1867," *Fortnightly Review*, October 1867, 464.

30. Sidney Colvin, *Memories and Notes of Persons and Places, 1852–1912* (New York: Charles Scribner's Sons, 1921), 90.

31. George Henry Lewes, *The History of Philosophy from Thales to Comte*, 3rd ed. (London: Longmans, Green, 1867), xxxv–xxxix.

32. Walter Pater, "Poems by William Morris," *Westminster Review*, October 1868, 312; and Pater, *The Renaissance: Studies in Art and Literature* (London: Macmillan, 1873), 210.

33. On Pater and science see Billie Andrew Inman, "The Intellectual Context of Walter Pater's 'Conclusion,'" in *Walter Pater: An Imaginative Sense of Fact*, ed. Philip Dodd (London: Frank Cass, 1981), 12–29; and Ian Small, "The Vocabulary of Pater's Criticism and the Psychology of Aesthetics," *British Journal of Aesthetics* 18, no. 1 (1978): 81–87. Both assert that Pater was familiar with developments in biology and psychology by the 1860s, including the work of Darwin, Spencer, and Alexander Bain. See also Gerald Monsman, *Walter Pater* (Boston: G. K. Hill, 1877); and Regenia Gagnier, "Production, Reproduction, and Pleasure in Victorian Aesthetics and Economics," in *Victorian Sexual Dissidence*, ed. Richard Dellamora (Chicago: University of Chicago Press, 1999), 127–46. Gagnier sees a cultural paradigm shift from the criticism of Ruskin as responding to the "productive" body of the artist to that of Pater as newly focused on the "pleasured" body of the observer.

34. Bain, *Emotions and the Will*, 571–72.

35. Inman, "Intellectual Context," 14.

36. Walter Pater, *Plato and Platonism: A Series of Lectures* (London: Macmillan, 1907), 19.

37. Pater, *Plato and Platonism*, 19–21. F. C. McGrath has argued that Pater was reading Darwin as early as 1865 and found in the scientist's ideas a basis for his developing interests in the determinism of the environment along with the relativism of observation. McGrath, *The Sensible Spirit: Walter Pater and the Modernist Paradigm* (Tampa: University of South Florida Press, 1986), 27.

38. Walter Pater, "Dante Gabriel Rossetti," in *The English Poets: Selections with Critical Introductions by Various Writers*, 2nd ed., ed. Thomas Ward, vol. 4, *Wordsworth to Rossetti* (London: Macmillan, 1883), 638.

39. Elizabeth Prettejohn, *Art for Art's Sake: Aestheticism in Victorian Painting* (New Haven, CT: Yale University Press, 2007), 215.

40. Letter to Thomas and Martha Combe, February 12, 1860, MS Eng. Lett. C. 296, folio 71, Bodleian Library. See also Virginia Surtees, *The Paintings and the Drawings of Dante Gabriel Rossetti, 1828–1882* (Oxford: Clarendon, 1971), 69.

41. Jane Munro, "More Like a Work of Art than of Nature: Darwin, Beauty and Sexual Selection," in *Endless Forms: Charles Darwin, Natural Science, and the Visual Arts*, ed. Diana Donald and Jane Munro (Cambridge: Fitzwilliam Museum, 2009), 269–72.

42. Laurence Shafe, "Why Is the Peacock's Tail So Beautiful?," in *Darwin and Theories of Aesthetics and Cultural History*, ed. Barbara Larson and Sabine Flach (Burlington, VT: Ashgate, 2013), 37–52; and Jonathan Smith, *Charles Darwin and Victorian Visual Culture* (Cambridge: Cambridge University Press, 2006), 125–36.

43. William Michael Rossetti, *Dante Gabriel Rossetti: His Family Letters*, vol. 1 (London: Ellis and Elvey, 1895), 203.

44. Tim Barringer, *The Pre-Raphaelites: Reading the Image* (London: Calmann and King, 1998), 150.

45. Nicola Humble, "Domestic Arts," in *The Cambridge Companion to Victorian Culture*, ed. Francis O'Gorman (Cambridge: Cambridge University Press, 2010), 229.

46. Robert Michael Brain, *The Pulse of Modernism: Physiological Aesthetics in Fin-de-Siècle Europe* (Seattle: University of Washington Press, 2015), 39–42. In 1850 Ferdinand Cohn and Robert Koch found the protoplasmic constituents in plants and animals to be identical.

47. John Ruskin, *Proserpina: Studies of Wayside Flowers*, vol. 1, in *The Works of John Ruskin*, edited by E. T. Cook and A. Wedderburn, vol. 25 (London: George Allen, 1875), 481.

48. John Ruskin to Rossetti 1865 (Denmark Hill), *Letters of John Ruskin*, in *The Complete Works of John Ruskin*, ed. E. T. Cook and A. Wedderburn, vol. 36 (London: George Allen, 1909), 491.

49. John Ruskin, *Of the Truth of Vegetation*, vol. 1 of *Modern Painters*, in *The Works of John Ruskin*, ed. E. T. Cook and A. Wedderburn, vol. 3 (London: George Allen, 1906), 574–605; Ruskin, *Of Leaf Beauty*, vol. 5 of *Modern Painters*, in *The Works of John Ruskin*, ed. E. T. Cook and A. Wedderburn, vol. 7 (London: George Allen, 1906), 3–132.

50. W. M. Rossetti, "The Royal Academy Exhibition," *Fraser's Magazine*, June 1863, 793.

51. Moore was also familiar with new theories of color that affected artists on the continent as well, such as M. E. Chevreul's *The Principles of Harmony and Contrast of Colours* (1839). See Robyn Asleson, *Albert Moore* (London: Phaidon, 2000), 89–92, 96–99, 114.

52. Algernon Swinburne, "Notes on Some Pictures of 1868," in Swinburne, *The Complete Works of Algernon Charles Swinburne*, edited by Edmund Gosse and Thomas James Wise, vol. 15 (New York: Russell and Russell, 1968), 198.

53. Christopher Newall, "Impression of the Effect," in *Pre-Raphaelite Vision: Truth to Nature*, ed. Allen Staley and Christopher Newall (London: Tate, 2004), 209. In her study of British landscape painting after the mid-1870s, Anne Helmreich has noted how the rise of psychology as a discipline coincided with a call by scientists such as physicist John Tyndall for artists to work from imagination and the emotions, and that this platform encouraged a paradigm shift for certain landscape artists like the former Pre-Raphaelite John Everett Millais to move away from observed detail toward affect and mental impressions. Helmreich, *Nature's Truth: Photography, Painting, and Science in Victorian Britain* (University Park: Pennsylvania State University Press, 2016).

54. See Caroline Arscott, "Subject and Object in Whistler: The Context of Physiological Aesthetics," in *Palaces of Art: Whistler and the Art Worlds of Aestheticism*, ed. Lee Glazer and Linda Merrill (Washington, DC: Smithsonian Institution Scholarly Press, 2013), 55–66.

55. Anna Robins posits that Whistler's interest in science dates mainly to the period after his trial of 1878 and was pursued in order to bolster his reputation, since he appeared to be "without a theoretical base." Robins, *A Fragile Modernism: Whistler and His Impressionist Followers* (New Haven, CT: Yale University Press, 2007), 17. See also Marc Simpson, ed., *Like Breath on Glass: Whistler, Inness, and the Art of Painting Softly* (Williamstown, MA: Sterling and Francine Clark Art Institute, 2008).

56. James Sully, "The Aesthetics of Human Character," *Fortnightly Review*, April 1, 1871, 507–8.

57. James Sully, "The Undefinable in Art," *Cornhill Magazine* 38, no. 227 (November 1878): 571.

58. Reproduced in Hermann von Helmholtz, *Science and Culture: Popular and Philosophical Essays*, edited by David Cahan (Chicago: University of Chicago Press, 1995), 279–308.

59. Helmholtz, *Science and Culture*, 279.

60. James Sully, "The Basis of Musical Sensation," *Fortnightly Review*, April 3, 1872, 428–43.

61. Elizabeth Prettejohn, *After the Pre-Raphaelites: Art and Aestheticism in Victorian England* (New Brunswick, N.J.: Rutgers University Press, 1999). 73.

62. Katharine A. Lochnan, *The Etchings of James McNeill Whistler* (New Haven, CT: Yale University Press, 1984), 98–100.

63. Mortimer Menpes, *Whistler as I Knew Him* (London: Charles Black, 1904), 22–23. Noted in Robins, *Fragile Modernism*, 18.

64. Hermann von Helmholtz, "The Eye as an Optical Instrument," in Helmholtz, *Popular Lectures on Scientific Subjects*, 213.

65. James Sully, *My Life and Friends: A Psychologist's Memories* (New York: Dutton, 1918), 100.

66. James Sully, *Sensation and Intuition: Studies in Psychology and Aesthetics* (London: Henry S. King, 1874), 368.

67. Sully acknowledged the intense delight "savages" may experience, from time to time, over color or decoration, but the complexity of mental development in the white European is such that there is "a vast increase in the *number* of pleasures" (*Sensation and Intuition*, 354). With mental development, there was also an increase in the quantity of pure enjoyment. An evolved mind can distinguish between different impressions and ideas and note points of resemblance. The artist was supremely gifted in aesthetics: "The artist's eye notes myriad points of diversity of resemblance among visual forms and shades of tint which wholly escape the attention of ordinary men" (357). With a cultivated mind there was greater subtlety in the meaning brought to a particular color, its emotional sign.

68. Sully, *Sensation and Intuition*, 367.

69. Sully, *Sensation and Intuition*, 361–62.

70. Sully, *Sensation and Intuition*, 347.

Chapter 8: Darwinian Aesthetics and Aestheticism in James McNeill Whistler's Peacock Room

1. [Andrew Lang], "In and Out of London: Mr. Whistler's Peacocks," *London*, February 17, 1877, 63.

2. The earliest use of peacock feathers associated with the movement that came to be known as aestheticism may be Frederic Leighton's *Pavonia* (1858–1859, held in private collection).

3. A brief overview on the global cultural history of the peacock is Christine Jackson, *Peacock* (London: Reaktion Books, 2005). See also the catalogue for the exhibition, Bartholomew F. Bland and Laura Vookles, eds., *Strut: The Peacock and Beauty in Art* (New York: Fordham University Press, 2014).

4. On peacocks, beauty, and mate selection in Victorian popular culture, see, for instance, Susan David Bernstein, "Designs after Nature: Evolutionary Fashions, Animals, and Gender," in *Victorian Animal Dreams: Representations of Animals in Victorian Literature and Culture*, ed. Deborah Denenholz Morse and Martin A. Danahay (Burlington, VT: Ashgate, 2007), 65–80.

5. Linda Nochlin, "The Darwin Effect," *Nineteenth-Century Art Worldwide* 2, no. 2 (2003), http://www.19thc-artworldwide.org/spring03/76-spring03/spring03article/218-the-darwin-effect-introduction. See bibliography for further notable sources on British art and Darwinism. Jane Munro, Richard A. Kaye, Linda Merrill, and Laurence Shafe have all mentioned the Peacock Room in conjunction with Darwinism. See Jane Munro, "'More like a Work of Art than of Nature': Darwin, Beauty and Sexual Selection," in *Endless Forms: Charles Darwin, Natural Science and the Visual Arts*, ed. Diana Donald and Jane Munro

(Cambridge: Fitzwilliam Museum, 2009), 280–82; Richard A. Kaye, *The Flirt's Tragedy: Desire without End in Victorian and Edwardian Fiction* (Charlottesville: University of Virginia Press, 2002), 116–17; Linda Merrill, *The Peacock Room: A Cultural Biography* (New Haven, CT: Yale University Press, 1998), 218; and Laurence Shafe, "Why is the Peacock's Tail So Beautiful?," in *Darwin and Theories of Aesthetics and Cultural History*, ed. Barbara Larson and Sabine Flach (Burlington, VT: Ashgate, 2013), 45.

6. Barbara Larson, introduction to *The Art of Evolution: Darwin, Darwinisms, and Visual Culture*, ed. Barbara Larson and Fae Brauer (Hanover, NH: University Press of New England, 2009), 2.

7. Alison Syme, *A Touch of Blossom: John Singer Sargent and the Queer Flora of Fin-de-Siècle Art* (University Park: Pennsylvania State University Press, 2010), 3–4.

8. Syme, *Touch of Blossom*, 4.

9. Merrill, *Peacock Room*, 19. Merrill's book remains the single most comprehensive and authoritative volume on the subject, though scholars in the last two decades have refined our understanding of the Peacock Room.

10. The room measures 33 2/3 feet long, 20 feet wide, and almost 14 feet in height.

11. For a biography of Leyland, see Merrill, *Peacock Room*, 109–45.

12. The intended fourth *Symphony* would succeed *Symphony in White, No. 1: The White Girl* (1860–1861, held at the National Gallery of Art, Washington, DC), *Symphony in White, No. 2: The Little White Girl* (1864, held at the Tate Britain, London), and *Symphony in White, No. 3* (1865–1867, held at the Barber Institute of Fine Arts, Birmingham, UK).

13. James McNeill Whistler, "Mr Whistler's 'Ten O'Clock'" (London: Chatto & Windus, 1888), 14.

14. Linda Merrill and Robyn Asleson, "The Lost Symphony: *The Three Girls* and Related *Projects*," *The Lost Symphony: Whistler & the Perfection of Art*, January 16–May 30, 2016, Freer Gallery of Art and Arthur M. Sackler Gallery, Smithsonian National Museum of Asian Art, Washington, DC, https://asia.si.edu/exhibition/related-projects/.

15. Anna Matilda Whistler to Frederick Leyland, March 11, 1869, *The Correspondence of James McNeill Whistler*, online ed., ed. Margaret F. MacDonald, Patricia de Montfort and Nigel Thorp, http://www.whistler.arts.gla.ac.uk/correspondence/exhibit/display/?rs=2&exhibid=LoRa-1869&sort=.

16. It was Leyland who suggested that Whistler call his atmospheric night landscapes "Nocturnes," in keeping with Whistler's musically titled "Harmonies," "Arrangements," and "Symphonies" (Merrill, *Peacock Room*, 142). The music in the frame of *The Three Girls* also alludes to Whistler's theory of aesthetic correspondences between painting and music.

17. The final canvas was almost complete in 1876, but Whistler cut it up when he fell out with Leyland over payment for the Peacock Room in 1877. The sole surviving section of the painting is now on long-term loan to the Courtauld Gallery, London.

18. Theodore Child, "A Pre-Raphaelite Mansion," *Harper's*, December 1890, 82.

19. On Thomas Jeckyll, see Susan Weber Soros and Catherine Arbuthnott, eds., *Thomas Jeckyll: Architect and Designer, 1827–1881* (New Haven, CT: Yale University Press, 2003).

20. On Jeckyll's artistic decisions, see Merrill, *Peacock Room*, 197–204; Soros and Arbuthnott, *Thomas Jeckyll*, 191–95; and Cynthia Fischer, "Thomas Jeckyll, James McNeill Whistler, and the Harmony in Blue and Gold: The Peacock Room: A Re-Examination" (PhD diss., Virginia Commonwealth University, 2013), chs. 2–3.

21. Though the leather hangings have been identified in many sources as priceless Tudor relics brought to England as part of Catherine of Aragon's dowry, Merrill concludes that they were probably Dutch, from the late eighteenth century (*Peacock Room*, 191–95).

22. The concept of this shelving had early modern European precedents, but its form more closely resembled popular Anglo-Asian furniture, and its surfaces were covered in motifs drawn from fashionable Japanese pattern books. See Merrill, *Peacock Room*, 197–204, and David Park Curry, *James McNeill Whistler at the Freer Gallery of Art* (New York: W. W. Norton, 1984), 62–67.

23. On Jeckyll's illness and the myths that have arisen around it, see Merrill, *Peacock Room*, 206–7, 257–60.

24. Merrill, *Peacock Room*, 210.

25. On the last stages of the break, see Merrill, *Peacock Room*, 271–74.

26. James McNeill Whistler, February 1877, "Harmony in Blue and Gold. The Peacock Room," *Correspondence of James McNeill Whistler*, http://www.whistler.arts.gla.ac.uk/correspondence/recno/display/?cid=01734.

27. Merrill's *Peacock Room*, again, contains the most thorough discussion of Whistler's Asian and European sources. See also Curry, *Whistler at the Freer*, 60–67, Fischer, "Thomas Jeckyll," and Sally-Anne Huxtable, "Whistler's Peacock Room and the Artist as Magus," in *Palaces of Art: Whistler and the Art Worlds of Aestheticism*, ed. Lee Glazer and Linda Merrill (Washington, DC: Smithsonian Institution Scholarly Press, 2013), 67–79.

28. See Jackson, *Peacock*, 142–47, for an introduction to the variety of symbolic roles played by peacocks in Asia.

29. The link with the Orientalist trope of the languorous beauty persisted, however—for instance in Whistler's *La Princesse*, whose subject is portrayed alongside Japanese and Chinese accoutrements. As Aileen Tsui has pointed out,

Whistler and his peers relocated Orientalist tropes and desires from North Africa, the Middle East, and India to Japan and China, but maintained the imbalanced power dynamics of earlier forms and continued to collapse multiple distinct cultures into one imagined "Orient." See Tsui, "Whistler's *La Princesse du pays de la porcelaine*: Painting Re-Oriented," *Nineteenth-Century Art Worldwide* 9, no. 2 (Autumn 2010), http://www.19thc-artworldwide.org/autumn10/whistlers-la-princesse-du-pays-de-la-porcelaine.

30. The *La Princesse* screen and the peacock scroll are both visible in a contemporary photograph (ca. 1867–78) of the drawing room of Whistler's home at what is now 96 Cheyne Walk, Chelsea, London.

31. Of the many examples that Whistler could have drawn upon, one of the most immediate would have been Albert Moore's peacock frieze of 1872–1873, created for the front drawing room of businessman Augustus Frederick Lehmann's Mayfair home. Like Whistler, Moore presents an assembly of stately, peaceful peacocks in an atmosphere of homosocial bonding and mutual admiration. Moore's peacocks are portrayed in highly naturalistic poses, very possibly drawn from life. Preparatory sketches remain, though the finished mural does not. See Shafe, "Why Is the Peacock's Tail So Beautiful?," 45, and Robyn Asleson, *Albert Moore* (London: Phaidon, 2000), 125–26.

32. On the history and varying fortunes of Darwin's theory of sexual selection, see Erika Lorraine Milam, *Looking for a Few Good Males: Female Choice in Evolutionary Biology* (Baltimore, MD: Johns Hopkins University Press, 2010).

33. Darwin, *On the Origin of Species*, 1st ed. (London: John Murray, 1859), 88.

34. Darwin, *On the Origin of Species*, 89.

35. Darwin, *On the Origin of Species*, 88.

36. Darwin, *On the Origin of Species*, 88.

37. [Henry Charles Fleeming Jenkin], "The Origin of Species," *North British Review*, no. 92 (June 1867): 293.

38. George Campbell, Duke of Argyll, *The Reign of Law* (London: Alexander Strahan, 1867), 184.

39. Charles Darwin to Asa Gray, April 3, 1860, letter no. 2743, *Darwin Correspondence Project*, http://www.darwinproject.ac.uk/DCP-LETT-2743.

40. On Darwin's search for persuasive illustrations of avian sexual selection for use in the *Descent*, see Jonathan Smith, *Charles Darwin and Victorian Visual Culture* (Cambridge: Cambridge University Press, 2006), 114–25, and Julia Voss, *Darwin's Pictures: Views of Evolutionary Theory, 1837–1874*, trans. Lori Lantz (New Haven, CT: Yale University Press, 2010), ch. 3.

41. Ruskin's fullest engagements with the sciences can be found in *The Eagles' Nest* (1872), *Proserpina* (2 vols., 1879 and 1886, mainly dealing with botany), *Deucalion* (1879, primarily on geology), and finally *Love's Meinie* (1881). These books collect lectures Ruskin gave as the Slade Professor of Fine Arts at Oxford University from 1870 to 1879.

42. These lectures were published in 1881. See John Ruskin, *Love's Meinie: Three Lectures on Greek and English Birds*, in *The Works of John Ruskin*, ed. E. T. Cook and Alexander Wedderburn, vol. 25 (London: George Allen, 1907). *Meinie* is an archaic English word for "many"; Ruskin presents birds as "Love's Many," or heralds of divine Love. Smith, *Charles Darwin and Victorian Visual Culture*, 126.

43. Ruskin, *Love's Meinie*, 36.

44. Scholarship focusing on Ruskin's reactions to the theories of natural and sexual selection includes George Levine, *Realism, Ethics and Secularism: Essays on Victorian Literature and Science* (Cambridge: Cambridge University Press, 2008), 75–98; Philip Prodger, "Ugly Disagreements: Darwin and Ruskin Debate Sex and Beauty," in *The Art of Evolution*, ed. Barbara Larson and Fae Brauer, 18–39; Shafe, "Why is the Peacock's Tail So Beautiful?," 37–52; Smith, *Charles Darwin and Victorian Visual Culture*, esp. chs. 1, 3, and 4; Rachel Teukolsky, *The Literate Eye: Victorian Art Writing and Modernist Aesthetics* (New York: Oxford University Press, 2009), 149–66; and Sharon Aronofsky Weltman, *Performing the Victorian: John Ruskin and Identity in Theater, Science, and Education* (Columbus: Ohio State University Press, 2007), 39–61.

45. Ruskin, *Love's Meinie*; see also Smith, *Charles Darwin and Victorian Visual Culture*, 125–36.

46. Ruskin, *The Stones of Venice*, vol. 1, in *The Works of John Ruskin*, ed. E. T. Cook and Alexander Wedderburn, vol. 9 (London: George Allen, 1903), 72.

47. Ruskin, *Stones of Venice*, 451.

48. Ruskin, *Stones of Venice*, 451.

49. On Grant Allen and Victorian visual culture, see Smith, "Grant Allen, Physiological Aesthetics, and the Dissemination of Darwin's Botany," in *Science Serialized: Representations of the Sciences in Nineteenth-Century Periodicals*, ed. Geoffrey N. Cantor and Sally Shuttleworth (Cambridge: MIT Press, 2004), 285–306, and Smith, *Charles Darwin and Victorian Visual Culture*, ch. 4. See also Teukolsky, *Literate Eye*, 151–84.

50. Grant Allen, *Physiological Aesthetics* (London: Henry S. King, 1877), vii. Allen is quoting a line from the first volume of Ruskin's *Modern Painters* (1843).

51. Smith, "Grant Allen," 291. Other venues included *Macmillan's Magazine* and *Longman's Magazine*.

52. Andrew Lang, "Grant Allen," *Argosy* 71 (May–August 1900): 412. Lang recalls that Allen was reduced to writing for *London* after his scientific articles and books on physiological aesthetics garnered applause but scant income. *London* lacked a strong unifying theme, found little audience, and was dissolved in 1879.

53. Carolyn Arscott, "Subject and Object in Whistler: The Context of Physiological Aesthetics," in Glazer and Merrill, *Palaces of Art*, 59.

54. The golden aureole can also be read as illumination from Whistler's butterfly signature, placed prominently along the shutter's upper edge.

55. On venues for homoerotic gazes in the Victorian art world, see Andrew Stephenson, "Precarious Poses: The Problems of Artistic Visibility and its Homosocial Performances in Late-Nineteenth-Century London," *Visual Culture in Britain* 8, no. 1 (Spring 2007): 74. Whistler's self-presentation as an artist-aesthete-dandy was fashioned within this context; see also Stephenson, "Refashioning Modern Masculinity: Whistler, Aestheticism, and National Identity," in *English Art, 1860–1914: Modern Artists and Identity*, ed. David Peters Corbett and Lara Perry (Manchester: Manchester University Press, 2000), 133–49.

56. Merrill, *Peacock Room*, 219–25, 238–42; Curry, *Whistler at the Freer*, 64–65.

57. Merrill, *Peacock Room*, 221.

58. See Bernstein, "Designs after Nature," 67–80.

59. Darwin, *The Descent of Man, and Selection in Relation to Sex* (London: John Murray, 1871), 2:46.

60. Melody Barnett Deusner, "Whistler, Aestheticism, and the Networked World," in Glazer and Merrill, *Palaces of Art*, 151.

61. Arthur Symons, *Studies in Seven Arts*, 2nd ed. (London: E. P. Dutton, 1907), 130.

62. The clearest distinguishing marks of green peacocks are the straight shaft of feathers on their heads (as opposed to blue peacocks' crownlike ornament) and the overlapping, disc-shaped emerald feathers, resembling scales, that cover their throat and breast. It was these scalelike feathers that Whistler adapted for his painted patterns in the Peacock Room. Unlike Whistler's birds, blue peacocks have smooth breasts and few scalelike feathers.

63. The cartoon, in pencil and charcoal on paper, still survives and is displayed in the Hunterian Museum at the University of Glasgow. On the timeline of the creation of *Art and Money*, see Merrill, *Peacock Room*, 245–46.

64. James McNeill Whistler, draft of an unsent letter to Frederic Leyland, July 18 or 25, 1877, *Correspondence of James McNeill Whistler*, http://www.whistler.arts.gla.ac.uk/correspondence/art/display/?cid=2599&filename=m0583&catno=M.0583&rs=1.

65. Metaphorical connections between miserliness and infertility have a long history in Western art. Giotto's Arena Chapel (1303–1305) in Padua, Italy, for instance, makes pointed contrasts between the fruitful womb of the Virgin Mary and the perverse growth of profit through usury, a sin of which the patron's family was accused and which the commission of the chapel was intended to expiate. See Anne Derbes and Mark Sandona, "Barren Metal and the Fruitful Womb: The Program of Giotto's Arena Chapel in Padua," *Art Bulletin* 80, no. 2 (1998): 274–91.

66. The banking heir Sir Coutts Lindsay and his wife Blanche established the Grosvenor Gallery as an alternative venue to the Royal Academy. See Susan P. Casteras and Colleen Denney, eds., *The Grosvenor Gallery: A Palace of Art in*

Victorian England (New Haven, CT: Yale University Press, 1996), and Colleen Denney, *At the Temple of Art: The Grosvenor Gallery, 1877–1890* (Cranbury, NJ: Fairleigh Dickinson University Press, 2000). On anxieties over gender, culture, and Darwinian evolution fostered by the Grosvenor's Aesthetic milieu, see Anne Anderson, "'Fearful Consequences . . . of Living up to One's Teapot': Men, Women, and 'Cultchah' in the English Aesthetic Movement, c. 1870–1900," *Victorian Literature and Culture* 37, no. 1 (2009): 219–54.

67. Linda Merrill, *A Pot of Paint: Aesthetics on Trial in Whistler v. Ruskin* (Washington, DC: Smithsonian Institution Press, 1992), 35.

68. John Ruskin, "Life Guards of New Life," letter 79 (July 1877), *Fors Clavigera: Letters to the Workmen and Labourers of Great Britain*, vol. 7, in *The Works of John Ruskin*, ed. E. T. Cook and Alexander Wedderburn, vol. 29 (London: George Allen, 1907), 146–69.

69. On Whistler's life from the court verdict to his bankruptcy, see Merrill, *Pot of Paint*, 272–85.

70. Syme, *Touch of Blossom*, 16.

71. *The Gold Scab* was one of three satirical paintings that Whistler left in his studio to be found by Leyland's men. However, the other two paintings no longer exist. They were destroyed at an unknown point before Whistler's death in 1903. Contemporary accounts gave one painting the title *The Loves of the Lobsters*, the other *Mount Ararat, or, an Arrangement in Rats*. The few surviving descriptions of the paintings suggest that they mocked Leyland in the same vein as *The Gold Scab*. See Merrill, *Peacock Room*, 280–81.

Bibliography

Abrams, M. H. *Natural Supernaturalism: Tradition and Revolution in Romantic Literature.* New York: Norton, 1971.

"Abstract of Discussion on Anaesthesia, before the Philadelphia Co. Medical Society, March 9, 1852." *New Jersey Medical Reporter* 6 (1852): 345–51.

Adams, George. *The Massachusetts Register: State Record for the Year 1852.* Boston: Office, No. 19, Washington Street, 1852.

Adams, George. *The Massachusetts Register: State Record for the Year 1853.* Boston: Office, No. 19, Washington Street, 1853.

"Address by the President on 'The Census of Hallucinations.'" *Proceedings of the Society for Psychical Research,* July 8, 1889, 7–12.

"Admiral Fitzroy's Weather-Book." *London Review of Politics, Society, Literature, Art, and Science* 6, no. 147 (April 25, 1863): 443–45.

Alcott, William A. *The House I Live In, or, The Human Body, for the Use of Families and Schools.* 13th ed. Boston: Charles H. Peirce, 1847.

Alexander, Sarah C. *Victorian Literature and the Science of the Imponderable.* New York: Routledge, 2015.

Allen, C. H. "Weather Charts and Storm Warnings." *Leisure Hour,* no. 1394 (September 14, 1878): 587–91.

Allen, Grant. *Physiological Aesthetics.* London: Henry S. King, 1877.

Allen, Jasmine. "Stained Glass and the Culture of Spectacle, 1780–1862." *Visual Culture in Britain* 13, no. 1 (2012): 1–23.

Allen, Virginia M. "'One Strangling Golden Hair': Dante Gabriel Rossetti's *Lady Lilith.*" *Art Bulletin* 66, no. 2 (June 1984): 285–94.

Alpers, Svetlana. *The Art of Describing: Dutch Art in the Seventeenth Century.* Chicago: University of Chicago Press, 1984.

Ambler, Louise Todd, and Melissa Banta, eds. *The Invention of Photography and Its Impact on Learning.* Cambridge, MA: Harvard University Press, 1989.

Amtower, Laurel. "Authorizing the Reader in Chaucer's *House of Fame.*" *Philological Quarterly* 79, no. 3 (Summer 2000): 273–91.

Anderson, Anne. "'Fearful Consequences . . . of Living Up to One's Teapot': Men, Women, and 'Cultchah' in the English Aesthetic Movement, c. 1870–1900." *Victorian Literature and Culture* 37, no. 1 (2009): 219–54.

Anderson, Katharine. *Predicting the Weather: Victorians and the Science of Meteorology*. Chicago: University of Chicago Press, 2005.

Anderson, Katharine. "The Weather Prophets: Science and Reputation in Victorian Meteorology." *History of Science* 37, no. 2 (June 1999): 179–216.

Anderson, Nancy A., and Michael R. Dietrich, eds. *The Educated Eye: Visual Culture and Pedagogy in the Life Sciences*. Hanover, NH: Dartmouth College Press, 2012.

Anduaga, Aitor. *Politics, Statistics and Weather Forecasting, 1840–1910: Taming the Weather*. London: Routledge, 2020.

Angell, G. M. "Recollections of the First Operation Performed under Anesthesia." *International Journal of Surgery* 14 (August 1901): 249–50.

Archibald, Diana C. "Beauty, Unity, and the Ideal: Wholeness and Heterogeneity in the Kelmscott *Chaucer*." *Studies in Medievalism* 7 (1995): 169–80.

Arjona, Jamie M. "Sublime Perversions: Capturing the Uncanny Affects of Queer Temporalities in Mississippian Ruins." *Journal of Social Archaeology* 16, no. 2 (2016): 189–215.

Armstrong, Isobel. *Victorian Glassworlds: Glass Culture and the Imagination, 1830–1880*. Oxford: Oxford University Press, 2008.

Arney, W. De W. "Colour Blindness." *English Mechanic and World of Science* 55 (May 27, 1892): 304–6.

Arscott, Caroline. "Subject and Object in Whistler: The Context of Physiological Aesthetics." In *Palaces of Art: Whistler and the Art Worlds of Aestheticism*, edited by Lee Glazer and Linda Merrill, 55–65. Washington, DC: Smithsonian Institution Scholarly Press, 2013.

Arscott, Caroline. *William Morris and Edward Burne-Jones: Interlacings*. New Haven, CT: Yale University Press, 2008.

"Art Chronicle." *Portfolio* (1879): 105–12.

"The Art of the Year." *Month: A Catholic Magazine and Review*, June 1895, 153–64.

"Artificial Light in Studios." *Lightning* 1, no. 12 (January 7, 1892): 268.

"Artificial Light in Studios." *Lightning* 1, no. 13 (January 14, 1892): 288–89.

Ashhurst, John. *The Principles and Practice of Surgery*. Philadelphia: Henry C. Lea, 1871.

Asleson, Robyn. *Albert Moore*. London: Phaidon, 2000.

"At the Back of the North Wind." *Art Journal* 33 (March 1871): 96.

"At the Banquet, Saturday, May 4." *Punch*, May 11, 1895, 221.

"Atmospheric Waves." *Chambers's Edinburgh Journal*, no. 394 (July 19, 1851): 42–44.

Aubin, David, Charlotte Bigg, and Heinz Otto Sibum, eds. *The Heavens on Earth: Observatories and Astronomy in Nineteenth-Century Science and Culture*. Durham, NC: Duke University Press, 2010.

Auerbach, Nina. *Woman and the Demon: The Life of a Victorian Myth*. Cambridge, MA: Harvard University Press, 1982.

Bain, Alexander. *The Emotions and the Will.* London: John W. Parker and Son, 1859.

Bain, Alexander. *The Senses and the Intellect.* London: J. W. Parker, 1855.

Baker, Robert. *Before Bioethics: A History of American Medical Ethics from the Colonial Period to the Bioethics Revolution.* Oxford: Oxford University Press, 2013.

Ball, Philip. *Bright Earth: Art and the Invention of Color.* Chicago: University of Chicago Press, 2003.

Banvard, John. *Description of Banvard's Panorama of the Mississippi River, with the Adventures of the Artist.* Boston: John Putnam, 1847.

Banvard, John. *Description of Banvard's Geographical Painting of the Mississippi, with Historical Additions of the War.* New York: L. H. Biglow, 1862.

Barger, M. Susan, and William B. White. *The Daguerreotype: Nineteenth-Century Technology and Modern Science.* Baltimore, MD: Johns Hopkins University Press, 2000.

Barnhart, Terry A. *American Antiquities: Revisiting the Origins of American Archaeology.* Lincoln: University of Nebraska Press, 2015.

Bartram, William. *Travels through North and South Carolina, Georgia, East and West Florida, the Cherokee Country, the Extensive Territories of the Muscogulges or Creek Confederacy, and the Country of the Chactaws.* Dublin: J. Moore, W. Jones, R. McAllister, and J. Rice, 1793.

Barringer, Tim. *The Pre-Raphaelites: Reading the Image.* London: Calmann and King, 1998.

Batchelor, David. *Chromophobia.* London: Reaktion Books, 2000.

Bate, Percy. *The English Pre-Raphaelite Painters: Their Associates and Successors.* London: George Bell and Sons, 1899.

Baudelaire, Charles. "Salon de 1859." In *Critique d'art suivi de critique musicale,* 268–342. Paris: Gallimard, 1992.

Baxandall, Michael. *Painting and Experience in Fifteenth-Century Italy: A Primer in the Social History of Pictorial Style.* Oxford: Oxford University Press, 1972.

Baudelaire, Charles. *Critique d'art suivi de critique musicale.* Paris: Gallimard, 1992.

Bauman, Zygmunt. *Liquid Modernity.* Cambridge: Polity, 2000.

Bedell, Rebecca Bailey. *The Anatomy of Nature: Geology and American Landscape Painting, 1825–1875.* Princeton, NJ: Princeton University Press, 2001.

Beer, Gillian. *Darwin's Plots: Evolutionary Narrative in Darwin, George Eliot and Nineteenth-Century Fiction.* Cambridge: Cambridge University Press, 1983.

Beer, Gillian. *Open Fields: Science in Cultural Encounter.* Oxford: Clarendon, 1999.

Behlmer, George K. "Grave Doubts: Victorian Medicine, Moral Panic, and the Signs of Death." *Journal of British Studies* 42, no. 2 (2003): 206–35.

Belden-Adams, Kris. "Harvard's Composite 'Class' Pictures." *Photographies* 8, no. 2 (2015): 125–36.

Bell, Charles. *Engravings of the Arteries; Illustrating the Second Volume of the Anatomy of the Human Body, and Serving as an Introduction to the Surgery of the Arteries.* 3rd ed. London: Longman, 1811.

Bell, Charles. *Engravings of the Arteries; Illustrating the Second Volume of the Anatomy of the Human Body, and Serving as an Introduction to the Surgery of the Arteries.* 1st American ed., from the 3rd London ed. Philadelphia: Anthony Finley, 1812.

Bell, Charles. *Idea of a New Anatomy of the Brain—Submitted for the Observations of His Friends.* London: Strahan and Preston, 1811.

Bellion, Wendy. *Citizen Spectator: Art, Illusion, and Visual Perception in Early National America.* Chapel Hill: University of North Carolina Press, 2011.

Bendiner, Kenneth. *An Introduction to Victorian Painting.* New Haven, CT: Yale University Press, 1985.

Bendiner, Kenneth. "John Brett's 'The Glacier of Rosenlaui.'" *Art Journal* 44, no. 3. (Autumn 1984): 241–48.

Bending, Lucy. *The Representation of Bodily Pain in Late Nineteenth-Century English Culture.* Oxford: Clarendon, 2004.

Bennett, Tony. *The Birth of the Museum: History, Theory, Politics.* London: Routledge, 2013.

Benson, Donald R. "'Catching Light': Physics and Art in Walter Pater's Cultural Context." In *One Culture: Essays in Science and Literature,* edited by George Levine with Alan Rauch, 143–63. Madison: University of Wisconsin Press, 1987.

Berkhofer, Robert F. *The White Man's Indian: Images of the American Indian from Columbus to the Present.* New York: Knopf, 1978.

Berkowitz, Carin. *Charles Bell and the Anatomy of Reform.* Chicago: University of Chicago Press, 2015.

Berkowitz, Carin, and Bernard Lightman. *Science Museums in Transition: Cultures of Display in Nineteenth-Century Britain and America.* Pittsburgh: University of Pittsburgh Press, 2017.

Berlo, Janet Catherine, ed. *The Early Years of Native American Art History: The Politics of Scholarship and Collecting.* Seattle: University of Washington Press, 1992.

Berman, Marshall. *All That Is Solid Melts into Air: The Experience of Modernity.* New York: Simon and Schuster, 1982.

Bernstein, Susan David. "Designs after Nature: Evolutionary Fashions, Animals, and Gender." In *Victorian Animal Dreams: Representations of Animals in Victorian Literature and Culture,* edited by Deborah Denenholz Morse and Martin A. Danahay, 65–80. Burlington, VT: Ashgate, 2007.

Biegel, Hermann. *The Human Hair: Its Structure, Growth, Diseases, and Their Treatment.* London: Henry Renshaw, 1869.

Bigelow, Henry Jacob. "Insensibility during Surgical Operations Produced by Inhalation." *Boston Medical and Surgical Journal* 35, no. 16 (1846): http://doi.org/10.1056/NEJM184611180351601.

Bigelow, Henry Jacob. *Medical Education in America: Being the Annual Address Read before the Massachusetts Medical Society.* Cambridge, MA: Welch, Bigelow, 1871.

Bigelow, Henry Jacob. *Surgical Anesthesia: Addresses and Other Papers.* Boston: Little, Brown, 1900.

Bindman, David. *Ape to Apollo: Aesthetics and the Idea of Race in the Eighteenth Century.* London: Reaktion Books, 2002.

Birt, William Radcliffe. "A Contribution to our Knowledge of Atmospheric Waves." *London, Edinburgh, and Dublin Philosophical Magazine and Journal of Science* 44, series 4, no. 291 (August 1872): 125–38.

Birt, William Radcliffe. *Handbook of the Law of Storms: being a Digest of the Principal Facts of Revolving Storms for the Use of Commanders in Her Majesty's Navy and Mercantile Marine.* London: George Philip and Son, 1856.

Birt, William Radcliffe. "On Certain Atmospheric or Barometric Waves which Traversed Europe during November 1842." *London, Edinburgh, and Dublin Philosophical Magazine and Journal of Science* 30, series 3, no. 204 (1847): 482–502.

Bjelajac, David. *American Art: A Cultural History.* Upper Saddle River, NJ: Prentice Hall, 2005.

Bland, Bartholomew F., and Laura Vookles, eds. *Strut: The Peacock and Beauty in Art.* New York: Fordham University Press, 2014.

Bleichmar, Daniela. *Visible Empire: Botanical Expeditions and Visual Culture in the Hispanic Enlightenment.* Chicago: University of Chicago Press, 2012.

Blix, Göran. *From Paris to Pompeii: French Romanticism and the Cultural Politics of Archaeology.* Philadelphia: University of Pennsylvania Press, 2009.

Blum, Ann Shelby. *Picturing Nature: Nineteenth-Century Zoological Illustration.* Princeton, NJ: Princeton University Press, 1993.

Boe, Roy A. "The Panoramas of the Mississippi." *Mississippi Quarterly* 16, no. 4 (Fall 1963): 203–11.

Boehm, Katharina, ed. *Bodies and Things in Nineteenth-Century Literature and Culture.* New York: Palgrave Macmillan, 2012.

Boime, Albert. *The Magisterial Gaze: Manifest Destiny and American Landscape Painting, c. 1830–1865.* Washington, DC: Smithsonian Institution Press, 1991.

Boos, Florence S., ed. "William Morris's 'Commercial War': A Critical Edition." *Journal of Pre-Raphaelite Studies* 19 (Fall 2010): 45–65.

Bourgeois, Brigitte. "Secure for Eternity: Assembly Techniques for Large Statuary in the Sixteenth to the Nineteenth Century." In *History of Restoration of Ancient Stone Sculptures*, edited by Janet Burnett Grossman, Jerry Podany, and Marion True, 149–62. Los Angeles: J. Paul Getty Museum, 2003.

Bourke, Joanna. *The Story of Pain: From Prayer to Painkillers*. Oxford: Oxford University Press, 2014.

Bourrit, Marc-Théodore. *A Relation of a Journey to the Glaciers, in the Dutchy of Savoy*. Translated by C. and F. Davy. Norwich: Richard Beatniffe, 1775.

Bowditch, Nathaniel J. *The Ether Controversy: Vindication of the Hospital Report of 1848*. Boston: J. Wilson, 1848.

Bowditch, Vincent Y. *Life and Correspondence of Henry Ingersoll Bowditch*. Boston: Houghton, Mifflin, 1902.

Bown, Nicola. *Fairies in Nineteenth-Century Art and Literature*. New York: Cambridge University Press, 2001.

Bown, Nicola, Carolyn Burdett, and Pamela Thurschwell, eds. *The Victorian Supernatural*. Cambridge: Cambridge University Press, 2004.

Brace, C. Loring. *"Race" Is a Four-letter Word: The Genesis of the Concept*. New York: Oxford University Press, 2005.

Brain, Robert Michael. "*Protoplasmania*: Huxley, Haeckel, and the Vibratory Organism in Late Nineteenth-Century Science and Art." In *The Art of Evolution: Darwin, Darwinisms, and Visual Culture*, edited by Barbara Larson and Fae Brauer, 92–122. Hanover, NH: University Press of New England, 2009.

Brain, Robert Michael. *The Pulse of Modernism: Physiological Aesthetics in Fin-de-Siècle Europe*. Seattle: University of Washington Press, 2015.

Bramwell, Valerie, and Robert McCracken Peck. *All in the Bones: A Biography of Benjamin Waterhouse Hawkins*. Philadelphia: Academy of Natural Sciences of Philadelphia, 2008.

Brauer, Fae. "The Transparent Body: Biocultures of Evolution, Eugenics, and Scientific Racism." In *A History of Visual Culture: Western Civilization from the 18th to the 21st Century*, edited by Jane Kromm and Susan Benforado Bakewell, 89–103. Oxford: Berg, 2009.

Bredekamp, Horst. *Darwins Korallen: Die frühen Evolutionsmodelle und die Tradition der Naturgeschichte*. Berlin: Klaus Wagenbach, 2005.

Bresson, Julie, Ngai Liu, Marc Fischler, and Alain Bresson. "Anesthesia, Sleep and Death: From Mythology to the Operating Room." *Anesthesia & Analgesia* 117, no. 5 (November 2013): 1257–59.

Breyer, John, and William Butcher. "Nothing New under the Earth: The Geology of Jules Verne's *Journey to the Centre of the Earth*." *Earth Sciences History* 22, no. 1 (2003): 36–54.

Broadsides Collection. American Philosophical Society Library. Philadelphia, Pennsylvania.

Brontë, Charlotte. *Jane Eyre*. London: J. M. Dent, 1908.

Browman, David L., and Douglas R. Givens. "Stratigraphic Excavation: The First 'New Archaeology.'" *American Anthropologist*, n.s., 98, no. 1 (March 1996): 80–95.

Browner, Stephanie. "Ideologies of the Anesthetic: Professionalism, Egalitarianism, and the Ether Controversy." *American Quarterly* 51, no. 1 (March 1999): 108–43.

Bruhm, Steven. *Gothic Bodies: The Politics of Pain in Romantic Fiction*. Philadelphia: University of Pennsylvania Press, 1994.

Brunt, David. "A Hundred Years of Meteorology (1851–1951)." *Advancement of Science* 8, no. 30 (1951): 114–24.

Buchan, Alexander. *Handy Book of Meteorology*. 2nd ed. Edinburgh: William Blackwood and Sons, 1868.

Buchan, Alexander. *Introductory Text-Book of Meteorology*. Edinburgh: William Blackwood and Sons, 1871.

Buckley, Arabella. *The Fairy-land of Science*. London: Edward Stanford, 1879.

Budge, Gavin. "The Hallucination of the Real: Pre-Raphaelite Vision as a Crisis of Romantic Masculinity." In *Pre-Raphaelite Masculinities: Constructions of Masculinity in Art and Literature*, edited by Amelia Yeates and Serena Trowbridge, 55–79. Baltimore, MD: Johns Hopkins University Press, 2015.

Bull, Webster, and Martha Bull. *Something in the Ether: A Bicentennial History of Massachusetts General Hospital, 1811–2011*. Beverly, MA: Memoirs Unlimited, 2011.

Bulwer-Lytton, Edward. *The Coming Race*. Santa Barbara, CA: Woodbridge, 1989.

Burgess, Michael David. "The Imperial Federation Movement in Great Britain, 1869–1893." PhD diss., University of Leicester, 1976.

Burne-Jones, Georgiana. *Memorials of Edward Burne-Jones*. 2 vols. London: Macmillan, 1904.

Burns, Sarah. *Painting the Dark Side: Art and the Gothic Imagination in Nineteenth-Century America*. Berkeley: University of California Press, 2004.

Buzard, James, Joseph W. Childers, and Eileen Gillooly, eds. *Victorian Prism: Refractions of the Crystal Palace*. Charlottesville: University of Virginia Press, 2007.

Bynum, William F. *Science and the Practice of Medicine in the Nineteenth Century*. Cambridge: Cambridge University Press, 1994.

Byrd, Vance. *A Pedagogy of Observation: Nineteenth-Century Panoramas, German Literature, and Reading Culture*. Lewisburg, PA: Bucknell University Press, 2017.

Cahan, David, ed. *From Natural Philosophy to the Sciences: Writing the History of Nineteenth-Century Science*. Chicago: University of Chicago Press, 2003.

Cahan, David. "Helmholtz and the British Scientific Elite: From Force Conservation to Energy Conservation." *Notes and Records of the Royal Society of London* 66, no. 1 (2012): 55–68.

Cahan, David. *Hermann von Helmholtz and the Foundations of Nineteenth-Century Science.* Berkeley: University of California Press, 1993.

Calhoun, Robert Dabney. "A History of Concordia Parish, Louisiana: Seventh Installment." *Louisiana Historical Quarterly* 16, no. 3 (July 1933): 454–78.

Campbell, George, Duke of Argyll. *The Reign of Law.* London: Alexander Strahan, 1867.

Cantor, Geoffrey, Gowan Dawson, Graeme Gooday, Richard Noakes, Sally Shuttleworth, and Jonathan R. Topham. *Science in the Nineteenth-Century Periodical: Reading the Magazine of Nature.* Cambridge: Cambridge University Press, 2004.

Carpenter, William. "Mesmerism, Odylism, Table-Turning, and Spiritualism." *Fraser's Monthly Magazine,* March 1877, 382–405.

Carpenter, William. *Principles of Human Physiology.* 5th ed. London: John Churchill, 1855.

Carpenter, William. *Principles of Mental Physiology.* 2nd ed. London: H. S. King, 1874.

Case, Thomas. *Physical Realism: Being an Analytical Philosophy from the Physical Objects of Science to the Physical Data of Sense.* London: Longmans, Green, 1888.

Casid, Jill H. *Shadows of Projection: Recasting the Enlightenment Subject.* Minneapolis: University of Minnesota Press, 2015.

Casteras, Susan P. *Images of Victorian Womanhood in English Art.* London: Associated University Presses, 1987.

Casteras, Susan, and Colleen Denner. *The Grosvenor Gallery: A Palace of Art in Victorian England.* New Haven, CT: Yale University Press, 1996.

Catlin, George. *Letters and Notes on the Manners, Customs, and Conditions of the North American Indians.* Vol. 1. New York: Wiley & Putnam, 1841.

Cawsey, Kathy. "'Alum de glas' or 'Alymed glass'? Manuscript Reading in Book III of *The House of Fame.*" *University of Toronto Quarterly* 73, no. 4 (Fall 2004): 972–79.

Chapman, John. *Chloroform and Other Anaesthetics: Their History, and Use during Childbirth.* London: Williams and Norgate, 1859.

Chaucer, Geoffrey. *The Works of Geoffrey Chaucer.* Edited by F. S. Ellis. Hammersmith: Kelmscott, 1896.

Child, Francis James. *English and Scottish Ballads.* Boston: Little, Brown, 1860.

Child, Theodore. "A Pre-Raphaelite Mansion." *Harper's,* December 1890, 81–99.

Clark, Elizabeth C. "'The Sacred Rights of the Weak': Pain, Sympathy, and the Culture of Individual Rights in Antebellum America." *Journal of American History* 82, no. 2 (September 1995): 463–93.

Clark, Kenneth. *The Nude: A Study in Ideal Form*. Princeton, NJ: Princeton University Press, 1956.

Clarke, Bruce. *Energy Forms: Allegory and Science in the Era of Classical Thermodynamics*. Ann Arbor: University of Michigan Press, 2001.

Claudet, A. "Production of Natural Colours by Photography." *British Journal of Photography* 14, no. 351 (1867): 40–42.

Claudet, Antoine. "The Progress and Present State of the Daguerreotype Art." *Journal of the Franklin Institute* 40, no. 1 (1845): 45–57.

Cobbe, Frances Power. "Dreams, as Illustrations of Unconscious Cerebration." *Macmillan's Magazine* 23, no. 138 (April 1871): 512–23.

Cobbe, Frances Power. *Hours of Work and Play*. London: N. Trübner, 1867.

Cobbe, Frances Power. "Unconscious Cerebration: A Psychological Study." *Macmillan's Magazine* 23, no. 133 (November 1870): 24–37.

Codell, Julie. "The Dilemma of the Artist in Millais's *Lorenzo and Isabella*: Phrenology, the Gaze and the Social Discourse." *Art History* 14, no. 1 (March 1991): 51–66.

Codell, Julie. "Empiricism, Naturalism and Science in Millais's Paintings." In *John Everett Millais: Beyond the Pre-Raphaelite Brotherhood*, edited by Deborah Mancoff, 119–47. New Haven, CT: Yale University Press, 2001.

Coley, David K. "'Withyn a temple ymad of glas': Glazing, Glossing, and Patronage in Chaucer's *House of Fame*." *Chaucer Review* 45, no. 1 (2010): 59–84.

"Collated Opinion upon the Exhibition of the Royal Academy: Sir John E. Millais, R. A." Supplement to *Artist*, June 1895, n.p.

Colley, Ann C. *Victorians in the Mountains: Sinking the Sublime*. Farnham, Surrey, UK: Ashgate, 2010.

Collins, Alan. "England." In *The Oxford Handbook of the History of Psychology: Global Perspectives*, edited by David B. Baker, 182–210. Oxford: Oxford University Press, 2012.

Collins, Wilkie. *Woman in White*. New York: Penguin, 1985. Originally serialized 1859–1860.

Colvin, Sidney. "English Painters and Painting in 1867." *Fortnightly Review*, October 1867, 464.

Colvin, Sidney. *Memories and Notes of Persons and Places, 1852–1912*. New York: Charles Scribner's Sons, 1921.

Colwell-Chanthaphonh, Chip. "Reconciling American Archaeology and Native America." *Daedalus* 138, no. 2 (Spring 2009): 94–104.

Conn, Steven. *Museums and American Intellectual Life, 1876–1926*. Chicago: University of Chicago Press, 1998.

Connor, Steven. *The Matter of Air: Science and the Art of the Ethereal*. London: Reaktion Books, 2010.

Cook, E. T. Introduction to *The Works of John Ruskin*, edited by E. T. Cook and Alexander Wedderburn, vol. 26. London: George Allen, 1906.

Cooper, Suzanne Fagence. *Effie: The Passionate Lives of Effie Gray, John Ruskin and John Everett Millais.* New York: St. Martin's, 2010.

Corton, Christine L. *London Fog: The Biography.* Cambridge, MA: Belknap Press of Harvard University Press, 2015.

Coryell, John R. "The Seals' Crystal Palace." *St. Nicholas,* November 1891 to April 1892, 366–67.

Cosgrove, Denis E. *Geography and Vision: Seeing, Imagining, and Representing the World.* London: I. B. Tauris, 2008.

Cotter, John L. "Update on Natchez Man." *American Antiquity* 56, no. 1 (1991): 36–39.

Cottle, E. Wyndham. *The Hair in Health and Disease.* London: J & A Churchill, 1877.

Cowling, Mary. *The Artist as Anthropologist: The Representation of Type and Character in Victorian Art.* Cambridge: Cambridge University Press, 1989.

Crary, Jonathan. "Attention and Modernity in the Nineteenth Century." In *Picturing Science, Producing Art,* edited by Carol A. Jones and Peter Galison, 475–99. New York: Routledge, 1998.

Crary, Jonathan. "Géricault, the Panorama, and Sites of Reality in the Early Nineteenth Century." *Grey Room* 1, no. 9 (2002): 5–25.

Crary, Jonathan. *Suspensions of Perception: Attention, Spectacle, and Modern Culture.* Cambridge: MIT Press, 1999.

Crary, Jonathan. *Techniques of the Observer: On Vision and Modernity in the Nineteenth Century.* Cambridge: MIT Press, 1990.

Cryle, Peter, and Elizabeth Stephens. *Normality: A Critical Genealogy.* Chicago: University of Chicago Press, 2017.

Culin, Stewart. "The Dickeson Collection of American Antiquities." *Bulletin of the Free Museum of Science and Art of the University of Pennsylvania* 2, no. 3 (1900): 113–69.

Curry, David Park. *James McNeill Whistler at the Freer Gallery of Art.* New York: W. W. Norton, 1984.

Dahlberg, Laurie Virginia. "The Material Ethereal: Photography and the Alchemical Ancestor." In *Art & Alchemy,* edited by Jacob Wamberg, 83–100. Copenhagen: Museum Tusculanum Press, 2006.

"Daily Weather Maps." *London Review* 3, no. 62 (September 7, 1861): 302.

Dallas, Eneas Sweetland. *The Gay Science.* Vol. 1. London: Chapman and Hall, 1866.

Daly, Nicholas. *The Demographic Imagination and the Nineteenth-Century City: Paris, London, New York.* Cambridge: Cambridge University Press, 2015.

Dames, Nicholas. *The Physiology of the Novel: Reading, Neural Science, and the Form of Victorian Fiction.* Oxford: Oxford University Press, 2007.

Darwin, Charles. *The Descent of Man, and Selection in Relation to Sex.* 1st ed. 2 vols. London: John Murray, 1871.

Darwin, Charles. *The Descent of Man, and Selection in Relation to Sex.* 2nd ed. London: John Murray, 1874.

Darwin, Charles. *On the Origin of Species by Means of Natural Selection, or the Preservation of Favoured Races in the Struggle for Life.* 1st ed. London: John Murray, 1859. Reprint, New York: Avenel Books, 1979.

Darwin, Charles. *On the Origin of Species by Means of Natural Selection, or the Preservation of Favoured Races in the Struggle for Life.* 5th ed. London: John Murray, 1869.

Darwin, Charles. *On the Origin of Species*, 6th ed. London: John Murray, 1872.

Darwin, Charles, ed. *Voyage of H.M.S. Beagle, Part 4 (Fish).* London: Smith, Elder, 1840–1842.

Daston, Lorraine. "Fear and Loathing of the Imagination in Science." *Daedalus* 134, no. 4 (2005): 16–30.

Daston, Lorraine, and Peter Galison. "The Image of Objectivity." In "Seeing Science," special issue, *Representations* 40 (Fall 1992): 81–128.

Daston, Lorraine, and Peter Galison. *Objectivity.* New York: Zone Books, 2007.

Davis, Elliot Bostwick. "Training the Eye and the Hand: Drawing Books in Nineteenth-Century America." PhD diss., Columbia University, 1992.

Davis, Jennifer. "The London Garotting Panic of 1862: A Moral Panic and the Creation of a Criminal Class in Mid-Victorian England." In *Crime and the Law: A Social History of Crime in Western Europe since 1500*, edited by Vic A. C. Gatrell, Bruce Lenman, and Geoffrey Parker, 190–213. London: Europa, 1980.

Davy, Humphry. *Consolations in Travel, or, The Last Days of a Philosopher.* London: John Murray, 1830.

Dawson, Gowan. "Intrinsic Earthliness: Science, Materialism, and the Fleshly School of Poetry." *Victorian Poetry* 41, no. 1 (Spring 2003): 113–29.

Dawson, Gowan. *Show Me the Bone: Reconstruction Prehistoric Monsters in Nineteenth-Century Britain and America.* Chicago: University of Chicago Press, 2016.

Dawson, Gowan, and Bernard Lightman. Introduction to *Victorian Science and Literature.* Vol. 1. London: Pickering and Chatto, 2011.

Dawson, Gowan, and Bernard Lightman, eds. *Victorian Scientific Naturalism: Community, Identity, Continuity.* Chicago: University of Chicago Press, 2014.

Dawson, Gowan, and Sally Shuttleworth. "Introduction: Science and Victorian Poetry." *Victorian Poetry* 41, no. 1 (Spring 2003): 1–10.

Dearborn, Kerry. *Baptized Imagination: The Theology of George MacDonald.* Aldershot, UK: Ashgate, 2006.

Dearinger, David. *Paintings and Sculpture in the Collection of the National Academy of Design, 1826–1925.* New York: Hudson Hills, 2004.

Debus, Allen A. *Prehistoric Monsters: The Real and Imagined Creatures of the Past That We Love to Fear.* Jefferson, NC: McFarland, 2010.

Debus, Allen A. "Re-framing the Science in Jules Verne's *Journey to the Centre of the Earth.*" *Science Fiction Studies* 33, no. 3 (2006): 405–20.

DeLucia, Christine. "Antiquarian Collecting and the Transits of Indigenous Material Culture: Rethinking 'Indian Relics' and Tribal Histories." *Commonplace* 17, no. 2, accessed January 29, 2020. http://commonplace.online/article/antiquarian-collecting-and-the-transits-of-indigenous-material-culture/.

DeLucia, Christine. "Fugitive Collections in New England Indian Country: Indigenous Material Culture and Early American History Making at Ezra Stiles's Yale Museum." *William and Mary Quarterly* 75, no. 1 (2018), 109–50.

Denney, Colleen. *At the Temple of Art: The Grosvenor Gallery, 1877–1890.* Cranbury, NJ: Fairleigh Dickinson University Press, 2000.

Derbes, Anne, and Mark Sandona. "Barren Metal and the Fruitful Womb: The Program of Giotto's Arena Chapel in Padua." *Art Bulletin* 80, no. 2 (June 1998): 274–91.

Descartes, René. *The Philosophical Writings of Descartes.* Vols. 1–3. Translated by John Cottingham, Robert Stoothoff, and Dugald Murdoch. Cambridge: Cambridge University Press, 1988.

Desmond, Adrian. *Archetypes and Ancestors: Palaeontology in Victorian London, 1850–1875.* London: Blond and Briggs, 1982.

Desmond, Adrian. *The Politics of Evolution: Morphology, Medicine, and Reform in Radical London.* Chicago: University of Chicago Press, 1989.

Dettelbach, Michael. "The Face of Nature: Precise Measurement, Mapping, and Sensibility in the Work of Alexander von Humboldt." *Studies in the History and Philosophy of Biological and Biomedical Sciences* 30, no. 4 (1999): 473–504.

Deusner, Melody Barnett. "Whistler, Aestheticism, and the Networked World." In *Palaces of Art,* edited by Lee Glazer and Linda Merrill, 149–64. Washington, DC: Smithsonian Institution Scholarly Press, 2013.

DeYoung, Ursula. *A Vision of Modern Science: John Tyndall and the Role of the Scientist in Victorian Culture.* New York: Palgrave Macmillan, 2011.

Díaz-Andreu, Margarita. *A World History of Nineteenth-Century Archaeology: Nationalism, Colonialism, and the Past.* Oxford: Oxford University Press, 2007.

Dickeson, Montroville Wilson. "Aboriginal Structures, upon the Premises of T. Benard, Esq." *Lotus* 1, no. 13 (June 24, 1848): 196–98.

Dickeson, Montroville Wilson. *The American Numismatic Manual of the Currency or Money of the Aborigines, and Colonial, State, and United States Coin.* Philadelphia: J. B. Lippincott, 1859.

Dickeson, Montroville Wilson. "Ancient Fortification, and Burial Place of the Beluxies, Pascagoula, Mississippi." *Lotus* 1, no. 11 (June 10, 1848): 161–63.

Dickeson, Montroville Wilson. "Exploration of an Ancient Burial Ground, on the Estate of Major General Quitman." *Lotus* 1, no. 8 (May 20, 1848): 113–15.

Dickeson, Montroville Wilson. "Monumental Grandeur of the Mississippi River Valley! Now Exhibiting for a Short Time Only, With Scientific Lectures on American Aerchiology [*sic*]." Newark, NJ: Mercury Office, ca. 1851. printed broadside, 15×19 cm. Western Americana Collection, Rare Books and Special Collections, Firestone Library, Princeton University.

Dickeson, Montroville Wilson. "Opening of Mounds on the Plantation of Col. A. L. Bingaman." *Lotus* 1, no. 9 (May 27, 1848): 129–31.

Dickeson, Montroville Wilson. "Opening of Mounds upon the Plantation of William Feriday, Concordia Parish, Louisiana." *Lotus* 1, no. 10 (June 3, 1848): 145–47.

Dickeson, Montroville Wilson. "White Apple Village." *Lotus* 1, no. 24 (September 9, 1848): 371–73.

Dingley, Robert. "The Ruins of the Future: Macaulay's New Zealander and the Spirit of the Age." In *Histories of the Future: Studies in Fact, Fantasy and Science Fiction*, edited by Alan Sandison and Robert Dingley, 15–33. Basingstoke, UK: Palgrave, 2000.

"Discussion on Anaesthesia." *Boston Medical and Surgical Journal* 46, no. 11 (1852): 224–25.

Dolman, Frederick. "Painters and their Pictures: Mr. Frank Dicksee." *Strand Magazine* 23, no. 137 (June 1902): 483–49.

Donald, Diana. *Picturing Animals in Britain, 1750–1850*. New Haven, CT: Yale University Press, 2007.

Donald, Diana, and Jane Munro, eds. *Endless Forms: Charles Darwin, Natural Science and the Visual Arts*. Cambridge: Fitzwilliam Museum, 2009.

Dove, Heinrich. *The Law of Storms Considered in Connection with the Ordinary Movements of the Atmosphere*. 2nd ed. Translated by Robert H. Scott. London: Longman, Roberts, Green, 1862.

Dowling, Linda. *The Vulgarization of Art: The Victorians and Aesthetic Democracy*. Charlottesville: University Press of Virginia, 1996.

Dubin, Nina. *Futures and Ruins: Eighteenth-Century Paris and the Art of Hubert Robert*. Los Angeles: Getty Research Institute, 2010.

Duchenne, G.-B. *Mécanisme de la physionomie humaine ou analyse électro-physiologique de l'expression des passions*. Paris: Librairie J.-B. Bailliere et Fils, 1862.

Dugatkin, Lee Alan. *Mr. Jefferson and the Giant Moose: Natural History in Early America*. Chicago: University of Chicago Press, 2009.

Duncum, Barbara M. *The Development of Inhalation Anaesthesia*. London: Oxford University Press, 1947.

Dyer, Richard. *White: Essays on Race and Culture*. New York: Routledge, 1997.

Eastlake, Elizabeth. "The Crystal Palace." *Quarterly Review* 96, no. 192 (March 1855): 303–54.

Edgerton, Samuel, Jr. "Art and Science." *Art and Science I, Life Sciences* (Summer 1984): 107–8.

Edina. "Some Recent Materializations: No. 3, Séance with Mrs. Davidson." *Light: A Journal of Psychical, Occult, and Mystical Research* 15 (June 15, 1895): 281.

"The Electric Light and Eyesight." *Electrician*, October 1, 1886, 430–31.

"Electricity at the World's Fair." *Electrical World* 20, no. 9 (1892): 131–33.

Ellis, Steve. *Chaucer at Large: The Poet in the Modern Imagination.* Minneapolis: University of Minnesota Press, 2000.

Encyclopaedia Britannica. 9th ed. Vol. 16. Edinburgh: Adam and Charles Black, 1883.

Espy, James Pollard. *Report on Meteorology.* Washington, DC: Government Printing Office, 1850.

Everett, Edwin Mallard. *The Party of Humanity: The* Fortnightly Review *and Its Contributors, 1865–1874.* Chapel Hill: University of North Carolina Press, 1939.

F. G. "Recent Weather." *Reader* 2, no. 51 (December 19, 1863): 730.

F., F. G. "Science is Measurement." *Leisure Hour* 31 (1882): 663–65.

Fagan, Brian M., and Nadia Durrani. *A Brief History of Archaeology: Classical Times to the Twenty-First Century.* 2nd ed. New York: Routledge, 2016.

"Faraday Exhibiting Ghosts." *Nature* 22, no. 583 (October 21, 1880): 583.

Faraday, Michael. *A Course of Six Lectures on the Chemical History of a Candle.* London: Griffin, Bohn, 1861.

Fenster, Julie M. *Ether Day: The Strange Tale of America's Greatest Medical Discovery and the Haunted Men Who Made It.* New York: HarperCollins, 2001.

Ferguson, T. J. "Native Americans and the Practice of Archaeology." *Annual Review of Anthropology* 25 (1996): 63–79.

Filmer, Kath. *The Victorian Fantasists: Essays on Culture, Society, and Belief in the Mythopoeic Fiction of the Victorian Age.* Houndmills, Basingstoke, UK: Macmillan, 1991.

Fink, Lois Marie, and Joshua C. Taylor. *Academy: The Academic Tradition in American Art.* Washington, DC: Smithsonian Institution Press, 1975.

"First Impressions of the R. A." *Punch*, May 11, 1895, 220–21.

Fischer, Cynthia. "Thomas Jeckyll, James McNeill Whistler, and the Harmony in Blue and Gold: The Peacock Room: A Re-Examination." PhD diss., Virginia Commonwealth University, 2013.

Fisher, Jean. "In Search of the 'Inauthentic': Disturbing Signs in Contemporary Native American Art." *Art Journal* 51, no. 3 (Autumn 1992): 44–50.

Fisher, Philip. *Making and Effacing Art: Modern American Art in a Culture of Museums.* New York: Oxford University Press, 1991.

Fitzroy, Robert. *Barometer and Weather Guide.* 4th ed. London: George E. Eyre and William Spottiswoode, 1861.

Fitzroy, Robert. *The Weather Book: A Manual of Practical Meteorology.* London: Longman, Green, Longman, Roberts and Green, 1863.

[Fleeming Jenkin, Henry Charles]. "The Origin of Species." *North British Review* no. 92 (June 1867): 277–318.

Fleming, Gordon H. *John Everett Millais, A Biography.* London: Constable, 1998.

Fleming, James Rodger, Vladimir Jankovic, and Deborah R. Coen, eds. *Intimate Universality: Local and Global Themes in the History of Weather and Climate.* Sagamore Beach, MA: Science History, 2006.

Flint, Kate. "Arthur Hughes as Illustrator for Children." In *Children and Their Books: A Celebration of the Work of Iona and Peter Opie,* edited by Gillian Avery and Julia Briggs, 201–20. Oxford: Clarendon, 1989.

Flint, Kate. *Flash! Photography, Writing and Surprising Illumination.* Oxford: Oxford University Press, 2017.

Flint, Kate. *The Victorians and the Visual Imagination.* Cambridge: Cambridge University Press, 2000.

Flower, Margaret Cameron Coss. *Victorian Jewellery.* North Chelmsford, MA: Courier, 2002.

Forbes, James D. *Travels through the Alps of Savoy and Other Parts of the Pennine Chain with Observations on the Phenomena of Glaciers.* Edinburgh: Adam & Charles Black, 1843.

Forrest, Derek William. *Francis Galton: The Life and Work of a Victorian Genius.* London: Paul Elek, 1974.

Forwood, W. Stump. *An Historical and Descriptive Narrative of the Mammoth Cave of Kentucky.* Philadelphia: J. B. Lippincott, 1870.

Foster, Hal, ed., *Vision and Visuality: Discussions in Contemporary Culture.* New York: Dia Art Foundation/New Press, 1999.

Fox, Daniel M., and Christopher Lawrence. *Photographing Medicine: Images and Power in Britain and America since 1840.* New York: Greenwood, 1988.

Foucault, Michel. *The Birth of the Clinic: An Archeology of Medical Knowledge.* New York: Vintage Books, 1994.

Foucault, Michel. *Discipline and Punish: The Birth of the Prison.* New York: Vintage Books, 1995.

Foucault, Michel. "Of Other Spaces." *Diacritics* 16, no. 1 (Spring 1986): 22–27.

Foucault, Michel. *The Order of Things: An Archaeology of the Human Sciences.* London: Routledge, 1989.

Franklin, J. Jeffrey. *Spirit Matters: Occult Beliefs, Alternative Religions, and the Crisis of Faith in Victorian Britain.* Ithaca, NY: Cornell University Press, 2018.

Fredeman, William E., ed. *A Pre-Raphaelite Gazette: The Penkill Letters of Arthur Hughes to William Bell Scott and Alice Boyd, 1886–97.* Manchester: John Rylands Library, 1967.

Freud, Sigmund. *On Sexuality: Three Essays on the Theory of Sexuality and Other Works.* Edited by Angela Richards. Translated by James Strachey Harmondsworth: Penguin, 1977.

Frigg, Roman, and Matthew C. Hunter. *Beyond Mimesis and Convention: Representation in Art and Science.* Dordrecht: Springer, 2010.

Frith, Henry. *How to Read Character in Features, Forms, and Faces: A Guide to the General Outlines of Physiognomy.* London: Ward, Lock, Bowden, 1891.

Fryd, Vivien Green, "Rereading the Indian in Benjamin West's 'Death of General Wolfe.'" *American Art* 9, no. 1 (Spring 1995): 73–85.

Fuller, Margaret. *At Home and Abroad, or, Things and Thoughts in America and Europe.* Boston: Roberts Brothers, 1895.

Fülop-Miller, René. "The Day Man Conquered Pain." *Rotarian*, September 1946, 41.

Fyfe, Aileen, and Bernard Lightman, eds. *Science in the Marketplace: Nineteenth-Century Sites and Experiences.* Chicago: University of Chicago Press, 2007.

Gaarden, Bonnie. *The Christian Goddess: Archetype and Theology in the Fantasies of George MacDonald.* Madison, NJ: Fairleigh Dickinson University Press, 2011.

Gagnier, Regenia. "Production, Reproduction, and Pleasure in Victorian Aesthetics and Economics." In *Victorian Sexual Dissidence*, edited by Richard Dellamora, 127–46. Chicago: University of Chicago Press, 1999.

Galison, Peter. "Images Scatter into Data, Data Gather into Images." In *Iconoclash: Beyond the Image Wars in Science, Religion and Art*, edited by Bruno Latour and Peter Weibel, 300–23. Cambridge: MIT Press, 2002.

Galton, Francis. "A Development of the Theory of Cyclones." *Proceedings of the Royal Society of London* 12, no. 54 (January 8, 1863): 385–87.

Galton, Francis. *Meteorographica, or Methods of Mapping the Weather.* London: Macmillan, 1863.

Gamwell, Lynn. *Exploring the Invisible: Art, Science, and the Spiritual.* Princeton, NJ: Princeton University Press, 2002.

Garratt, Peter. "The Eye (In)Exact: G. H. Lewes and the Problems of Perception." In *Victorian Empiricism: Self, Knowledge, and Reality in Ruskin, Bain, Lewes, Spencer, and George Eliot*, 102–26. Madison, NJ: Fairleigh Dickinson University Press, 2010.

Gatty, Mrs. Alfred [Margaret]. *Parables from Nature.* London: Bell and Daldy, 1858.

Geller, Pamela L., and Christopher M. Stojanowski. "The Vanishing Black Indian: Revisiting Craniometry and Historic Collections." *American Journal of Physical Anthropology* 162, no. 2 (February 2017): 267–84.

Gentile, Phillip. "Viewing the Iconic Mississippi: Strategies of Reenactment in River Panoramas and Bill Morrison's *The Great Flood* (2013)." *Southern Quarterly* 52, no. 3 (Spring 2015): 121–36.

Gibbons, Carey. "The Limits of the Body in Victorian Illustration: Arthur Hughes and Frederick Sandys." PhD diss., Courtauld Institute of Art, University of London, 2016.

Giberne, Agnes. *The Ocean of Air: Meteorology for Beginners*. London: Seeley, 1890.

Gilbert, Pamela. *Cholera and Nation: Doctoring the Social Body in Victorian England*. Albany: State University of New York Press, 2008.

Gillespie, Sarah Kate. *The Early American Daguerreotype: Cross-Currents in Art and Technology*. Cambridge: MIT Press, 2016.

Gillooly, Eileen. "Rhetorical Remedies for Taxonomic Troubles." In *Victorian Prism: Refractions of the Crystal Palace*, edited by James Buzard, Joseph W. Childers, and Eileen Gillooly, 23–39. Charlottesville: University of Virginia Press, 2007.

Ginzburg, Carlo. "Morelli, Freud and Sherlock Holmes: Clues and Scientific Method." *History Workshop*, no. 9 (Spring 1980): 5–36.

Gitter, Elisabeth G. "The Power of Women's Hair in the Victorian Imagination." *PMLA* 99, no. 5 (October 1984): 936–54.

Glazer, Lee, and Linda Merrill, eds. *Palaces of Art: Whistler and the Art Worlds of Aestheticism*. Washington, DC: Smithsonian Institution Scholarly Press, 2013.

Gliddon, George R. *Hand-Book to the American Panorama of the Nile*. London: James Madden, 1849.

Good, Gregory A. "A Shift of View: Meteorology in John Herschel's Terrestrial Physics." In *Intimate Universality: Local and Global Themes in the History of Weather and Climate*, edited by James Rodger Fleming, Vladimir Jankovic, and Deborah R. Coen. Sagamore Beach, MA: Science History, 2006.

Good, John Mason. *A Physiological System of Nosology*. London: Printed by Bensley for E. Cox, 1817.

Goodrum, Matthew R., and Cora Olson. "The Quest for an Absolute Chronology in Human Prehistory: Anthropologists, Chemists, and the Fluorine Dating Method in Palaeoanthropology." *British Society for the History of Science* 42, no. 1 (March 2009): 95–114.

Godfrey, Helen. *Submarine Telegraphy and the Hunt for Gutta Percha: Challenge and Opportunity in a Global Trade*. Leiden: Brill, 2018.

Gold, Barri. *Thermopoetics: Energy in Victorian Literature and Science*. Cambridge: MIT Press, 2010.

Gooday, Graeme. "Electricity and the Sociable Circulation of Fear and Fearlessness." In *Geographies of Nineteenth-Century Science*, edited by David N. Livingstone and Charles W. J. Withers, 203–28. Chicago: University of Chicago, 2011.

Gould, Stephen Jay. *The Mismeasure of Man*. Rev. ed. New York: Norton, 1996.

Gould, Stephen Jay. *The Structure of Evolutionary Theory*. Cambridge, MA: Belknap Press of Harvard University Press, 2002.

Graves, Algernon. *The Royal Academy of Arts; a Complete Dictionary of Contributors and Their Work from Its Foundation in 1769 to 1904*. London: Henry Graves, 1905.

Gregory, Derek. *Geographical Imaginations*. Oxford: Blackwell, 2000.

Green, David. "Veins of Resemblance: Photography and Eugenics." *Oxford Art Journal* 7, no. 2 (1984): 3–16.

Green, Thomas. "On Death from Chloroform." *British Medical Journal* 2 (1872): 628–29.

Griffith, Frederick. "A Consideration of General Anesthesia with a Presentation of Instruments." *American Medicine* 6 (July–December 1903): 826.

Griffiths, Alison. "'Shivers Down Your Spine': Panoramas and the Origins of the Cinematic Reenactment." *Screen* 44, no. 1 (Spring 2003): 1–37.

Griffiths, Thomas. *Chemistry of the Four Ancient Elements, Fire, Air, Earth, and Water: An Essay Founded upon Lectures Delivered Before Her Most Gracious Majesty the Queen*. London: Samuel Highley, 1842.

Grigsby, Darcy Grimaldo. "Food Chains: French Abolitionism and Human Consumption (1787–1819)." In *An Economy of Colour: Visual Culture and the Atlantic World, 1600–1830*, edited by Geoff Quilley and Kay Dian Kriz, 153–75. Manchester: Manchester University Press, 2003.

Grilli, Stephanie. "Pre-Raphaelitism and Phrenology." In *Pre-Raphaelite Papers*, edited by Leslie Parris, 44–60. London: Tate, 1984.

"Grosvenor Gallery Winter Exhibition." *Academy* 454 (January 15, 1881): 50.

Groth, Helen, and Natalya Lusty. *Dreams and Modernity: A Cultural History*. Abingdon: Routledge, 2013.

Gura, Philip F. *The American Antiquarian Society, 1812–2012: A Bicentennial History*. Worcester, MA: American Antiquarian Society, 2012.

Gurney, George, and Therese Thau Heyman, eds. *George Catlin and His Indian Gallery*. Washington, DC: Smithsonian American Art Museum, 2002.

Gurney, Edmund. "Hallucinations." *Proceedings of the Society for Psychical Research* 3 (1885): 151–89.

Gurney, Edmund. *Phantasms of the Living*. 2 vols. London: Trübner, 1886.

Gurney, Peter. "'A Palace for the People'? The Crystal Palace and Consumer Culture in Victorian England." In *Victorian Prism: Refractions of the Crystal Palace*, edited by James Buzard, Joseph W. Childers, and Eileen Gillooly, 138–50. Charlottesville: University of Virginia Press, 2007.

Gutta Percha: Its Discovery, History, and Manifold Uses. London: Benjamin L. Green, 1851.

Gutta Percha Company. *Pattern Book of Ornaments for the Use of the Trade*. London: Gutta Percha Works, 1850.

Hain, Mark, Michael J. Lewis, and Stephen May. *Pennsylvania Academy of the Fine Arts, 1805–2005: 200 Years of Excellence*. Philadelphia: Pennsylvania Academy of the Fine Arts, 2005.

Hall, Thomas Wright. *Correlation Theory of Chemical Action and Affinity*. London: Remington, 1888.

Halpine, Susana Maria. "Real Scientists Do It with Models: The Art of Science Visualization." *Teaching Artist Journal* 6, no. 1 (2008): 5–19.

Hamerton, Philip Gilbert. *The Graphic Arts: A Treatise on Varieties of Drawing, Painting, and Engraving in Comparison with Each Other and with Nature*. London: Seeley, Jackson, and Halliday, 1882.

Hamerton, Philip Gilbert. "Notes on Aesthetics (1879–80)." *Portfolio Papers*. London: Seeley, 1889.

Hammerschlag, Keren Rosa. *Frederick Leighton: Death, Mortality, Resurrection*. Aldershot, UK: Ashgate, 2015.

Hammerschlag, Keren Rosa. "The Gentleman Artist-Surgeon in Late Victorian Group Portraiture." *Visual Culture in Britain* 14, no. 2 (2013): 154–78.

Hansen, Bert. "The Early History of Glacial Theory in British Geology." *Journal of Glaciology*, 9, no. 55 (1970): 135–41.

Hantman, Jeffrey L., and Gary Dunham. "The Enlightened Archaeologist." *Archaeology* 46, no. 3 (May/June 1993): 44–49.

Hares-Stryker, Carolyn. "'Slight Channels': Arthur Hughes and the Illustration of Children's Books." In *Haunted Texts: Studies in Pre-Raphaelitism in Honour of William E. Fredeman*, edited by David Latham, 91–118. Toronto: University of Toronto Press, 2003.

Hariot, Thomas. *A Briefe and True Report of the Newfoundland of Virginia*. Frankfurt-am-Main: Ioannis Wecheli and Theodori de Bry, 1590.

Harrington, Mark. "History of the Weather Map." *Report of the International Meteorological Congress, Chicago, Ill., August 21–24, 1893, U. S. Weather Bureau Bulletin* no. 11, pt. 2 (1895): 327–35.

Harrison, Frederic. *The Choice of Books, and Other Literary Pieces*. London: Macmillan, 1886.

Hart, Ernest Abraham. *Hypnotism, Mesmerism and the New Witchcraft*. New York: Appleton, 1893.

Hartman, Saidiya V. *Scenes of Subjection: Terror, Slavery, and Self-Making in Nineteenth-Century America*. New York; Oxford: Oxford University Press, 1997.

Harvey, David. *Justice, Nature and the Geography of Difference*. London: Blackwell, 1996.

Hatzenbuehler, Ronald. "Questioning Whether Thomas Jefferson Was the 'Father' of American Archaeology." *History and Anthropology* 22, no. 1 (2011): 121–29.

Hawkins, Benjamin Waterhouse. *Comparative Anatomy as Applied to the Purposes of the Artist*. London: Winsor & Newton, 1860.

Hawkins, Benjamin Waterhouse. *The Science of Drawing Simplified*. London: Smith, Elder, 1843.

Hawthorne, Nathaniel. *The House of the Seven Gables.* New York: Charles Merrill, 1907.

Heart, Jonathan. "A letter from Major Jonathan Heart, to Benjamin Smith Barton, M. D. . . . . containing observations on the Ancient Works of Art, the Native Inhabitants, &c. of the Western-Country, Fort Harmar, 5th January, 1791." In *Transactions of the American Philosophical Society*, vol. 3, 214–22. Philadelphia: American Philosophical Society, 1793.

Hecimovich, Gregg. *Puzzling the Reader: Riddles in Nineteenth-Century British Literature.* New York: Peter Lang, 2008.

Heilbron, Bertha L. "Making a Motion Picture in 1848: Henry Lewis on the Upper Mississippi," *Minnesota History* 17, no. 2 (June 1936): 131–58.

Heilbron, Bertha L. "A Mississippi Panorama." *Minnesota History* 23, no. 4 (December 1942): 349–54.

Hein, Rolland. *George MacDonald: Victorian Mythmaker.* Whitehorn, CA: Johannesen, 1993.

Hein, Rolland. *The Harmony Within: The Spiritual Vision of George MacDonald.* Eureka, CA: Sunrise Books, 1982.

Helmholtz, Hermann. *Popular Lectures on Scientific Subjects.* Translated by E. Atkinson; introduction by John Tyndall. London: Longmans, Green, 1873.

Helmholtz, Hermann von. *Science and Culture: Popular and Philosophical Essays.* Edited by David Cahan. Chicago: University of Chicago Press, 1995.

Helmreich, Anne. *Nature's Truth: Photography, Painting, and Science in Victorian Britain.* University Park: Pennsylvania State University Press, 2016.

Henderson, Andrea. "Magic Mirrors: Formalist Realism in Victorian Physics and Photography." *Representations* 117, no. 1 (Winter 2012): 120–50.

Hentschel, Klaus. *Mapping the Spectrum: Techniques of Visual Representation in Research and Teaching.* Oxford: Oxford University Press, 2009.

Hentschel, Klaus. *Visual Cultures in Science and Technology: A Comparative History.* Oxford: Oxford University Press, 2014.

Herschel, John F. W. *Meteorology from the Encyclopaedia Britannica.* Edinburgh: Adam and Charles Black, 1861.

Herschel, John F. W. "On Barometric Waves, February 5th, 1867." *Proceedings of the Literary and Philosophical Society of Manchester* 6 (1867): 91–93.

Herschel, John F. W. "The Weather and Weather Prophets." *Good Words* 5 (January 1864): 57–64.

Hevly, Bruce. "The Heroic Science of Glacier Motion." *Osiris* 11 (1996): 66–86.

History of Wages in the United States from Colonial Times to 1928. Washington: Government Printing Office, 1934.

Holmes, Abiel. "Meeting of October 24, 1814. Address." In *Proceedings of the American Antiquarian Society, 1812–1849*, 57–70. Worcester, MA: American Antiquarian Society, 1912.

Holmes, John. *The Pre-Raphaelites and Science.* New Haven, CT: Yale University Press, 2018.

Holmes, Oliver Wendell. *An Introductory Lecture, delivered at the Massachusetts Medical College, November 3, 1847.* Boston: Ticknor, 1847.

Honig, Edith Lazaros. *Breaking the Angelic Image: Woman Power in Victorian Children's Fantasy.* New York: Greenwood, 1988.

Hooper, Robert. *Lexicon Medicum, or, Medical Dictionary.* New York: J. & J. Harper, 1829.

Hooper-Greenhill, Eileen. *Museums and the Shaping of Knowledge.* London: Routledge, 1992.

Housman, Laurence. "The Illustrations of Arthur Hughes." *Bibliophile* 1, no. 5 (July 1908): 231–37.

Howe, Catherine. "Average Joes and Mean Girls: The Representation and Transformation of the Average American, 1890–1945." PhD diss., University of California, Santa Barbara, 2012.

Hudson, Nicholas. "The 'Hottentot Venus,' Sexuality, and the Changing Aesthetics of Race, 1650–1850." *Mosaic* 41, no. 1 (2008): 19–41.

Huhtamo, Erkki. *Illusions in Motion: Media Archaeology of the Moving Panorama and Related Spectacles.* Cambridge: MIT Press, 2013.

Humble, Nicola. "Domestic Arts." In *The Cambridge Companion to Victorian Culture,* edited by Francis O'Gorman, 219–35. Cambridge: Cambridge University Press, 2010.

Humboldt, Alexander von. "Account of the Great Cavern of the Guacharo." *Edinburgh Philosophical Journal* 3 (1820): 83–92.

Humboldt, Alexander von. *Cosmos: A Sketch of a Physical Description of the Universe.* Vol. 1. Translated by E. C. Otté. London: Henry G. Bohn, 1864.

Humboldt, Alexander von. *Florae Fribergensis specimen: plantas cryptogamicas praesertim subterraneas exhibens.* Berolini: H. A. Rottmann, 1793.

Humboldt, Alexander von. *Views of Nature.* Edited by Stephen T. Jackson and Laura Dassow Walls. Translated by Mark W. Person. Chicago: University of Chicago Press, 2014.

Hunter, Mary. "Intern, Orderly, Artist, Corpse: Emerging Masculinities in Henri Gervex's *Autopsy at Hôtel-Dieu.*" *Oxford Art Journal* 38, no. 3 (2015): 405–26.

Hunter, Matthew C. *Wicked Intelligence: Visual Art and the Science of Experiment in Restoration London.* Chicago: University of Chicago Press, 2013.

Hutchinson, Elizabeth. "From Pantheon to Indian Gallery: Art and Sovereignty on the Early Nineteenth-Century Cultural Frontier." *Journal of American Studies* 47, no. 2 (May 2013): 313–37.

Hutchinson, Elizabeth. *The Indian Craze: Primitivism, Modernism, and Transculturation in American Art, 1890–1915.* Durham, NC: Duke University Press, 2009.

Hutton, Muriel. "The George MacDonald Collection." *Yale University Library Gazette* 51, no. 2 (1976): 74–75.

Hutton, Ronald. *The Triumph of the Moon: A History of Modern Pagan Witchcraft.* Oxford: Oxford University Press, 1999.

Huxley, Thomas H. *Evidence as to Man's Place in Nature.* London: Williams and Norgate, 1863.

Huxley, Thomas H. "On the Physical Basis of Life." 1868. In *Select Works of Thomas H. Huxley,* 451–62. New York: J. B. Alden, 1886.

Huxtable, Sally-Anne. "Whistler's Peacock Room and the Artist as Magus." In *Palaces of Art: Whistler and the Art Worlds of Aestheticism,* edited by Lee Glazer and Linda Merrill, 67–81. Washington, DC: Smithsonian Institution Scholarly Press, 2013.

Hyman, Stanley Edgar. *The Tangled Bank: Darwin, Marx, Frazer and Freud as Imaginative Writers.* New York: Atheneum, 1962.

Impey, Oliver, and Arthur MacGregor, eds. *The Origins of Museums: The Cabinet of Curiosities in Sixteenth- and Seventeenth-Century Europe.* Oxford: Ashmolean Museum, 2017.

Ingold, Tim. *Lines: A Brief History.* London: Routledge, 2007.

Inman, Billie Andrew. "The Intellectual Context of Walter Pater's 'Conclusion.'" In *Walter Pater: An Imaginative Sense of Fact,* edited by Philip Dodd, 12–29. London: Frank Cass, 1981.

Jackson, Christine. *Peacock.* London: Reaktion Books, 2005.

Jackson, Joe. *A World on Fire: A Heretic, An Aristocrat, and the Race to Discover Oxygen.* New York: Viking, 2005.

Jackson, Roland. *The Ascent of John Tyndall: Victorian Scientist, Mountaineer, and Public Intellectual.* Oxford: Oxford University Press, 2018.

James, William. *Psychology.* New York: H. Holt, 1892.

James, William. "The Sentiment of Rationality." In *Essays in Popular Philosophy,* 63–110. London: Longmans, Green, 1896.

Jankovic, Vladimir. "Ideological Crests versus Empirical Troughs: John Herschel's and William Radcliffe Birt's Research on Atmospheric Waves, 1843–50." *British Journal for the History of Science* 31, no. 1 (March 1998): 21–40.

Jardine, Nicholas, and Isla Fay. *Observing the World through Images: Diagrams and Figures in the Early-Modern Arts and Sciences.* Boston: Brill, 2014.

Jefferies, Richard. *After London; or Wild England.* Edited by Mark Frost. Edinburgh: Edinburgh University Press, 2017.

Jefferson, Thomas. *Notes on the State of Virginia.* Richmond, VA: J. W. Randolph, 1853. Originally published 1787.

Jenkins, Alice. *Space and the 'March of Mind': Literature and the Physical Sciences in Britain, 1815–1850.* Oxford: Oxford University Press, 2007.

Jerrold, Blanchard, and Gustave Doré. *London: A Pilgrimage.* London: Grant, 1872.

Johnson, Joseph. *George MacDonald: A Biographical and Critical Appreciation.* London: Sir Isaac Pitman and Sons, 1906.

Johnson, Lucetta. "Beyond the Hair-Line: The Representation of Hair in the Work of Dante Gabriel Rossetti." PhD diss., Courtauld Institute of Art, University of London, 2010.

Johnston, Matthew N. *Narrating the Landscape: Print Culture and American Expansion in the Nineteenth Century.* Norman: University of Oklahoma Press, 2016.

Jones, Caroline A., and Peter Galison. *Picturing Science, Producing Art.* London: Routledge, 1998.

Jones, David J. *Gothic Machine: Textualities, Pre-cinematic Media and Film in Popular Visual Culture, 1670–1910.* Cardiff: University of Wales Press, 2011.

Jones, David. "To the Editor of the Royal American Magazine." *Royal American Magazine; or, The Universal Repository of Instruction and Amusement* 2, no. 1 (January 1775): 29–30.

Jordanova, Ludmilla. *Sexual Visions: Images of Gender in Science and Medicine between the Eighteenth and Twentieth Centuries.* Madison: University of Wisconsin Press, 1989.

Kaalund, Nanna Katrine Lüders. "A Frosty Disagreement: John Tyndall, James David Forbes, and the Early Formation of the X-Club." *Annals of Science* 74, no. 4 (2017): 282–98.

Karp, Ivan, and Stephen Lavine, eds. *Exhibiting Cultures: The Poetics and Politics of Museum Display.* Washington, DC: Smithsonian Institution Press, 1991.

Kastor, John A. *Mergers of Teaching Hospitals in Boston, New York, and Northern California.* Ann Arbor: University of Michigan Press, 2001.

Kaye, Richard A. *The Flirt's Tragedy: Desire without End in Victorian and Edwardian Fiction.* Charlottesville: University of Virginia Press, 2002.

Keene, Melanie. *Science in Wonderland: Scientific Fairy Tales of Victorian Britain.* Oxford: Oxford University Press, 2015.

Kehoe, Alice Beck. *The Land of Prehistory: A Critical History of American Archaeology.* New York: Routledge, 1998.

Keller, Corey. *Brought to Light: Photography and the Invisible, 1840–1900.* New Haven, CT: Yale University Press, 2008.

Kember, Joe, John Plunkett, and Jill Sullivan. *Popular Exhibitions, Science and Showmanship 1840–1910.* London: Pickering and Chatto, 2013.

Kemp, Martin. *The Human Animal in Western Art and Science.* Chicago: University of Chicago Press, 2007.

Kemp, Martin. *The Science of Art: Optical Themes in Western Art from Brunelleschi to Seurat.* New Haven, CT: Yale University Press, 1990.

Kemp, Martin. *Seen and Unseen: Art, Science, and Intuition from Leonardo to the Hubble Telescope.* New York: Oxford, 2006.

Kemp, Martin, and Marina Wallace. *Spectacular Bodies: The Art and Science of the Human Body from Leonardo to Now.* London: Hayward Gallery, 2000.

Kennedy, Roger. "Jefferson and the Indians." *Winterthur Portfolio* 27, nos. 2–3 (Summer–Autumn 1992): 105–12.

Kenny, Stephen C. "'I Can Do the Child No Good': Dr. Sims and the Enslaved Infants of Montgomery, Alabama." *Social History of Medicine* 20, no. 2 (2007): 223–41.

Kenny, Stephen C. "Power, Opportunism, Racism: Human Experiments under American Slavery." *Endeavour* 39, no. 1 (2015): 10–20.

Kleiner, Fred S. *Gardner's Art through the Ages.* Boston: Cengage Learning, 2013.

Kohl, Philip L., Irina Podgorny, and Stefanie Gänger, eds. *Nature and Antiquities: The Making of Archaeology in the Americas.* Tucson: University of Arizona Press, 2014.

Kohlstedt, Sally Gregory. "Nature by Design: Masculinity and Animal Display in Nineteenth-Century America." In *Figuring It Out: Science, Gender, and Visual Culture*, edited by Ann B. Shteir and Bernard V. Lightman, 110–39. Hanover, NH: University Press of New England, 2006.

Kohlstedt, Sally Gregory. "Place and Museum Space: The Smithsonian Institution, National Identity, and the American West, 1846–1896." In *Geographies of Nineteenth-Century Science*, edited by David N. Livingstone and Charles W. J. Withers, 399–437. Chicago: Chicago University Press, 2011.

Kolodny, Annette. "Fictions of American Prehistory: Indians, Archeology, and National Origin Myths." *American Literature* 75, no. 4 (2004): 693–721.

Kontou, Tatiana, ed. *The Ashgate Research Companion to Nineteenth-Century Spiritualism and the Occult.* London: Routledge, 2016.

Kooistra, Lorraine Janzen. *Christina Rossetti and Illustration: A Publishing History.* Athens: Ohio University Press, 2002.

Kooistra, Lorraine Janzen. *Poetry, Pictures, and Popular Publishing: The Illustrated Gift Book and Victorian Visual Culture, 1855–1875.* Athens: Ohio University Press, 2011.

Kort, Pamela, and Max Hollein, eds. *Darwin: Art and the Search for Origins.* Cologne: Wienand, 2009.

"La Fantasmagorie." *Magasin Pittoresque* (1849): 51–55.

[Lang, Andrew]. "In and Out of London: Mr. Whistler's Peacocks." *London*, February 17, 1877, 63.

Lang, Andrew. "Grant Allen." *Argosy* 71 (May–August 1900): 410–15.

Larson, Barbara. "Evolution and Victorian Art." In *Evolution and Victorian Culture*, edited by Bernard Lightman and Bennett Zon, 121–48. Cambridge: Cambridge University Press, 2014.

Larson, Barbara, and Fae Brauer, eds. *The Art of Evolution: Darwin, Darwinisms, and Visual Culture.* Hanover, NH: University Press of New England, 2009.

Larson, Barbara, and Sabine Flach, eds. *Darwin and Theories of Aesthetics and Cultural History*. Aldershot, UK: Ashgate, 2013.

Latour, Bruno. "What Is Iconoclash?" In *Iconoclash: Beyond the Image Wars in Science, Religion and Art*, edited by Bruno Latour and Peter Weibel, 14–37. Cambridge: MIT Press, 2002.

Laughton, J. K. "Weather Forecasting." *Fraser's Magazine*, August 1879, 242–54.

Lawrence, Christopher. "Medical Minds, Surgical Bodies: Corporeality and the Doctors." In *Science Incarnate: Historical Embodiments of Natural Knowledge*, edited by Christopher Lawrence and Steven Shapin, 156–201. Chicago: University of Chicago Press, 1998.

Laycock, Thomas. *Mind and Brain, or, the Correlations of Consciousness and Organisation*. Vol. 2. Edinburgh: Sutherland and Knox, 1860.

Le Brun, Charles. *Expressions des passions de l'âme*. Augsburg, 1732.

Lederer, Susan. *Subjected to Science: Human Experimentation in American before the Second World War*. Baltimore, MD: Johns Hopkins University Press, 1995.

Leighton, Frederic. *Addresses Delivered to the Students of the Royal Academy by the Late Lord Leighton*. London: Kegan, Paul, Trench and Tubner, 1896.

Lempfert, R. G. K. "British Weather Forecasts: Past and Present." *Quarterly Journal of the Royal Meteorological Society* 39, no. 167 (1913): 173–84.

Levere, Trevor H. *Chemists and Chemistry in Nature and Society, 1770–1878*. London: Routledge, 1994.

Levere, Trevor H. "The Rich Economy of Nature: Chemistry in the Nineteenth Century." In *Nature and the Victorian Imagination*, edited by U. C. Knoepflmacher and G. B. Tennyson, 189–200. Berkeley: University of California Press, 1977.

Levine, Caroline. *Form: Whole, Rhythm, Hierarchy, Network*. Princeton, NJ: Princeton University Press, 2017.

Levine, George. *Darwin and the Novelists: Patterns of Science in Victorian Fiction*. Cambridge, MA: Harvard University Press, 1988.

Levine, George. *Dying to Know: Scientific Epistomology and Narrative in Victorian England*. Chicago: University of Chicago Press, 2002.

Levine, George. *Realism, Ethics and Secularism: Essays on Victorian Literature and Science*. Cambridge: Cambridge University Press, 2008.

Levinson, Marjorie. "What Is New Formalism?" *PMLA* 122, no. 2 (2007): 558–69.

Lewes, George Henry. *The History of Philosophy from Thales to Comte*, 3rd ed. London: Longmans, Green, 1867.

Lewes, G. H. *Physiology of Common Life*. Vol. 2. Edinburgh: W. Blackwood, 1860.

Lewes, George Henry. *Problems of Life and Mind. Third Series. Problem the First, The Study of Psychology, Its Object, Scope, and Method*. London: Trübner, 1879.

Lewes, George. "Mr. Darwin's Hypotheses, Part I." *Fortnightly Review*, April 1, 1868, 353–73.

Lewes, George. "Mr. Darwin's Hypotheses, Part II." *Fortnightly Review*, June 1, 1868, 611–28.

Lewes, George. "Mr. Darwin's Hypotheses, Part III." *Fortnightly Review*, July 1, 1868, 61–80.

Lewes, George. "Mr. Darwin's Hypotheses, Conclusion." *Fortnightly Review*, November 1, 1868, 492–509.

Liebig, Justus. *Familiar Letters on Chemistry*. London: Walton and Maberly, 1859.

Lightman, Bernard. "Science and Culture." In *The Cambridge Companion to Victorian Culture*, edited by Francis O'Gorman, 12–42. Cambridge: Cambridge University Press, 2010.

Lightman, Bernard. "Depicting Nature, Defining Roles: The Gender Politics of Victorian Illustration." In *Figuring It Out: Science, Gender, and Visual Culture*, edited by Ann B. Shteir and Bernard V. Lightman, 214–39. Hanover, NH: University Press of New England, 2006.

Lightman, Bernard. "Refashioning the Spaces of London Science: Elite Epistemes in the Nineteenth Century." In *Geographies of Nineteenth-Century Science*, edited by David N. Livingstone and Charles W. J. Withers, 25–50. Chicago: University of Chicago Press, 2011.

Lightman, Bernard. *Victorian Popularizers of Science: Designing Nature for New Audiences*. Chicago: University of Chicago Press, 2009.

Lightman, Bernard. "Victorian Science and Popular Visual Culture." *Early Popular Visual Culture* 10, no. 1 (February 2012): 1–5.

Lightman, Bernard, ed. *Victorian Science in Context*. Chicago: University of Chicago Press, 2009.

Lightman, Bernard, and Bennett Zon, eds. *Evolution and Victorian Culture*. Cambridge: Cambridge University Press, 2014.

Lindberg, David. *Theories of Vision from al-Kindi to Kepler*. Chicago: University of Chicago Press, 1976.

"Literature." *Aberdeen Weekly Journal*, December 9, 1886, 2.

Litvack, Leon. "Continental Art and the 'Cockneyfied Corkonian': German and French Influences on Daniel Maclise." In *Daniel Maclise 1806–1870: Romancing the Past*, edited by Peter Murray, 198–214. Cork: Crawford Art Gallery and Gandon Editions, 2008.

Livingstone, David N., and Charles W. J. Withers, eds. *Geographies of Nineteenth-Century Science*. Chicago: University of Chicago Press, 2011.

Lochnan, Katharine A. *The Etchings of James McNeill Whistler*. New Haven, CT: Yale University Press, 1984.

Lochnan, Katharine A., ed. *Turner, Whistler, Monet*. London: Tate, 2004.

"London Letter." *Critic*, May 25, 1895, 387.

Looby, Christopher. "The Constitution of Nature: Taxonomy as Politics in Jefferson, Peale, and Bartram." *Early American Literature* 22, no. 3 (1987): 252–73.

"The Looker On." *Blackwood's Magazine*, 1895, 902–29.

Lowry, Bates, and Isabel Barrett Lowry. *Silver Canvas: Daguerreotype Masterpieces from the J. Paul Getty Museum*. Los Angeles: J. Paul Getty Museum, 1998.

Lowry, Bates, and Isabel Barrett Lowry. "Simultaneous Developments: Documentary Photography and Painless Surgery." In *Young America: The Daguerreotypes of Southworth & Hawes*, edited by Grant B. Romer and Brian Wallis, 75–88. New York: International Center of Photography, 2005.

Luarca-Shoaf, Nanette. "Currents of Time on the Lower Mississippi: M. W. Dickeson and *Monumental Grandeur of the Mississippi Valley*." In "The Mississippi River in Antebellum Visual Culture," 170–239. PhD diss., University of Delaware, 2012.

Luarca-Shoaf, Nanette. "Excavating a Nineteenth-Century Mass Medium." *American Art* 27, no. 2 (Summer 2013): 15–20.

Ludmerer, Kenneth M. *Learning to Heal: The Development of American Medical Education*. New York: Basic Books, 1985.

Lupacchini, Rosella, and Annarita Angelini, eds. *The Art of Science: From Perspective Drawing to Quantum Randomness*. Cham, Switzerland: Springer, 2014.

Macaulay, T. B. Review of *The Ecclesiastical and Political History of the Popes, during the Sixteenth and Seventeenth Centuries*, by Leopold van Ranke, translated by Sarah Austin. *Edinburgh Review* 72, no. 145 (October 1840): 227–58.

MacCarthy, Fiona. *The Last Pre-Raphaelite: Edward Burne-Jones and the Victorian Imagination*. London: Faber and Faber, 2011.

MacDonald, George. *At the Back of the North Wind*. London: Strahan, 1871.

MacDonald, George. Collection. Beinecke Rare Book and Manuscript Library, Yale University.

MacDonald, George. *A Dish of Orts: Chiefly Papers on the Imagination, and on Shakespeare*. London: Sampson, Low, Marston, 1893.

MacDonald, George. *Unspoken Sermons: Series I, II, and III*. Radford, VA: Wilder, 2008.

MacDonald, Greville. *George MacDonald and His Wife*. London: G. Allen & Unwin, 1924.

Mackenzie, Morell. *Diseases of the Pharynx, Larynx, and Trachea*. 2 vols. New York: William Wood, 1880.

MacLachlan, C. J. M., John Patrick Pazdziora, and Ginger Stelle. *Rethinking George MacDonald: Contexts and Contemporaries*. Glasgow: Scottish Literature International, 2013.

MacLeod, Alexander. *Scripture, Meteorology and Modern Science*. Glasgow: George Gallie, 1867.

Maclise, Joseph. *Comparative Osteology: Being Morphological Studies to Demonstrate the Archetype Skeleton of Vertebrated Animals*. London: Taylor and Walton, 1847.

Maclise, Joseph. *Surgical Anatomy*. Philadelphia: Blanchard and Lea, 1851.

Maclise, Joseph. *Surgical Anatomy*. London: Churchill, 1851.

Maclise, Joseph. *Surgical Anatomy*. 2nd ed. Philadelphia: Blanchard & Lea, 1859.

Maclise, Joseph. *Surgical Anatomy*. 2nd ed. London: Churchill, 1856.

Maclise, Joseph. *Surgical Anatomy*. Philadelphia: Henry C. Lea, 1870.

Madison, James. "A Letter on the Supposed Fortifications of the Western Country, from Bishop Madison of Virginia to Dr. Barton." *Transactions of the American Philosophical Society* 6, no. 1 (1804): 132–42.

Manlove, Colin. *The Impulse of Fantasy Literature*. London: Macmillan, 1983.

"Mapping the Weather." *London Review of Politics, Society, Literature, Art, and Science* 7, no. 174 (October 31, 1863): 469–70.

Markovits, Stefanie. "Form Things: Looking at Genre through Victorian Diamonds." *Victorian Studies* 52, no. 4 (Summer 2010): 591–619.

Marriott, William. "The Earliest Telegraphic Daily Meteorological Reports and Weather Maps." *Quarterly Journal of the Royal Meteorological Society* 29, no. 126 (1903): 123–31.

Marsh, Jan, ed. *Black Victorians: Black People in British Art, 1800–1900*. Aldershot, UK: Lund Humphries, 2006.

Marshall, Nancy Rose. *City of Gold and Mud: Painting Victorian London*. New Haven, CT: Yale University Press, 2012.

Marshall, Nancy Rose. "The Many Shades of Shakespeare: Representations of Othello and Desdmona in Victorian Visual Culture." In *Transculturation in British Art, 1770–1930*, edited by Julie F. Codell, 73–92. New York: Routledge, 2017.

Martin, Geoffrey J. *American Geography and Geographers: Toward Geographical Science*. Oxford: Oxford University Press, 2015.

Marx, Karl, and Frederick Engels. *Manifesto of the Communist Party*. Translated by Samuel Moore with Frederick Engels. London: William Reeves, 1888.

[Masson, David]. "The Academy II." *Spectator* 74 (May 18, 1895): 683–85.

Maudsley, Henry. *Body and Mind*. London: Macmillan, 1873.

Maudsley, Henry. *Body and Will*. London: Kegan, Paul, Trench, 1883.

Maudsley, Henry. *The Physiology and Pathology of Mind*. London: Macmillan, 1867.

Maxwell, Anne. *Picture Imperfect: Photography and Eugenics, 1870–1940*. Brighton: Sussex Academic, 2008.

May, Rebecca E. "'This Shattered Prison': Bodily Dissolution, *Wuthering Heights*, and Joseph Maclise's Dissection Manuals." *Nineteenth-Century Contexts* 33, no. 5 (2011): 415–36.

McClees, James E. *Elements of Photography: Being a Brief Detail of the Principles upon Which the Different Styles of Heliographic Pictures Are Produced*. Philadelphia: J. H. Jones, 1855.

McClintock, Anne. *Imperial Leather: Race, Gender, and Sexuality in the Colonial Contest*. New York: Routledge, 1995.

McCorristine, Shane. *Spectres of the Self: Thinking about Ghosts and Ghost-Seeing in England, 1750–1920.* Cambridge: Cambridge University Press, 2010.

McDermott, John Francis. *The Lost Panoramas of the Mississippi.* Chicago: University of Chicago Press, 1958.

McGann, Jerome J. "The Beauty of Medusa: A Study in Romantic Literary Iconology." *Studies in Romanticism* 11, no. 1 (Winter 1972): 3–25.

McGillis, Roderick. *For the Childlike: George MacDonald's Fantasies for Children.* West Lafayette, IN: Children's Literature Association, 1992.

McGrath, F. C. *The Sensible Spirit: Walter Pater and the Modernist Paradigm.* Tampa: University of South Florida Press, 1986.

"Meeting of June 1, 1814." In *Proceedings of the American Antiquarian Society, 1812–1849,* 43–44.

"Meeting of June 2, 1818." In *Proceedings of the American Antiquarian Society, 1812–1849,* 133.

"Meeting of October 23, 1816." In *Proceedings of the American Antiquarian Society, 1812–1849,* 99–100.

"Meeting of October 23, 1817." In *Proceedings of the American Antiquarian Society, 1812–1849,* 126–27.

Meigs, J. Aitken. *Catalogue of Human Crania, in the Collection of the Academy of Natural Sciences.* Philadelphia: J. B. Lippincott, 1857.

Meisel, Martin. *Chaos Imagined: Literature, Art, Science.* New York: Columbia University Press, 2016.

Meneley, Tobias. "Anthropocene Air." *Minnesota Review* 83 (2014): 93–101.

Menpes, Mortimer. *Whistler as I Knew Him.* London: Charles Black, 1904.

Merleau-Ponty, Maurice. *The Visible and the Invisible.* Edited by Claude Lefort. Translated by Alphonso Lingis. Evanston, IL: Northwestern University Press, 1968.

Merrill, Linda. *The Peacock Room: A Cultural Biography.* New Haven, CT: Yale University Press, 1998.

Merrill, Linda. *A Pot of Paint: Aesthetics on Trial in Whistler v. Ruskin.* Washington, DC: Smithsonian Institution Press, 1992.

Messer, Peter C. "Republican Animals: Politics, Science, and the Birth of Ecology." *Journal for Eighteenth-Century Studies* 33, no. 4 (December 2010): 599–613.

Milam, Erica. *Looking for a Few Good Males: Female Choice in Evolutionary Biology.* Baltimore, MD: Johns Hopkins University Press, 2010.

"The Millais and Rossetti Exhibitions." *Fortnightly Review,* February 1, 1898, 189–96.

Miller, Angela. "The Panorama, the Cinema, and the Emergence of the Spectacular." *Wide Angle* 18, no. 2 (April 1996): 34–69.

Miller, D. A. *The Novel and the Police.* Berkeley: University of California Press, 1989.

Miller, Elizabeth Carolyn. *Slow Print: Literary Radicalism and Late Victorian Print Culture.* Stanford, CA: Stanford University Press, 2013.

Miller, Jay. *Ancestral Mounds: Vitality and Volatility of Native America.* Lincoln: University of Nebraska Press, 2015.

Miller, Mimi. "Art and Society in Antebellum Natchez." In *Picturing Mississippi: Land of Plenty, Pain, and Promise, 1817–2017,* edited by Jochen Wierich, 20–31. Jackson: Mississippi Museum of Art, 2017.

Miller, Robert J. *Native America, Discovered and Conquered: Thomas Jefferson, Lewis and Clark, and Manifest Destiny.* Lincoln: University of Nebraska Press, 2008.

Millais, John Guille. *Life and Letters of John Everett Millais.* 2 vols. London: Methuen, 1899.

Mirzoeff, Nicholas. "Visualizing the Anthropocene." *Public Culture* 26, no. 2 (Spring 2014): 213–32.

Muir, Matthew Moncrieff Pattison. *Chemistry of Fire.* London: Methuen, 1893.

Monkhouse, Cosmo. "Wright of Derby." *Magazine of Art* 6 (1883): 383–89.

Monmonier, Mark. *Air Apparent: How Meteorologists Learned to Map, Predict, and Dramatize Weather.* Chicago: University of Chicago Press, 1999.

Monsman, Gerald. *Walter Pater.* Boston: G. K. Hill, 1977.

Morgan, Benjamin. *The Outward Mind: Materialist Aesthetics in Victorian Science and Literature.* Chicago: University of Chicago Press, 2017.

Morris, Christopher. "The Intellectual Lives of Natchez and Concord, and the Legacies of Slavery." *Southern Quarterly* 54, no. 2 (Winter 2017): 49–70.

Morris, William. *Stories in Prose, Stories in Verse, Shorter Poems, Lectures and Essays,* edited by G. D. H. Cole. London: Nonesuch, 1934.

Morris, William. *Icelandic Journals.* London: Mare's Nest, 1996.

Morris, William. *The Ideal Book: Essays and Lectures on the Arts of the Book.* Edited by William S. Peterson. Berkeley: University of California Press, 1982.

Morse, Deborah Denenholz, and Martin A. Danahay, eds. *Victorian Animal Dreams: Representations of Animals in Victorian Literature and Culture.* Aldershot, UK: Ashgate, 2007.

Morton, A. L. "Morris, Marx, and Engels." *Journal of William Morris Studies* 7, no. 1 (Autumn 1986): 45–54.

Morus, Iwan Rhys. *Frankenstein's Children: Electricity, Exhibition, and Experiment in Early-Nineteenth-Century London.* Princeton, NJ: Princeton University Press, 2014.

Morus, Iwan Rhys. "Illuminating Illusions, or, the Victorian Art of Seeing Things." *Early Popular Visual Culture* 10, no. 1 (2012): 37–50.

Morus, Iwan Rhys. "Seeing and Believing Science." *Isis* 97, no. 1 (March 2006): 101–10.

Moscoso, Javier. *Pain: A Cultural History.* New York: Palgrave Macmillan, 2012.

Müller, Johannes. *Elements of Physiology*. Translated by William Baly. London: Taylor and Walton, 1842.

Müller, Johannes. *Über die phantastischen Gesichtserscheinungen*. Coblenz: J. Hölscher, 1826.

Müller, Johannes. *Zür vergleichenden Physiologie des Gesichtsinnes des Menschen und Thiere*. Liepzig: C. Cnobloch, 1826.

Mulvey, Laura. "Visual Pleasure and Narrative Cinema." *Screen* 16, no. 3 (October 1975): 6–18.

Munro, Jane. "'More like a Work of Art than of Nature': Darwin, Beauty and Sexual Selection." In *Endless Forms*, edited by Diana Donald and Jane Munro, 253–91. New Haven: Yale University Press, in association with the Fitzwilliam Museum and the Yale Center for British Art, 2009.

Murray, Peter, ed. *Daniel Maclise, 1806–1870: Romancing the Past*. Cork: Crawford Art Gallery and Gandon Editions, 2008.

Myers, Frederic. "On Recognised Apparitions Occurring More than a Year after Death." *Proceedings of the Society for Psychical Research* 6 (1889): 13–65.

National Meteorological Library and Archive. Met Office. Exeter, UK.

Nead, Lynda. *Myths of Sexuality: Representations of Women in Victorian Britain*. Oxford: B. Blackwell, 1988.

Nead, Lynda. *Victorian Babylon: People, Streets, and Images in Nineteenth-Century London*. New Haven, CT: Yale University Press, 2000.

Neill, Anna. *Primitive Minds: Evolution and Spiritual Experience in the Victorian Novel*. Columbus: Ohio State University Press, 2013.

"New Books and New Editions." *Pall Mall Gazette*, November 28, 1868, 11.

Newall, Christopher. "Impression of the Effect." In *Pre-Raphaelite Vision: Truth to Nature*, edited by Allen Staley and Christopher Newall, 209–17. London: Tate, 2004.

Newall, Christopher. "'Understanding the Landscape." In *Pre-Raphaelite Vision: Truth to Nature*, edited by Allen Staley and Christopher Newall, 133–72. London: Tate, 2004.

Nickel, Douglas. "Talbot's Natural Magic." *History of Photography* 26, no. 2 (Summer 2002): 132–40.

"A Night Terror in Africa." *Temple Bar: A London Magazine for Town and Country Readers* 43 (December 1874): 68–78.

Noakes, Richard. "Spiritualism, Science, and the Supernatural in Mid-Victorian Britain." In *The Victorian Supernatural*, edited by Nicola Bown, Carolyn Burdett, and Pamela Thurschwell, 23–43. Cambridge: Cambridge University Press, 2004.

Nochlin, Linda. "The Darwin Effect." *Nineteenth-Century Art Worldwide* 2, no. 2 (Summer 2003). http://www.19thc-artworldwide.org/spring03/76-spring03/spring03article/218-the-darwin-effect-introduction.

"Notes." *Telegraphic Journal and Electrical Review* 12 (January 6, 1883):17.

"Notes on the Royal Academy Exhibition." *Bow Bells: A Magazine of General Literature and Art*, May 10, 1895, 466.

Nott, J. C., and Geo. R. Gliddon, *Types of Mankind; or, Ethnological Researches.* 6th ed. Philadelphia: Lippincott, Grambo, 1854.

Nova, Alessandro. *The Book of the Wind: The Representation of the Invisible.* Montreal: McGill-Queen's University Press, 2011.

Nussbaumer, Samuel U., and Heinz J. Zumbühl. "The Little Ice Age History of the Glacier des Bossons (Mont Blanc massif, France): A New High-Resolution Glacier Length Curve Based on Historical Documents." *Climatic Change* 111, no. 2 (2012): 301–34.

Oakley, Maroussia. "A Spark of Genius: An Investigation into the Publishing Context of the Book and Periodical Illustrations of Arthur Hughes (1832–1915)." PhD diss., University of London, 2012.

Oettermann, Stephan. *The Panorama: History of a Mass Medium.* Translated by Deborah Lucas Schneider. New York: Zone Books, 1997.

Ofek, Galia. *Representations of Hair in Victorian Literature and Culture.* Farnham, UK: Ashgate, 2009.

Ojanuga, Durrenda. "The Medical Ethics of the 'Father of Gynaecology,' Dr. J. Marion Sims." *Journal of Medical Ethics* 19, no. 1 (1993): 28–31.

Ono, Ayako. "Networks of Modernism: A New Look at Whistler and Japan." In *Palaces of Art: Whistler and the Art Worlds of Aestheticism*, edited by Lee Glazer and Linda Merrill, 165–72. Washington, DC: Smithsonian Institution Scholarly Press, 2013.

O'Sullivan, Patrick. "'Morris the Red, Morris the Green'—A Partial Review." *Journal of the William Morris Society* 19, no. 3 (Winter 2011): 22–38.

Otter, Chris. *The Victorian Eye: A Political History of Light and Vision in Britain, 1800–1910.* Chicago: University of Chicago Press, 2008.

"Owen and Maclise on the *Archetype Skeleton*." *British and Foreign Medico-chirurgical Review or Quarterly Journal of Practical Medicine and Surgery* 2 (July 1848): 107–21.

Pamboukian, Sylvia. "'Looking Radiant': Science, Photography and the X-Ray Craze of 1896." *Victorian Review* 27, no. 2 (2001): 56–74.

Pang, Alex Soojung-Kim. *Empire and the Sun: Victorian Solar Eclipse Expeditions.* Stanford, CA: Stanford University Press, 2002.

Pang, Alex Soojung-Kim. "Visual Representation and Post-constructivist History of Science." *Historical Studies in the Physical and Biological Sciences* 28, no. 1 (1997): 139–71.

Parish, Edmund. *Hallucinations and Illusions: A Study of the Fallacies of Perception.* London: W. Scott, 1897.

Parkinson, Gavin. *Surrealism, Art and Modern Science: Relativity, Quantum Mechanics, Epistomology.* New Haven, CT: Yale University Press, 2008.

Pater, Walter. "Dante Gabriel Rossetti." In *The English Poets: Selections with Critical Introductions by Various Writers*, 2nd ed., ed. Thomas Ward, vol. 4, *Wordsworth to Rossetti*, 631–41. London: Macmillan, 1883.

Pater, Walter. *Plato and Platonism: A Series of Lectures*. London: Macmillan, 1907.

Pater, Walter. "Poems by William Morris." *Westminster Review*, 1868, 300–312.

Pater, Walter. *The Renaissance: Studies in Art and Literature*. London: MacMillan, 1873.

Patterson, Thomas C. *Toward the Social History of Archaeology in the United States*. Belmont, CA: Wadsworth/Thomson Learning, 2002.

Patterson, Thomas C. "Who Did Archaeology in the United States before There Were Archaeologists and Why? Preprofessional Archaeologies in the Nineteenth Century." In *Processual and Postprocessual Archaeologies: Multiple Ways of Knowing the Past*, edited by Robert W. Preucel, 242–50. Carbondale: Center for Archaeological Investigations, Southern Illinois University at Carbondale, 1991.

Paul, Carole. *The First Modern Museums of Art: The Birth of an Institution in Eighteenth- and Early Nineteenth-Century Europe*. Los Angeles: J. Paul Getty Museum, 2012.

Pauly, Daniel. *Darwin's Fishes: An Encyclopedia of Ichthyology, Ecology, and Evolution*. Cambridge: Cambridge University Press, 2007, 107.

Pauwels, Luc, ed. *Visual Cultures of Science: Rethinking Representational Practices in the Production of a Scientific Reality*. Lebanon, NH: University Press of New England, 2006.

Payne, Christiana. *John Brett: Pre-Raphaelite Landscape Painter*. New Haven, CT: Yale University Press, 2010.

Pennington, John, and Roderick McGillis, eds. *Behind the Back of the North Wind: Critical Essays on George Macdonald's Classic Book*. Hamden, CT: Winged Lion, 2011.

Pepper, John Henry. *True History of a Ghost and All about Metempsychosis*. London: Cassell, 1890.

Pernick, Martin S. *A Calculus of Suffering: Pain, Professionalism, and Anesthesia in Nineteenth-Century America*. New York: Columbia University Press, 1985.

Peterson, William S. *The Kelmscott Press: A History of William Morris's Typographical Adventure*. Berkeley: University of California Press, 1991.

Petherbridge, Deanna, and Ludmilla Jordanova. *The Quick and the Dead: Artists and Anatomy*. Berkeley: University of California Press, 1997.

Pettitt, Claire. *Patent Inventions: Intellectual Property and the Victorian Novel*. Oxford: Oxford University Press, 2004.

Phillips, Claude. "Fine Art: The Royal Academy I." *Academy*, May 11, 1895, 407–8.

Phillips, Claude. "Pictures of the Year." *Fortnightly Review*, n.s., 57, no. 342 June 1, 1895, 925–39.

Phillips, Helen. "Chaucer and the Nineteenth-Century City." *Chaucer and the City*, edited by Ardis Butterfield, 193–210. Cambridge: D. S. Brewer, 2006.

Phillips, Ruth B. *Trading Identities: The Souvenir in Native North American Art from the Northeast, 1700–1900*. Seattle: University of Washington Press, 1998.

Pia Di Bella, Maria, and James Elkins, eds. *Representations of Pain in Art and Visual Culture*. New York: Routledge, 2013.

Pickstone, John V. *Ways of Knowing: A New History of Science, Technology and Medicine*. Chicago: University of Chicago Press, 2001.

Pomian, Krzysztof. *Collectors and Curiosities: Paris and Venice, 1500–1800*. Oxford: Polity, 1990.

Postle, Martin. "Leighton's Lost Model: The Rediscovery of Mary Lloyd." *Apollo*, February 1996, 27–29.

Pratt-Smith, Stella. "The Poetic Science of Nineteenth-Century Electricity." *Nineteenth-Century Contexts* 34, no. 1 (2012): 51–62.

Prettejohn, Elizabeth. *After the Pre-Raphaelites: Art and Aestheticism in Victorian England*. New Brunswick, NJ: Rutgers University Press, 1999.

Prettejohn, Elizabeth. *Art for Art's Sake: Aestheticism in Victorian Painting*. New Haven, CT: Yale University Press, 2007.

Proceedings of the American Antiquarian Society, 1812–1849. Worcester, MA: American Antiquarian Society, 1912.

Prodger, Phillip. *Darwin's Camera: Art and Photography in the Theory of Evolution*. Oxford: Oxford University Press, 2009.

Prodger, Phillip. "Ugly Disagreements: Darwin and Ruskin Debate Sex and Beauty." In *The Art of Evolution: Darwin, Darwinisms, and Visual Culture*, edited by Barbara Larson and Fae Brauer, 40–58. Hanover, NH: Dartmouth College Press, 2009.

Prum, Richard O. *The Evolution of Beauty: How Darwin's Forgotten Theory of Mate Choice Shapes the Animal World—and Us*. New York: Doubleday, 2017.

Pullman, Bernard. *The Atom in the History of Human Thought*. Oxford: Oxford University Press, 2001.

Pyne, Kathleen. *Art and the Higher Life: Painting and Evolutionary Thought in Late Nineteenth-Century America*. Austin: University of Texas Press, 1996.

Quain, Richard. *The Anatomy of the Arteries of the Human Body with Its Applications to Pathology and Operative Surgery in Lithographic Drawings with Practical Commentaries. The Drawings from Nature and on Stone by Joseph Maclise*. London: Taylor and Walton, 1844.

Quinn, William A. "Chaucer's Recital of Presence in the *House of Fame* and the Embodiment of Authority." *Chaucer Review* 43, no. 2 (2008): 171–96.

Raeper, William. *George MacDonald*. Tring, UK: Lion, 1987.

Rampley, Matthew. *The Seductions of Darwin: Art, Evolution, Neuroscience*. University Park: Pennsylvania State University Press, 2017.

Raue, Charles Godlove. *Psychology as a Natural Science Applied to the Solution of Occult Psychic Phenomena.* Philadelphia: Porter and Coates, 1889.

Redman, Samuel J. *Bone Rooms: From Scientific Racism to Human Prehistory in Museums.* Cambridge, MA: Harvard University Press, 2016.

Regis, Pamela. *Describing Early America: Bartram, Jefferson, Crèvecoeur, and the Influence of Natural History.* Philadelphia: University of Pennsylvania Press, 1999.

Reid, J. Eadie. *Sir J. E. Millais, P.R.A.* London: Walter Scott, 1909.

Reilly, Kara. "Two Venuses: Historicizing the Anatomical Female Body." *Performance Research* 19, no. 4 (2014): 111–21.

"Report on the Census of Hallucinations." *Proceedings of the Society for Psychical Research.* London: Kegan, Paul, Trench, Trübner, 1894.

Review of *The Life and Works of Joseph Wright, A.R.A., commonly called "Wright of Derby"*, by William Bemrose. *Art Journal* 48 (1886): 95–96.

Review of *Rudimentary Electricity*, by W. Snow Harris, *Regulations of the Electric Telegraph Company*, and *Traité de télégraphie electrique*, by François Moigno, *Edinburgh Review*, American ed., 90 (October 1849): 227–40.

Reynolds, Joshua. *The Discourses of Joshua Reynolds.* Edited by John Burnet. London: James Carpenter, 1842.

Richmond, Velma Bourgeois. "Edward Burne-Jones's Chaucer Portraits in the Kelmscott 'Chaucer.'" *Chaucer Review* 40, no. 1 (2005): 1–38.

Risse, Guenter B. *Mending Bodies, Saving Souls: A History of Hospitals.* New York: Oxford University Press, 1999.

Rivett, Sarah. *Unscripted America: Indigenous Languages and the Origins of a Literary Nation.* New York: Oxford University Press, 2017.

Robb, David S. *God's Fiction: Symbolism and Allegory in the Works of George MacDonald.* Eureka, CA: Sunrise Books, 1987.

Roberts, Leonard. *Arthur Hughes: His Life and Works.* Woodbridge, UK: Antique Collectors' Club, 1997.

Robins, Anna. *A Fragile Modernism: Whistler and His Impressionist Followers.* New Haven, CT: Yale University Press, 2007.

Robinson, Duncan. *William Morris, Edward Burne-Jones and the Kelmscott Chaucer.* Mount Kisco, NY: Moyer Bell, 1986.

Romer, Grant B. "'A High Reputation with All True Artists and Connoisseurs': The Daguerrian Careers of A.S. Southworth and J.J. Hawes." In *Young America: The Daguerreotypes of Southworth & Hawes*, edited by Grant B. Romer and Brian Wallis, 20–55. New York: International Center of Photography, 2005.

Root-Bernstein, Robert Scott. "On Paradigms and Revolutions in Science and Art: The Challenge of Interpretation." *Art Journal* 44, no. 2(Summer 1984): 109–19.

Rosaldo, Renato. "Imperialist Nostalgia." *Representations* 26 (Spring 1989): 107–22.

Rose, Gillian. "Making Space for the Female Subject of Feminism: The Spatial Subversions of Holzer, Kruger and Sherman." In *Mapping the Subject: Geographies of Cultural Transformation*, edited by Steve Pile and Nigel Thrift, 332–54. London: Routledge, 1995.

Rosenberg, Charles E. *The Care of Strangers: The Rise of America's Hospital System*. New York: Basic Books, 1987.

Rosenberg, Charles E. *No Other Gods: On Science and American Social Thought*. Baltimore, MD: Johns Hopkins University Press, 1976.

Rosenberg, Gary D., and Renee M. Clary, eds. *Museums at the Forefront of the History and Philosophy of Geology: History Made, History in the Making*. Special Paper 535. Boulder: Geological Society of America, 2018.

Rosenfeld, Jason. *John Everett Millais*. London: Phaidon, 2012.

Ross, William Alexander. *Pyrology, or, Fire Chemistry*. London: E. & F. N. Spon, 1875.

Rossetti, William Michael. *Dante Gabriel Rossetti: His Family Letters*. Vol. 1. London: Ellis and Elvey, 1895.

Rossetti, W. M. "The Royal Academy Exhibition." *Fraser's Magazine*, June 1863, 783–95.

Routhier, Jessica Skwire, Kevin J. Avery, and Thomas Hardiman, Jr. *The Painters' Panorama: Narrative, Art, and Faith in the Moving Panorama of Pilgrim's Progress*. Hanover, NH: University Press of New England, 2015.

"The Royal Academy II." *Academy* 47 (May 25, 1895): 449–50.

"The Royal Academy of Arts 1895." *Art Journal* 57 (June 1895): 161–81.

Royal Academy Illustrated. London: Cassell, 1895.

Rudwick, Martin J. S. *Earth's Deep History: How It Was Discovered and Why It Matters*. Chicago: University of Chicago Press, 2014.

Rudwick, Martin J. S. "The Emergence of a Visual Language for Geological Science, 1760–1830." *History of Science* 14 (1976): 149–95.

Rudwick, Martin J. S. *Scenes from Deep Time: Early Pictorial Representations of the Pre-historic World*. Chicago: University of Chicago Press, 1992.

Rue, Warren de la. "Eclipse of the Sun, July 1860, Rivabellosa, Observed at Near Miranda de Ebro, Spain." *Philosophical Transactions of the Royal Society of London* 152 (1862): 333–416.

Ruiz-Gómez, Natasha. "Shaking the Tyranny of the Cadaver: Doctor Paul Richer and the 'Living Écorché.'" In *Bodies beyond Borders. Moving Anatomies, 1750–1950*, edited by Kaat Wils, Raf De Bont, and Sokhieng Au, 231–57. Leuven, Belgium: University of Leuven Press, 2017.

Rupke, Nicolaas A. *Richard Owen: Biology without Darwin*. Chicago: University of Chicago Press, 2009.

Rupke, Nicolaas A. "Richard Owen's Vertebrate Archetype." *Isis* 84 (1993): 231–51.

Ruskin, John. *Deucalion: Collected Studies of the Lapse of Waves and Life of Stones*. In *The Works of John Ruskin*, edited by E. T. Cook and Alexander Wedderburn, vol. 26. London: George Allen, 1906.

Ruskin, John. *Fors Clavigera: Letters to the Workmen and Labourers of Great Britain*. Vol. 3. In *The Works of John Ruskin*, edited by E. T. Cook and Alexander Wedderburn, vol. 27. London: George Allen, 1907.

Ruskin, John. *Fors Clavigera: Letters to the Workmen and Labourers of Great Britain*. Vol. 7. In *The Works of John Ruskin*, edited by E. T. Cook and Alexander Wedderburn, vol. 29. London: George Allen, 1907.

Ruskin, John. "The Laws of Fésole." In *The Works of John Ruskin*, edited by E. T. Cook and Alexander Wedderburn, vol. 15. London: George Allen, 1904.

Ruskin, John. *Letters of John Ruskin*. In *The Complete Works of John Ruskin*, edited by E. T. Cook and A. Wedderburn, vol. 36. London: George Allen, 1909.

Ruskin, John. *Love's Meinie: Three Lectures on Greek and English Birds*. In *The Works of John Ruskin*, edited by E. T. Cook and Alexander Wedderburn, vol. 25. London: George Allen, 1907.

Ruskin, John. *Proserpina: Studies of Wayside Flowers*. Vol. 1. In *The Works of John Ruskin*, edited by E. T. Cook and A. Wedderburn, vol. 25. London: George Allen, 1875.

Ruskin, John. *Of Leaf Beauty*. Vol. 5 of *Modern Painters*. In *The Works of John Ruskin*, edited by E. T. Cook and A. Wedderburn, vol. 7. London: George Allen, 1906.

Ruskin, John. *Of Mountain Beauty*. Vol. 4 of *Modern Painters*. 1856. New York: John Wiley & Sons, 1880.

Ruskin, John. *Of the Truth of Vegetation*. Vol. 1 of *Modern Painters*. In *The Works of John Ruskin*, edited by E. T. Cook and A. Wedderburn, vol. 3. London: George Allen, 1906.

Ruskin, John. *Queen of the Air, Being a Study of the Greek Myths of Storm and Cloud*. 1869. London: George Allen, 1906.

Ruskin, John. *The Stones of Venice*. Vol. 1. In *The Works of John Ruskin*, edited by E. T. Cook and Alexander Wedderburn, vol. 9. London: George Allen, 1903.

Rylance, Rick. *Victorian Psychology and British Culture, 1850–1880*. Oxford: Oxford University Press, 2000.

Ryves, G. T. "The Times Daily Weather Chart." *Time: A Monthly Miscellany of Interesting and Amusing Literature* 9, no. 54 (September 1883): 345–50.

Sadler, Glenn Edward, ed. *An Expression of Character: The Letters of George MacDonald*. Grand Rapids, MI: W. B. Eerdmans, 1994.

Sappol, Michael. *Dream Anatomy*. Bethesda, MD: National Library of Medicine, 2002.

Sappol, Michael., ed. *Hidden Treasure: The National Library of Medicine*. New York: Blast Books, 2012.

Sappol, Michael. *A Traffic of Dead Bodies: Anatomy and Embodied Social Identity in Nineteenth-Century America.* Princeton, NJ: Princeton University Press, 2002.

Saunders, Gill. *Picturing Plants: An Analytical History of Botanical Illustration.* Berkeley: University of California Press, 1995.

Savage, Kirk. *Standing Soldiers, Kneeling Slaves: Race, War, and Monument in Nineteenth-Century America.* Princeton, NJ: Princeton University Press, 2018.

Sawyer, Paul L. "Ruskin and Tyndall: The Poetry of Matter and the Poetry of Spirit." *Annals of the New York Academy of Sciences* 360, no. 1 (April 1981): 217–46.

Saxe, Burton. *The Cave Secret, or, The Mystery of Night Island.* New York: American News, 1865.

Sayre, Gordon M. *Les Sauvages Américains: Representations of Native Americans in French and English Colonial Literature.* Chapel Hill: University of North Carolina Press, 1997.

Sayre, Gordon M. "The Mound Builders and the Imagination of American Antiquity in Jefferson, Bartram, and Chateaubriand." *Early American Literature* 33, no. 3 (1998): 225–49.

Schaaf, Larry J. *Out of the Shadows: Herschel, Talbot, and the Invention of Photography.* New Haven, CT: Yale University Press, 1992.

Schatz, Stephanie L. "Lewis Carroll's Dream-Child and Victorian Child Psychopathology." *Journal of the History of Ideas* 76, no. 1 (January 2015): 93–114.

Scheckel, Susan. *The Insistence of the Indian: Race and Nationalism in Nineteenth-Century American Culture.* Princeton, NJ: Princeton University Press, 1998.

Schickore, Jutta. *The Microscope and the Eye: A History of Reflections, 1740–1870.* Chicago: University of Chicago Press, 2007.

Schiebinger, Londa. "Skeletons in the Closet: The First Illustration of a Female Skeleton in Eighteenth-Century Anatomy." *Representations* 14 (1986): 42–82.

Schimmel, Julie. "Inventing the 'Indian.'" In *The West as America: Reinterpreting Images of the Frontier, 1820–1920*, edited by William H. Truettner, 149–89. Washington, DC: Smithsonian Institution Press, 1991.

Schnapp, Alain. "French Archaeology: Between National Identity and Cultural Identity." In *Nationalism and Archaeology in Europe*, edited by Margarita Díaz-Andreu and Timothy Champion, 48–67. New York: Routledge, 1996.

Schofield, Robert E. *The Enlightened Joseph Priestley: A Study of his Life and Work from 1773 to 1804.* University Park: Pennsylvania State University Press, 2010.

Schoolcraft, Henry Rowe. *An Address, Delivered before the Was-Ah Ho-De-No-Son-Ne, or New Confederacy of the Iroquois, at its Third Annual Council.* Rochester, NY: Jerome and Brothers, 1846.

Scott, Robert H. "Weather Charts in Newspapers." *Journal of the Society of Arts* 23, no. 1183 (July 23, 1875): 776–82.

Scott, Robert H. "A Word about Weather." *Chambers's Journal of Popular Literature, Science, and* Art 51, no. 554 (August 8, 1874): 511–12.

Secord, James. *Visions of Science: Books and Readers at the Dawn of the Victorian Age*. Chicago: University of Chicago Press, 2014.

Sekula, Alan. "The Body and the Archive." *October* 39 (Winter 1986): 3–64.

"A Septette of Favourite Paintings." *Strand* 33 (May 1907): 483–90.

Serres, Michel. "Turner Translates Carnot." In *Hermes: Literature, Science, Philosophy*, by Michel Serres, edited by Josue V. Harari and David F. Bell, 52–64. Baltimore, MD: Johns Hopkins University Press, 1982.

Shaberman, R. B. "George MacDonald and Lewis Carroll." *North Wind: A Journal of George MacDonald Studies* 1, no. 1 (1982): 10–30. https://digitalcommons.snc.edu/northwind/vol1/iss1/1.

Shakespeare, William. *The Tempest*. In *The Norton Shakespeare*, 3rd ed., edited by Stephen Greenblatt et al. New York: W. W. Norton, 2016.

Shaw, Napier. *Manual of Meteorology*. Vol. 1, *Meteorology in History*. Cambridge: Cambridge University Press, 1932.

Shafe, Laurence. "Why Is the Peacock's Tail so Beautiful?" In *Darwin and Theories of Aesthetics and Cultural History*, edited by Barbara Larson and Sabine Flach, 37–52. Burlington, VT: Ashgate, 2013.

Sheehan, Tanya. *Doctored: The Medicine of Photography in Nineteenth-Century America*. University Park: Pennsylvania State University Press, 2011.

Shires, Linda M. *Perspectives: Modes of Viewing and Knowing in Nineteenth-Century England*. Columbus: Ohio State University Press, 2009.

Shteir, Ann B., and Bernard V. Lightman, eds. *Figuring It Out: Science, Gender, and Visual Culture*. Hanover, NH: University Press of New England, 2006.

Shteir, Ann B. "Iconographies of Flora: The Goddess of Flowers in the Cultural History of Botany." In *Figuring It Out: Science, Gender, and Visual Culture*, edited by Ann B. Shteir and Bernard V. Lightman, 3–27. Hanover, NH: University Press of New England, 2006.

Shuttleworth, Sally. *Charlotte Brontë and Victorian Psychology*. Cambridge: Cambridge University Press, 1996.

Shuttleworth, Sally. *The Mind of the Child: Child Development in Literature, Science, and Medicine, 1840–1900*. Oxford: Oxford University Press, 2010.

Siegel, Daniel M. "Thomson, Maxwell, and the Universal Ether in Victorian Physics." In *Conceptions of Ether: Studies in the History of Ether Theories, 1740–1900*, edited by G. N. Cantor and M. J. S. Hodge, 239–68. Cambridge: Cambridge University Press, 1981.

Sienkewicz, Julia A. "Beyond the Mohawk Warrior: Reinterpreting Benjamin West's Evocations of American Indians." *19: Interdisciplinary Studies in the Long Nineteenth Century* (November 2009). http://doi.org/10.16995/ntn.515.

Silverberg, Robert. *Mound Builders of Ancient America: The Archaeology of a Myth*. Greenwich, CT: New York Graphic Society, 1968.

Silver, Carole G. *Strange and Secret Peoples: Fairies and Victorian Consciousness.* New York: Oxford University Press, 1999.

Simmons, John E. *Museums: A History.* Lanham, MD: Rowman & Littlefield, 2016.

Simonin, Louis. *Underground Life, or, Mines and Miners.* Translated by Henry W. Bristow. London: Chapman and Hall, 1869.

Simpson, James Young. *History of Modern Anaesthetics: A Second Letter to Dr. J. Bigelow.* Edinburgh: Edmonston and Douglas, 1870.

Simpson, Marc, ed. *Like Breath on Glass: Whistler, Inness, and the Art of Painting Softly.* Williamstown, MA: Sterling and Francine Clark Art Institute, 2008.

"Sir John Everett Millais, Bart., R.A., A Character Sketch." *Review of Reviews* 11 (1895): 543–51.

Skilton, David John. "Tourists at the Ruins of London: The Metropolis and the Struggle for Empire." *Cercles: revue pluridisciplinaire du monde anglophone* 17 (2007): 93–119.

Slipp, Naomi. "For the Edification of All: Nineteenth-Century American Medicine, Art, and the Role of the Classical Cast in Cultural Life." *Athanor* 32 (2014): 65–73.

Slipp, Naomi. "'The Secret Figure': Artistic Anatomy and the Medical Body in Nineteenth-Century American Culture." PhD diss., Boston University, 2015.

Smajić, Srdjan. *Ghost-Seers, Detectives, and Spiritualists: Theories of Vision in Victorian Literature and Science.* Cambridge: Cambridge University Press, 2010.

Small, Ian. "The Vocabulary of Pater's Criticism and the Psychology of Aesthetics." *British Journal of Aesthetics* 18, no. 1 (1978): 81–87.

Smith, Alison, and Jason Rosenfeld. *Millais.* London: Tate, 2008.

Smith, John R. *Descriptive Book of the Tour of Europe, the Largest Moving Panorama in the World.* New York: Pettiner and Gray, 1855.

Smith, John R. *Descriptive Pamphlet of Smith's Leviathan Panorama of the Mississippi River!* Philadelphia: Wm. S. Young, 1848.

Smith, Jonathan. *Charles Darwin and Victorian Visual Culture.* Cambridge: Cambridge University Press, 2006.

Smith, Jonathan. "Grant Allen, Physiological Aesthetics, and the Dissemination of Darwin's Botany." In *Science Serialized: Representations of the Sciences in Nineteenth-Century Periodicals,* edited by Geoffrey N. Cantor and Sally Shuttleworth, 285–306. Cambridge: MIT Press, 2004.

Smith, Roger. "The Physiology of the Will: Mind, Body, and Psychology in the Periodical Literature, 1855–1875." In *Science Serialized: Representations of the Sciences in Nineteenth-Century Periodicals,* edited by Geoffry N. Cantor and Sally Shuttleworth, 81–110. Cambridge: MIT Press, 2004.

Smith, Shawn Michelle. *At the Edge of Sight: Photography and the Unseen.* Durham, NC: Duke University Press, 2013.

Smith, Truman. *An Examination of the Question of Anaesthesia, Arising on the Memorial of Charles Thomas Wells.* New York: John A. Gray, 1859.

Sobieszek, Robert A. *The Spirit of Fact: The Daguerreotypes of Southworth & Hawes.* Boston: D. R. Godine, 1976.

Solomon-Godeau, Abigail. *Photography at the Dock: Essays on Photographic History, Institutions, and Practices.* Minneapolis: University of Minnesota Press, 2009.

Sontag, Susan. *Regarding the Pain of Others.* New York: Farrar, Straus and Giroux, 2003.

Soros, Susan Weber, and Catherine Arbuthnott, eds. *Thomas Jeckyll: Architect and Designer, 1827–1881.* New Haven, CT: Yale University Press, 2003.

Southworth, Albert S. "An Address to the National Photographic Association of the United States, Delivered at Cleveland, Ohio, June 1870." *Philadelphia Photographer* 8, no. 94 (October 1871): 315–23.

Spencer, Herbert. *The Principles of Psychology.* 2nd ed. Vol. 1. London: Williams and Norgate, 1870.

Spencer, Herbert. *The Principles of Psychology.* 2nd ed. Vol. 2. 1872. Reprint, New York: Appleton, 1897.

Spencer, Herbert. *Essays: Scientific, Political and Speculative.* Vol. 2. London: Williams and Norgate, 1891.

Spielmann, M. H. *Millais and His Works.* London: Blackwood, 1898.

Spielmann, Marion. "The Royal Academy Exhibition 2." *Magazine of Art* (1895): 281–84.

Squier, Ephraim George, and Edwin Hamilton Davis. *Ancient Monuments of the Mississippi Valley,* edited by David J. Meltzer. Washington, DC: Smithsonian Institution Press, 1998.

Stafford, Barbara Maria. *Artful Science: Enlightenment Entertainment and the Eclipse of Visual Education.* Cambridge: MIT Press, 1996.

Stafford, Barbara Maria. *Body Criticism: Imaging the Unseen in Enlightenment Art and Medicine.* Cambridge: MIT Press, 1993.

Staley, Allen, and Christopher Newall. *The Pre-Raphaelite Vision: Truth to Nature.* London: Tate, 2004.

Starr, Paul. *The Social Transformation of American Medicine: The Rise of a Sovereign Profession and the Making of a Vast Industry.* New York: Basic Books, 1982.

"Stated Meeting, October 6, 1846." *Proceedings of the Academy of Natural Sciences of Philadelphia* 3, no. 5 (September–October 1846): 105–8.

"Stated Meeting, October 20, 1846." *Proceedings of the Academy of Natural Sciences of Philadelphia* 3, no. 5 (September–October 1846): 109–10.

Stepan, Nancy. *The Idea of Race in Science: Great Britain, 1800–1960.* Hamden, CT: Archon Books, 1982.

Stepan, Nancy. *Picturing Tropical Nature.* Ithaca, NY: Cornell University Press, 2001.

Stephens, Frederic George. "Modern Giants." *Germ* 4 (May 1850): 169–72.

Stephenson, Andrew. "Precarious Poses: The Problems of Artistic Visibility and Its Homosocial Performances in Late-Nineteenth-Century London." *Visual Culture in Britain* 8, no. 1 (Spring 2007): 73–103.

Stephenson, Andrew. "Refashioning Modern Masculinity: Whistler, Aestheticism, and National Identity." In *English Art, 1860–1914: Modern Artists and Identity*, edited by David Peters Corbett and Lara Perry, 133–49. Manchester: Manchester University Press, 2000.

S., B. [Balfour Stewart]. "Meteorology." *Encyclopaedia Britannica*, 9th ed. Vol. 16. Edinburgh: A. and C. Black, 1883.

Stoeckle, John D., and Guillermo C. Sanchez. "On Seeing Medicine's Science and Art: Cure and Care, Body and Patient." In *The Invention of Photography and Its Impact on Learning*, edited by Louise Todd Ambler and Melissa Banta, 72–89. Cambridge, MA: Harvard University Press, 1989.

Strickland, R. A., and J. F. Butterworth. "Sexual Dreaming during Anesthesia: Early Case Histories (1849–1888) of the Phenomenon." *Anesthesiology* 106, no. 6 (June 2007): 1232–36.

Sully, James. "The Aesthetics of Human Character." *Fortnightly Review*, April 1, 1871, 505–20.

Sully, James. "The Basis of Musical Sensation." *Fortnightly Review*, April 3, 1872, 428–43.

Sully, James. *My Life and Friends: A Psychologist's Memories.* New York: Dutton, 1918.

Sully, James. *Sensation and Intuition: Studies in Psychology and Aesthetics.* London: Henry S. King, 1874.

Sully, James. "The Undefinable in Art." *Cornhill Magazine* 38, no. 227 (November 1878): 559–72.

"The Supernatural in Art." *Borderland: A Quarterly Review and Index* (July 1895): 195.

Surtees, Virginia. *The Paintings and the Drawings of Dante Gabriel Rossetti, 1828–1882.* Oxford: Clarendon, 1971.

Sweeney, Gray. "Racism, Nationalism, and Nostalgia." In *Race-ing Art History: Critical Readings in Race and Art History*, edited by Kymberly N. Pinder, 155–68. New York: Routledge, 2002.

Swinburne, Algernon. *The Complete Works of Algernon Swinburne*, edited by Edmund Gosse and Thomas James Wise. New York: Russell and Russell, 1968.

Syme, Alison. *A Touch of Blossom: John Singer Sargent and the Queer Flora of Fin-de-Siècle Art.* University Park: Pennsylvania State University Press, 2010.

Symons, Arthur. *Studies in Seven Arts.* 2nd ed. London: E. P. Dutton, 1907.

Tauber, A. I. *Elusive Synthesis: Aesthetics and Science.* Boston: Springer, 1997.

Taylor, Hilary. *James McNeill Whistler.* London: Studio Vista, 1978.

Taylor, Jenny Bourne. "Fallacies of Memory in Nineteenth-Century Psychology: Henry Holland, William Carpenter and Frances Power Cobbe." *Victorian Review* 26, no. 1 (2000): 98–118.

Taylor, Jenny Bourne, and Sally Shuttleworth, eds. *Embodied Selves: An Anthology of Psychological Texts, 1830–1890.* Oxford: Clarendon, 1998.

Taylor, Jesse Oak. "The Novel as Climate Model: Realism and the Greenhouse Effect in *Bleak House.*" *Novel: A Forum on Fiction* 46, no. 1 (Spring 2013): 1–25.

Tenneriello, Susan. *Spectacle Culture and American Identity, 1815–1940.* New York: Palgrave Macmillan, 2013.

Terranova, Charissa N., and Meredith Tromble, eds. *The Routledge Companion to Biology in Art and Architecture.* New York: Routledge, 2017.

Teukolsky, Rachel. *The Literate Eye: Victorian Art Writing and Modernist Aesthetics.* New York: Oxford University Press, 2009.

Thomas, Isaiah. "Communication from the President." In *Proceedings of the American Antiquarian Society, 1812–1849,* 51–56.

Thomas, Isaiah. "Meeting of September 29, 1813. An Account of the American Antiquarian Society." In *Proceedings of the American Antiquarian Society, 1812–1849,* 14–19.

Thompson, John D., and Grace Goldin. *The Hospital: A Social and Architectural History.* New Haven, CT: Yale University Press, 1975.

Thomson, D. C. "Realism in Painting." *Art Journal* 19 (1880): 282–84.

Thomson, Keith. *Jefferson's Shadow: The Story of His Science.* New Haven, CT: Yale University Press, 2012.

Townsend, Joyce. *Pre-Raphaelite Painting Techniques.* London: Tate, 2004.

Trachtenberg, Alan. "The Daguerreotype and Antebellum America." In *American Photography: Local and Global Contexts,* edited by Bettina Gockel, with Patrizia Munforte, 1–14. Berlin: Akademie, 2012.

Trachtenberg, Alan. *Reading American Photographs: Images as History, Mathew Brady to Walker Evans.* New York: Hill & Wang, 1989.

Trachtenberg, Alan. "Seeing and Believing: Hawthorne's Reflections on the Daguerreotype in *The House of the Seven Gables.*" *American Literary History* 9, no. 3 (Autumn 1997): 460–81.

Tresch, John. *The Romantic Machine: Utopian Science and Technology after Napoleon.* Chicago: University of Chicago Press, 2012.

Treves, Frederick. "The Case of Sir John Millais: Passages from the Notebooks of Sir Frederick Treves." *British Royal College Surgeons Annual* (1961): 284–388.

Tribble, Scott. "Mounds, Myths, and Grave Mistakes: Wills De Hass and the Growing Pains of Nineteenth-Century Archaeology." *West Virginia History: A Journal of Regional Studies* 7, no. 1 (Spring 2013): 23–37.

Trigger, Bruce G. "Anglo-American Archaeology." *World Archaeology* 13, no. 2 (October 1981): 138–55.

Trigger, Bruce G. "Archaeology and the Image of the American Indian." *American Antiquity* 45, no. 4 (October 1980): 662–76.

Truettner, William H. *The Natural Man Observed: A Study of Catlin's Indian Gallery* Washington, DC: Smithsonian Institution Press, 1979.

Tseng, Shao-Chien. "Nadar's Photography of Subterranean Paris: Mapping the Urban Body." *History of Photography* 38, no. 3 (2014): 233–54.

Tsui, Aileen. "Whistler's *La Princesse du pays de la porcelaine*: Painting Re-Oriented." *Nineteenth-Century Art Worldwide* 9, no. 2 (Autumn 2010). http://www.19thc-artworldwide.org/autumn10/whistlers-la-princesse-du-pays-de-la-porcelaine.

Tucker, Jennifer. *Nature Exposed: Photography as Eyewitness in Victorian Science.* Baltimore, MD: Johns Hopkins University Press, 2005.

Tupper, John Lucas. "Progress of the Species." In *Poems*, by John Lucas Tupper, edited by William Michael Rossetti, 70–73. London: Longmans, Green, 1897.

Turpin, John. "Maclise, Daniel." In *Oxford Dictionary of National Biography*, edited by H. C. G. Matthew and Brian Harrison. Oxford University Press, 2004. https://doi.org/10.1093/ref:odnb/17682.

Tyndall, John. *Forms of Water in Clouds and Rivers, Ice and Glaciers.* New York: D. Appleton, 1872.

Tyndall, John. *Heat Considered as a Mode of Motion; Being a Course of Twelve Lectures Delivered at the Royal Institution of Great Britain in the Season of 1862.* London: Longman, Green, Longman, Roberts, & Green, 1863.

Tyndall, John. *The Scientific Use of the Imagination: A Discourse.* London: Longmans Green, 1872.

Underwood, Ted. "How Did the Conservation of Energy Become 'The Highest Law in All Science'?" In *Repositioning Victorian Sciences: Shifting Centres in Nineteenth-Century Scientific Thinking*, edited by David Clifford, Elisabeth Wadge, Alex Warwick, and Martin Willis, 119–30. London: Anthem, 2006.

Valance, Hélène. *Nocturne: Night in American Art, 1890–1917*, translated by Jane Marie Todd. New Haven, CT: Yale University Press, 2018.

Veit, Richard. "A Case of Archaeological Amnesia: A Contextual Biography of Montroville Wilson Dickeson (1810–1882), Early American Archaeologist." *Anthropology of Eastern North America* 25 (1997): 97–123.

Vera, Francisco, Rodrigo Rivera, and César Núñez. "Burning a Candle in a Vessel, a Simple Experiment with a Long History." *Science & Education* 20, no. 9 (September 2011): 881–93.

Verne, Jules. *Journey to the Centre of the Earth.* Mineola, NY: Dover, 2005.

Viets, Henry. "The Earliest Printed References in Newspapers and Journals to the First Public Demonstration of Ether Anesthesia in 1846." *Journal of the History of Medicine and Allied Sciences* 4, no. 2 (March 1949): 149–69.

Voss, Julia. *Darwin's Pictures: Views of Evolutionary Theory, 1837–1874*. Translated by Lori Lantz. New Haven, CT: Yale University Press, 2010.

Waegner, Cathy Covell, ed. *Mediating Indianness*. East Lansing: Michigan State University Press, 2015.

Wailoo, Keith. *Pain: A Political History*. Baltimore, MD: Johns Hopkins University Press, 2014.

Wales, J. *Discovery by the Late Dr. Horace Wells of the Applicability of Nitrous Oxyd Gas, Sulphuric Ether and other Vapors in Surgical Operations*. Hartford: Elihu Geer, 1852.

Walker, Malcolm. *History of the Meteorological Office*. New York: Cambridge University Press, 2012.

Walker, William. *Handbook of Drawing*. London: Seeley, Jackson & Halliday, 1880.

Wall, L. L. "The Medical Ethics of Dr J Marion Sims: A Fresh Look at the Historical Record." *Medical Ethics* 32, no. 6 (2006): 346–50.

Wallach, Alan. *Exhibiting Contradiction: Essays on the Art Museum in the United States*. Amherst: University of Massachusetts Press, 1998.

Wallen, Martin. *City of Health, Fields of Disease: Revolutions in the Poetry, Medicine, and Philosophy of Romanticism*. Aldershot, UK: Ashgate, 2004.

Waller, David. *The Perfect Man: The Muscular Life and Times of Eugen Sandow, Victorian Strongman*. Brighton, UK: Victorian Secrets, 2011.

Ward, James. "Mechanism and Morals: The World of Science and the World of History." *Hibbert Journal* (1906): 79–99.

Warner, John Harley. *Against the Spirit of the System: The French Impulse in Nineteenth-Century American Medicine*. Princeton, NJ: Princeton University Press, 1997.

Warner, John Harley. *The Therapeutic Perspective: Medical Perspective, Knowledge, and Identity in America, 1820–1885*. Princeton, NJ: Princeton University Press, 1997.

Warner, John Harley. "Witnessing Dissection: Photography, Medicine, and American Culture." In *Dissection: Photographs of a Rite of Passage in American Medicine, 1880–1930*, edited by J. H. Warner and James M. Edmonson, 7–29. New York: Blast Books, 2009.

Warner, Malcolm. "Millais in Reproduction." In *Writing the Pre-Raphaelites: "Text, Context, Subtext"*, edited by Michaela Giebelhausen and Tim Barringer, 215–36. London: Routledge, 2017.

Warren, Edward. *The Life of John Collins Warren*. Vol. 1. Boston: Ticknor and Fields, 1860.

Warren, Edward. *Some Account of the Letheon, or, Who Was the Discover?* Boston: Dutton and Wentworth, 1847.

Warren, John Collins. *Etherization: with Surgical Remarks*. Boston: William D. Ticknor, 1848.

Waterfield, Giles. *The People's Galleries: Art Museums and Exhibitions in Britain, 1800–1914*. New Haven, CT: Yale University Press, 2015.

"Weather Charts and Storm Warnings." *Leisure Hour*, September 7, 1878, 573–76.

"Weather Forecasts." *Chamber's Journal*, May 17, 1862, 307–10.

Webb, Jean. *"A Noble Unrest": Contemporary Essays on ιhe Work of George MacDonald*. Newcastle, UK: Cambridge Scholars, 2007.

Weiss, Marta. *Julia Margaret Cameron*. London: Mack, 2015.

Weltman, Sharon Aronofsky. *Performing the Victorian: John Ruskin and Identity in Theater, Science, and Education*. Columbus: Ohio State University Press, 2007.

Weston, Nancy. *Daniel Maclise: Irish Artist in Victorian London*. Dublin: Four Courts, 2001.

Whistler, James McNeill. "Mr Whistler's 'Ten O'Clock.'" London: Chatto & Windus, 1888.

Whistler, James McNeill. *The Correspondence of James McNeill Whistler, 1855–1903*. Edited by Margaret F. MacDonald, Patricia de Montfort and Nigel Thorp. http://www.whistler.arts.gla.ac.uk/correspondence/.

Whitaker, Robert D. "An Historical Note on the Conservation of Mass." *Journal of Chemical Education* 52, no. 10 (1975): 658–59.

Whytt, Robert. *An Essay on the Vital and Other Involuntary Motions of Animals*. Edinburgh: Hamilton, Balfour, & Neill, 1751.

Wickham, Annette, and Mark Murray-Flutter, eds. *Daniel Maclise: The Waterloo Cartoon*. London: Royal Academy, 2015.

Willey, Gordon R., and Jeremy A. Sabloff. *A History of American Archaeology*. London: Thames and Hudson, 1974.

Williams, Catherine. *Ruskin's Philosophy*. Penzance: Triton, 1975.

Williams, Rosalind. *Notes on the Underground: An Essay on Technology, Society, and the Imagination*. New ed. Cambridge: MIT Press, 2008.

Williams, Stephen. *Fantastic Archaeology: The Wild Side of North American Prehistory*. Philadelphia: University of Pennsylvania Press, 1991.

Willis, Deborah, ed. *Black Venus 2010: They Called Her "Hottentot"*. Philadelphia: Temple University Press, 2010.

Wilson, Andrew. "Human Resemblances to Lower Life." *Longman's Magazine* (1883): 44–66.

Wilson, Andrew. *Studies in Life and Sense*. London: Chatto & Windus, 1887.

Wilson, Stephen. "Introduction: Art, Science + Technology." In *Art + Science Now: How Scientific Research and Technological Innovation are Becoming Key to 21st-Century Aesthetics*. New York: Thames and Hudson, 2010."The Winds." *Cornhill Magazine* 11, no. 63 (March 1865, 346–53.

Winter, Alison. *Mesmerized: Powers of Mind in Victorian Britain*. Chicago: University of Chicago Press, 1998.

Wise, M. Norton. "Making Visible." *Isis* 97, no. 1 (March 2006): 75–82.

Wolf, Bryan J. "Of Air Pumps and Teapots: Joseph Wright of Derby, John Singleton Copley, and the Technology of Seeing." *Art History* 39, no. 2 (April 2016): 356–75.

Wolfe, Richard J. *Tarnished Idol: William Thomas Green Morton and the Introduction of Surgical Anesthesia: A Chronicle of the Ether Controversy.* San Anselm, CA: Norman, 2001.

Wolff, Robert Lee. *The Golden Key: A Study of the Fiction of George MacDonald.* New Haven, CT: Yale University Press, 1961.

Wood, John George. *The Common Objects of the Country.* London: G. Routledge, 1858.

Woolgar, Steve. *Science: The Very Idea.* London: Tavistock, 1988.

World of Wonders: A Record of Things Wonderful in Nature, Science, and Art. London: Cassell, Petter, and Galpin, 1873.

Wylde, James. *The Magic of Science: A Manual of Easy and Instructive Scientific Experiments.* London: Richard Griffin, 1861.

Yandell, Lunsford P. "A Plea for Anaesthetics in Midwifery." *Transactions of the Indiana State Medical Society* 21 (1871): 41–50.

Yanni, Carla. *Nature's Museums: Victorian Science and the Architecture of Display.* New York: Princeton Architectural Press, 2006.

Yonan, Michael. "The Man behind the Mask? Looking at Franz Xaver Messerschmidt." *Eighteenth-Century Studies* 42, no. 3 (Spring 2009): 431–51.

Young, Edward. *Pre-Raffaellitism, or, a Popular Enquiry into Some Newly-Asserted Principles.* London: Longman, Brown, Green, Longmans, & Roberts, 1857.

Young, Robert. *Darwin's Metaphor: Nature's Place in Victorian Culture.* Cambridge: Cambridge University Press, 1985.

Zigarovich, J. *Writing Death and Absence in the Victorian Novel: Engraved Narratives.* New York: Palgrave Macmillan, 2012.

Contributors

Rachael Z. DeLue, the Christopher Binyon Sarofim '86 Professor in American Art at Princeton University, studies American art and visual culture, focusing on intersections between art and science and the theory and practice of knowledge. She is the author of *George Inness and the Science of Landscape* (2004) and *Arthur Dove: Always Connect* (2016) and coeditor of *Landscape Theory* (2008) with James Elkins. She serves as editor-in-chief for the Terra Foundation Essays, a series of multiauthor volumes on salient concepts in American art, including picturing, scale, intermedia, circulation, and experience. She is currently working on a book about impossible images.

Carey Gibbons is a visiting assistant professor in the history of art and design at the Pratt Institute. She recently finished her PhD from the Courtauld Institute of Art in London. Her dissertation, "The Limits of the Body in Victorian Illustration: Arthur Hughes and Frederick Sandys," crosses disciplines, engaging with the illustrations of Hughes and Sandys in relation to Victorian literature, science, religion, and gender constructions. She has taught at a variety of institutions and held positions at the Museum of Modern Art, Fisk University Galleries, and Cooper Hewitt, Smithsonian Design Museum. She has a particular interest in illustration and recently curated an exhibition on Arthur Rackham at Columbia University's Rare Book and Manuscript Library.

Keren Rosa Hammerschlag is a lecturer in art history and curatorship in the Centre for Art History and Art Theory in the School of Art and Design at the Australian National University. From 2013 to 2018 she taught in art history, women's and gender studies, and disability studies at Georgetown University. She is the author of *Frederic Leighton: Death, Mortality, Resurrection* (2015), and numerous articles on Victorian and Edwardian art and visual culture. Her current research focuses on representations of racial difference and hybridity in Victorian painting.

Barbara Larson is a professor of modern European art at the University of West Florida. She has published extensively on many topics in science and visual culture. She is author of *The Dark Side of Nature: Science, Society, and the Fantastic in the Work of Odilon Redon* (2005) and lead editor of two books on Darwin, *The Art of Evolution: Darwin, Darwinisms, and Visual Culture* (2009) and *Darwin and Theories of Aesthetics and Cultural History* (2013). Also the editor of two series for Routledge/Taylor & Francis, Science and the Arts after 1750 and Science and Visual Culture in the Long Nineteenth Century: Sources and Documents, Larson is the recipient of many grants.

Nancy Rose Marshall is a professor of Victorian art and visual culture at the University of Wisconsin–Madison. Her books, catalogues, and articles seek to creatively illuminate nineteenth-century British ways of thinking and making through close examinations of form and materiality, embedded in Victorian textual evidence. Marshall's interest in exploring the intersections of art and science arose from her current project on the shifting representations and meanings of fire. Her previous book, *City of Gold and Mud* (2012), explored the formation of modernity through the gendered and raced images of Victorian London.

Caitlin Silberman completed a PhD in art history from the University of Wisconsin–Madison in 2017. Her dissertation, "Thinking with Birds in British Art, 1840–1900," considered the rhetorical strategies through which Victorian art and visual culture negotiated emergent challenges to interspecies difference. Her doctoral research was supported by the Paul Mellon Centre, the Yale Center for British Art, the Huntington Library, and the Smithsonian Institution, among others. An archivist as well as an art historian, Caitlin has held positions at London's Natural History Museum, the Stanley Kubrick Archive, and the University of Minnesota's Bell Museum.

Naomi Slipp is an associate professor of art history at Auburn University at Montgomery and holds a PhD from Boston University. Her research in American art focuses on nineteenth-century intersections between art, science, and identity. She has published in *Sculpture Journal* and the edited volumes *Beyond the Battlefield: New England and the Civil War* (2017), *Bodies beyond Borders: Moving Anatomies, 1750–1950* (2017), and *Ecocriticism and the Anthropocene in Nineteenth-Century Art and Visual Culture* (2019).

Alison Syme is an associate professor of modern art at the University of Toronto. Her publications, which explore intersections between the histories of art, literature, and science, focus on the role of metaphors in artistic practice and poetics. They include *A Touch of Blossom: John Singer Sargent and the Queer Flora of Fin-de-Siècle Art* (2010) and *Willow* (2014). She is currently completing a book on Edward Burne-Jones, the research for which has been funded by the Social Sciences and Humanities Council of Canada.

Index